Analysis and Stochastics of Growth Processes and Interface Models

Analysis and Stochastics of Growth Processes and Interface Models

Edited by

Peter Mörters, Roger Moser, Mathew Penrose,
Hartmut Schwetlick, and Johannes Zimmer

OXFORD
UNIVERSITY PRESS

Great Clarendon Street, Oxford OX2 6DP

Oxford University Press is a department of the University of Oxford.
It furthers the University's objective of excellence in research, scholarship,
and education by publishing worldwide in

Oxford New York

Auckland Cape Town Dar es Salaam Hong Kong Karachi
Kuala Lumpur Madrid Melbourne Mexico City Nairobi
New Delhi Shanghai Taipei Toronto

With offices in

Argentina Austria Brazil Chile Czech Republic France Greece
Guatemala Hungary Italy Japan Poland Portugal Singapore
South Korea Switzerland Thailand Turkey Ukraine Vietnam

Oxford is a registered trade mark of Oxford University Press
in the UK and in certain other countries

Published in the United States
by Oxford University Press Inc., New York

© Oxford University Press 2008

The moral rights of the authors have been asserted
Database right Oxford University Press (maker)

First Published 2008

All rights reserved. No part of this publication may be reproduced,
stored in a retrieval system, or transmitted, in any form or by any means,
without the prior permission in writing of Oxford University Press,
or as expressly permitted by law, or under terms agreed with the appropriate
reprographics rights organization. Enquiries concerning reproduction
outside the scope of the above should be sent to the Rights Department,
Oxford University Press, at the address above

You must not circulate this book in any other binding or cover
and you must impose the same condition on any acquirer

British Library Cataloguing in Publication Data

Data available

Library of Congress Cataloging in Publication Data

Data available

Typeset by Newgen Imaging Systems (P) Ltd., Chennai, India
Printed in Great Britain
on acid-free paper by
Biddles Ltd., King's Lynn, Norfolk

ISBN 978–0–19–923925–2

1 3 5 7 9 10 8 6 4 2

PREFACE

A regional meeting of the London Mathematical Society, followed by a workshop on 'Analysis and Stochastics of Growth Processes', was held at the University of Bath on 11–15 September 2006. The aim of these events was to bring together analysts and probabilists working on the mathematical description of growth phenomena, with models based on the physics of individual particles discussed alongside models based on the continuum description of large collections of particles.

Convinced by positive feedback from the participants that this exercise in interdisciplinary exchange was worthwhile, we invited the speakers of the meeting and the workshop to contribute to this volume with an article in the same spirit. We hope that the resulting collection will help bridge the gap between researchers studying phenomena of the same type with different approaches. The meeting and the workshop were funded by the London Mathematical Society and the Bath Institute for Complex Systems, whose generous support is gratefully acknowledged.

The Editors, Bath, 26th October 2007.

CONTENTS

List of Contributors ix

Introduction 1

PART I QUANTUM AND LATTICE MODELS

A Quantum and Lattice Models 7

1 Directed random growth models on the plane 9
   Timo Seppäläinen

2 The pleasures and pains of studying the two-type
   Richardson model 39
   Maria Deijfen and Olle Häggström

3 Ballistic phase of self-interacting random walks 55
   Dmitry Ioffe and Yvan Velenik

B Microscopic to Macroscopic Transition 81

4 Stochastic homogenization and energy of infinite sets of points 83
   Xavier Blanc

5 Validity and non-validity of propagation of chaos 101
   Karsten Matthies and Florian Theil

C Applications in Physics 121

6 Applications of the lace expansion to statistical-mechanical
   models 123
   Akira Sakai

7 Large deviations for empirical cycle counts of integer
   partitions and their relation to systems of Bosons 148
   Stefan Adams

8 Interacting Brownian motions and the Gross–Pitaevskii formula 173
   Stefan Adams and Wolfgang König

9 A short introduction to Anderson localization 194
   Dirk Hundertmark

PART II MICROSCOPIC MODELS

A Nucleation and Growth   221

10 Effective theories for Ostwald ripening   223
    Barbara Niethammer

11 Switching paths for Ising models with long-range
    interaction   244
    Nicolas Dirr

12 Nucleation and droplet growth as a stochastic process   265
    Oliver Penrose

    B Applications in Physics   279

13 On the stochastic Burgers equation with some applications to
    turbulence and astrophysics   281
    A. D. Neate and A. Truman

14 Liquid crystals and harmonic maps in polyhedral domains   306
    Apala Majumdar, Jonathan Robbins, and Maxim Zyskin

Index   327

LIST OF CONTRIBUTORS

Stefan Adams
Mathematics Institute
Zeeman Building
University of Warwick
Coventry CV4 7AL
United Kingdom

Xavier Blanc
Laboratoire Jacques-Louis Lions
Université Pierre et Marie Curie
  - Paris 6
Boîte courrier 187
4, Place Jussieu
75252 Paris cedex 05
France

Maria Deijfen
Department of Mathematics
Stockholm University
106 91 Stockholm
Sweden

Nicolas Dirr
Department of Mathematical
  Sciences
University of Bath
Bath BA2 7AY
United Kingdom

Olle Häggström
Mathematical Statistics
Chalmers University of Technology
S-412 96 Göteborg
Sweden

Dirk Hundertmark
Department of Mathematics
University of Illinois at
  Urbana-Champaign
1409 W. Green Street (MC-382)
Urbana
Illinois 61801
United States of America

Dmitry Ioffe
William Davidson Faculty of
  Industrial Engineering and
  Management
Technion - Israel Institute
  of Technology
Haifa 3200
Israel

Wolfgang König
Fakultät für Mathematik
  und Informatik
Mathematisches Institut
Abteilung Wirtschaftsmathematik/
  Stochastik
Postfach 10 09 20
D-04009 Leipzig
Germany

Apala Majumdar
Mathematical Institute
University of Oxford
24-29 St Giles'
Oxford OX1 3LB
United Kingdom

List of Contributors

Karsten Matthies
Department of Mathematical
 Sciences
University of Bath
Bath BA2 7AY
United Kingdom

Andrew Neate
Department of Mathematics
University of Wales Swansea
Singleton Park
Swansea SA2 8PP
United Kingdom

Barbara Niethammer
Mathematical Institute
University of Oxford
24-29 St Giles'
Oxford OX1 3LB
United Kingdom

Oliver Penrose
Heriot-Watt University
School of Mathematical and
 Computer Sciences
Edinburgh EH14 4AS
United Kingdom

Jonathan Robbins
Department of Mathematics
University of Bristol
University Walk
Bristol BS8 1TW
United Kingdom

Akira Sakai
Creative Research Initiative Sousei
Hokkaido University
North 21, West 10, Kita-ku
Sapporo 001-0021
Japan

Timo Seppäläinen
419 Van Vleck Hall
University of Wisconsin-Madison
Madison WI 53706-1388
United States of America

Florian Theil
Mathematics Institute
Zeeman Building
University of Warwick
Coventry CV4 7AL
United Kingdom

Aubrey Truman
Department of Mathematics
University of Wales Swansea
Singleton Park
Swansea SA2 8PP
United Kingdom

Yvan Velenik
Section de Mathématiques
Université de Genève
2-4, rue du Lièvre
Case postale 64
1211 Genève 4
Switzerland

Maxim Zyskin
Department of Mathematics
University of Texas
SETB 2.454 - 80 Fort Brown
Brownsville
TX 78520
USA

INTRODUCTION

There has been a significant increase recently in activities on the interface between analysis and probability. Considering the potential of a combined approach to the study of various physical systems, it seems likely that this trend will continue. Yet any attempt to cross the divide between different communities can be impeded by the lack of a common vocabulary, or more fundamentally, by a lack of awareness of developments in each other's fields.

Against this background, the invited speakers of the London Mathematical Society South West and South Wales regional meeting on 'Analysis and Stochastics of Growth Processes', held at the University of Bath on 11–15 September 2006, provided an excellent example of how stimulating the interaction between different communities can be. Many of them agreed to follow up this occasion with a contribution to these proceedings, and were joined in some cases by co-authors not present at the workshop. The result is a collection of articles, mainly of survey character, covering a range of topics in deterministic and stochastic analysis. In some of them the theories are motivated by a model with an underlying lattice structure, in others by macroscopic models.

Quantum and lattice models

Random growth models

Random growth models describing the evolution of an interface in the plane are discussed by *Seppäläinen*. For specific models, three basic questions are discussed. First, under appropriate scaling, what is the limiting shape of the interface and what is the partial differential equation governing its evolution? Second, how can random fluctuations around the limit behaviour be described? Third, how can atypical behaviour be characterized? The power of probabilistic tools is demonstrated by employing laws of large numbers, central limit theorems, and large deviation techniques to answer these questions, respectively.

The two-type Richardson model discussed by *Deijfen and Häggström* is concerned with the competition of two infectious entities and how they spread over a lattice; they are assumed mutually exclusive at each site. One of the main questions for this model is under what conditions both entities can grow infinitely with positive probability. This problem is much more difficult than one would intuitively expect, and despite various partial results remains open.

The article by *Ioffe and Velenik* presents a unified approach to a study of the ballistic phase for a large family of self-interacting random walks with a drift and self-interacting polymers with an external stretching force. The approach is based on a recent version of the Ornstein–Zernike theory.

Microscopic to macroscopic transition

Blanc discusses recent work on homogenization of an elliptic partial differential equation under certain periodic or random assumptions. The coefficients are nonconstant but are a stationary random deformation of a periodic set of coefficients; a limit is taken where the period (in d-space) of the periodicity shrinks to zero. He also describes related work on average energies of nonperiodic infinite sets of points.

Matthies and Theil survey a novel rigorous approach to analyse the validity of continuum approximations for deterministic interacting particle systems. In particular, they look at the Boltzmann–Grad limit of ballistic annihilation, a topic which has has received considerable attention in the physics literature. In this model, due to the deterministic nature of the evolution, it is possible that correlations build up and the mean field approximation by the Boltzmann equation breaks down. They find a sharp condition on the initial distribution which ensures the validity of the Boltzmann equation and demonstrate the failure of the mean-field theory if the condition is violated.

Applications in physics

The lace expansion is the subject of the article by *Sakai*. It is one of the few approaches for a rigorous investigation of critical behaviour for various statistical-mechanical models. The article summarizes some of the most intriguing lace-expansion results for self-avoiding walk, percolation, and the Ising model.

Quantum models pose many challenges in probability and analysis alike. One area is interacting many-particle systems, in particular the peculiar effect of Bose–Einstein condensation; it is predicted that, under certain conditions (in particular extremely low temperature), all particles will condense into one state. Some of the physical background is surveyed in the article by *Adams and König*. They also discuss the Gross–Pitaevskii approximation for dilute systems. Variational problems appear here naturally, as the quantum mechanical ground state is of interest. In connection with positive temperature, related probabilistic models, based on interacting Brownian motions in a trapping potential, are introduced. Again, large deviation techniques are used to determine the mean occupation measure, both for vanishing temperature and large particle number.

Also motivated by Bose gases is the article by *Adams*. Here the focus is on the analysis of symmetrized systems of interacting Brownian motions. A cycle structure approach is introduced for the symmetrized distributions of empirical path measures, and this leads to a phase transition in the mean path measure.

Anderson localization is another physical problem that has spurred much mathematical research. The issue here is how disorder, such as random changes in the spacing of a crystal, influences the movement of electrons and thus the crystal's conductivity. In 1977, Ph. Anderson was awarded the Nobel prize for his investigations on this subject. *Hundertmark* introduces the physical model, based on a random Schrödinger operator, and carefully reviews different notions of localization as well as rigorous proofs of localization. A very readable introduction to finite-volume criteria for localization via percolation arguments is followed by an elegant proof of localization for large disorder.

Macroscopic models

Nucleation and growth

In her article, *Niethammer* discusses the derivation and analysis of reduced models for a coarsening process known as Ostwald ripening, which is a paradigm for statistical self-similarity in coarsening systems. The underlying physical phenomenon appears in the late stage of phase transitions, when – due to a change in temperature or pressure for example – the energy of the underlying system becomes nonconvex and prefers two different phases of the material. Consequently a homogeneous mixture is unstable and, in order to minimize the energy, it separates into the two stable phases. Typical examples are the condensation of liquid droplets in a supersaturated vapour and phase separation in binary alloys after rapid cooling.

Dirr studies a multiscale model for a two-phases material, which is on the microscopic scale a stochastic process. Due to the stochasticity on the microscopic scale, deviations from the limiting deterministic evolution arise with small probability. These are described in two illustrative examples.

O. Penrose proposes a stochastic differential equation as the putative limit for a birth-and-death Markov chain model for the size of a metastable droplet, and uses the large deviations theory of Freidlin and Wentzell to give a variational analysis of the path properties of the solution to this stochastic differential equation, relating these results to the classical theory of Becker and Döring.

Applications in physics

The Burgers equation, first introduced as a model for pressureless gas dynamics, has more recently been used as a tool in the context of various theories such as hydrodynamic turbulence, statistical mechanics, or cosmology. *Neate and Truman* summarize a selection of results on the inviscid limit of the stochastic Burgers equation. They discuss geometric properties of the caustic, Hamilton–Jacobi level surfaces, and the Maxwell set, and they show that for small viscosity, there exists a vortex filament structure near the Maxwell set. Furthermore, they explain how the theory is related to the adhesion model for the formation of the early universe.

Majumdar, Robbins, and Zyskin investigate harmonic maps from a polyhedron to the the unit two-sphere, motivated by the study of liquid crystals. They look at the Dirichlet energy of homotopy classes of such harmonic maps, subject to tangent boundary conditions, and investigate lower and upper bounds for this Dirichlet energy on each homotopy class.

PART I

QUANTUM AND LATTICE MODELS

A
QUANTUM AND LATTICE MODELS

1

DIRECTED RANDOM GROWTH MODELS ON THE PLANE

Timo Seppäläinen

Abstract

This is a brief survey of laws of large numbers, fluctuation results and large deviation principles for asymmetric interacting particle systems that represent moving interfaces on the plane. We discuss the exclusion process, the Hammersley process, and the related last-passage growth models.

1.1 Introduction

This article is a brief overview of recent results for a class of stochastic processes that represent growth or motion of an interface in two-dimensional Euclidean space. The models discussed have in a sense rather orderly evolutions, and the word 'directed' is included in the title to evoke this feature.

Let us begin with generalities about these stochastic processes. The state at time $t \in [0, \infty)$ is of the form $h(t) = (h_i(t) : i \in \mathbb{Z})$ with the interpretation that the integer- or real-valued random variable $h_i(t)$ represents the height of the interface over site i of the substrate $\mathbb{Z}$. We call the state $h = (h_i)$ a *height function* on $\mathbb{Z}$. The interface on the plane is then represented by the graph $\{(i, h_i) : i \in \mathbb{Z}\}$. Each particular process has a state space that defines the set of admissible height functions. The state space will be defined by putting restrictions on the increments (discrete derivatives) $h_i - h_{i-1}$ of the height functions.

The random dynamics of the state are specified by the *jump rates* of the individual height variables h_i. The rates are functions of the current state h. For the sake of illustration, suppose that only jumps ± 1 are permitted for each variable h_i. Then the model is defined by giving two functions $p(h)$ and $q(h)$. If the current state is h, then the height value h_0 at the origin jumps down to a new value $h'_0 = h_0 - 1$ with rate $p(h)$, and jumps up to a new value $h'_0 = h_0 + 1$ with rate $q(h)$. In the spatially homogeneous case the rates for h_i are $p(\theta_i h)$ and $q(\theta_i h)$ where the spatial translations are defined by $(\theta_i h)_j = h_{i+j}$.

The quantity $p(h)$ is the rate for the jump $h_0 \curvearrowright h_0 - 1$ in an instantaneous sense: in the current state h, the probability that this jump happens in the next infinitesimal time interval $(0, dt)$ is $p(h)dt + o(dt)$. Rigorous constructions of the processes utilize Poisson processes or 'Poisson clocks'. A rate λ Poisson process $N(t)$ is a simple continuous time Markov chain: it starts at $N(0) = 0$, runs through the integers $0, 1, 2, 3, \ldots$ in increasing order, and waits for a rate λ exponential random time between jumps. A rate λ exponential random time is defined by its density $\varphi(t) = \lambda e^{-\lambda t}$ on $\mathbb{R}_+$. The number of jumps $N(s+t) - N(s)$ in time interval $(s, s+t]$ has the mean λt Poisson distribution:

$$P\{N(s+t) - N(s) = k\} = \frac{e^{-\lambda t}(\lambda t)^k}{k!} \quad (k \geq 0).$$

If the overall rates are bounded, say by $p(h) \leq \lambda$, then a Poisson clock with a time-varying rate $p(h(t))$ can be obtained from $N(t)$ by randomly accepting a jump at time t with probability $p(h(t))/\lambda$.

Later, we mention in passing rigorous constructions of some processes. In each case the outcome of the construction is that all the random variables $\{h_i(t) : i \in \mathbb{Z}, t \geq 0\}$ are defined as measurable functions on an underlying probability space $(\Omega, \mathcal{F}, \mathbf{P})$. Since these processes evolve through jumps, the appropriate path regularity is that with probability 1, the path $t \mapsto h(t)$ is right-continuous with left limits (cadlag for short). The use of Poisson clocks makes the stochastic process $h(t)$ a *Markov process*. This means that if the present state $h(t)$ is known, the future evolution $(h(s) : s > t)$ is statistically independent of the past $(h(s) : 0 \leq s < t)$. This is a consequence of the 'forgetfulness property' of the exponential distribution. For a complete discussion of these foundational matters we must refer to textbooks on probability theory and stochastic processes.

This article covers only asymmetric systems. Asymmetry in this context means that the height variables $h_i(t)$ on average tend to move more in one direction than the other. For definiteness, we define the models so that the downward direction is the preferred one. In fact, the great majority of the paper is concerned with *totally asymmetric systems* for which $q(h) \equiv 0$, in other words only downward jumps are permitted. Symmetric systems behave quite differently from asymmetric systems, hence restricting treatment to one or the other is natural.

Stochastic processes with a large number of interacting components, such as the height process $h(t) = (h_i(t) : i \in \mathbb{Z})$, belong in an area of probability theory called *interacting particle systems.* (Spitzer 1970) is one of the seminal papers of this subject. Here is a selection of books and lecture notes on the topic: De Masi and Presutti (1991), Durrett (1988), Kipnis and Landim (1999), Liggett (1985), Liggett (1999), Liggett (2004), Varadhan (2000). Krug and Spohn (1992) and Spohn (1991) are sources that combine mathematics and the theoretical physics side.

Our treatment is organized around three basic questions posed about stochastic models: (i) laws of large numbers, (ii) fluctuations, and (iii) large deviations.

(i) Laws of large numbers give deterministic limit shapes and evolutions under appropriate space and time scaling. A parameter $n \nearrow \infty$ gives the ratio of macroscopic and microscopic scales. A sequence of processes $h^n(t)$ indexed by n is considered. Under appropriate hypotheses the height process satisfies this

type of result: for $(t, x) \in \mathbb{R}_+ \times \mathbb{R}$

$$n^{-1} h^n_{[nx]}(nt) \to u(x, t) \quad \text{as } n \to \infty, \tag{1.1}$$

and the limit function u satisfies a Hamilton–Jacobi equation $u_t + f(u_x) = 0$.

(ii) Fluctuations. After a law of large numbers the next question concerns the random fluctuations around the large scale behavior. One seeks an exponent α that describes the magnitude of these fluctuations, and a precise description of them in the limit. A typical statement would be:

$$\frac{h^n_{[nx]}(nt) - nu(x,t)}{n^\alpha} \longrightarrow Z(t, x) \tag{1.2}$$

where $Z(t, x)$ is a random variable whose distribution would be described as part of the result. The convergence is of a weak type, where it is the probability distribution of the random variable on the left that converges.

(iii) Large deviations. The vanishing probabilities of atypical behaviour fall under this rubric. Often these probabilities decay as e^{-Cn^β} to leading order, with another exponent $\beta > 0$. The precise constant $C \in (0, \infty)$ is also of interest and comes in the form of a *rate function*. When all the ingredients are in place the result is called a large deviation principle (LDP). An LDP from the law of large numbers (1.1) with rate function $I : \mathbb{R} \to [0, \infty]$ could take this form:

$$\lim_{\varepsilon \searrow 0} \lim_{n \to \infty} n^{-\beta} \log \mathbf{P}\{h^n_{[nx]}(nt) \in (z - \varepsilon, z + \varepsilon)\} = -I(z) \tag{1.3}$$

valid for points z in some range. Positive values $I(z) > 0$ represent atypical behavior, while limit (1.1) would force $I(u(x, t)) = 0$.

Example. For classical examples of these statements let us consider one-dimensional nearest-neighbour random walk. Fix a parameter $0 < p < 1$. Let $\{X_k\}$ be independent, identically distributed (IID) ± 1-valued random variables with common distribution $\mathbf{P}\{X_k = 1\} = p = 1 - \mathbf{P}\{X_k = -1\}$. Define the random walk by $S_0 = 0$, $S_n = S_{n-1} + X_n$ for $n \geq 1$. Then the *strong law of large numbers* gives the long term velocity:

$$\lim_{n \to \infty} n^{-1} S_n = v \quad \text{where } v = \mathbf{E}(X_1) = 2p - 1.$$

The convergence in the limit above is almost sure (a.s.), that is, almost everywhere (a.e.) on the underlying probability space of the variables $\{X_k\}$.

The order of nontrivial fluctuations around the limit is $n^{1/2}$ ('diffusive') and in the limit these fluctuations are Gaussian. That is the content of the *central limit theorem*:

$$\lim_{n \to \infty} \mathbf{P}\left\{ \frac{S_n - nv}{\sigma n^{1/2}} \leq s \right\} = \Phi(s) \equiv \frac{1}{\sqrt{2\pi}} \int_{-\infty}^{s} e^{-z^2/2} \, dz.$$

The parameter $\sigma^2 = \mathbf{E}[(X_1 - v)^2]$ is the variance.

Random walk satisfies this LDP:

$$\lim_{\varepsilon \searrow 0} \lim_{n \to \infty} \frac{1}{n} \log \mathbf{P}\{|S_n - nx| \le n\varepsilon\} = -I(x) \qquad (1.4)$$

where the rate function $I \colon \mathbb{R} \to [0, \infty]$ is identically ∞ outside $[-1, 1]$ and:

$$I(x) = \frac{1-x}{2} \log \frac{1-x}{2(1-p)} + \frac{1+x}{2} \log \frac{1+x}{2p} \quad \text{for } x \in [-1, 1]. \qquad (1.5)$$

$I(x)$ can be interpreted as an entropy. Convex analysis plays a major role in large deviation theory. Part of the general theory behind this simple case is that I is the convex dual of the logarithmic moment generating function $\Lambda(\theta) = \log \mathbf{E}(e^{\theta X_1})$.

Results for random walk are covered in graduate probability texts such as (Durrett 2004) and (Kallenberg 2002).

At the outset we delineated the class of models discussed. Important models left out include *diffusion limited aggregation* (DLA) and *first-passage percolation*. Their interfaces are considerably more complicated than interfaces described by height functions. But even for the models discussed our treatment is not a complete representation of the mathematical progress of the past decade. In particular, this article does not delve into the recent work on Tracy–Widom fluctuations, Airy processes, and determinantal point processes. These topics are covered by many authors, and we give a number of references to the literature in Sections 1.3.1 and 1.3.2. Overall, the best hope for this article is that it might inspire the reader to look further into the references.

Recurrent notation. The set of nonnegative integers is $\mathbb{Z}_+ = \{0, 1, 2, \dots\}$, while $\mathbb{N} = \{1, 2, 3, \dots\}$. The integer part of a real x is $[x] = \max\{n \in \mathbb{Z} : n \le x\}$. $a \vee b = \max\{a, b\}$ and $a \wedge b = \min\{a, b\}$.

1.2 Limit shape and evolution

We begin with the much studied *corner growth model* and a description that is not directly in terms of height variables. Attach nonnegative weights $\{Y_{i,j}\}$ to the points (i, j) of the positive quadrant $\mathbb{N}^2$ of $\mathbb{Z}^2$, as in Fig. 1.1. $Y_{i,j}$ represents the time it takes to occupy point (i, j) *after* the points to its left and below have been occupied. Assume that everything outside the positive quadrant is occupied at the outset so the process can start. Once occupied, a point remains occupied. Thus this is a totally asymmetric growth model, for the growing cluster never loses points, only adds them.

Let $G(k, \ell)$ denote the time when point (k, ℓ) becomes occupied. The above explanation is summarized by these rules: $G(k, \ell) = 0$ for $(k, \ell) \notin \mathbb{N}^2$, and:

$$G(k, \ell) = G(k-1, \ell) \vee G(k, \ell-1) + Y_{k,\ell} \quad \text{for } (k, \ell) \in \mathbb{N}^2. \qquad (1.6)$$

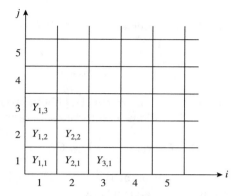

FIG. 1.1: Each point $(i,j) \in \mathbb{N}^2$ has a weight $Y_{i,j}$ attached to it.

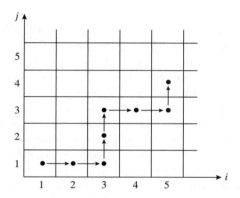

FIG. 1.2: An admissible path from $(1,1)$ to $(5,4)$.

The last equality can be iterated until the corner $(1,1)$ is reached, resulting in this last-passage formula for G:

$$G(k,\ell) = \max_{\pi \in \Pi_{k,\ell}} \sum_{(i,j) \in \pi} Y_{i,j} \qquad (1.7)$$

where $\Pi_{k,\ell}$ is the collection of nearest-neighbour up-right paths π from $(1,1)$ to (k,ℓ). Fig. 1.2 represents one such path for $(k,\ell) = (5,4)$.

This model and others of its kind are called *directed last-passage percolation models*. 'Directed' refers to the restrictions on admissible paths, and 'last-passage' to the feature that the occupation time $G(k,\ell)$ is determined by the slowest path to (k,ℓ). (By contrast, in first-passage percolation occupation times are determined by quickest paths.)

Our first goal is to argue the existence of a limit for $n^{-1}G([nx],[ny])$ as $n \to \infty$. Assume now that the weights $\{Y_{i,j}\}$ are i.i.d. non-negative random variables.

The idea is to exploit sub(super)additivity. Generalize the definition of $G(k, \ell)$ to:

$$G((k,\ell),(m,n)) = \max_{\pi \in \Pi_{(k,\ell),(m,n)}} \sum_{(i,j) \in \pi} Y_{i,j},$$

where $\Pi_{(k,\ell),(m,n)}$ is the collection of nearest-neighbour up-right paths π from $(k+1, \ell+1)$ to (m, n). The definitions lead to the superadditivity:

$$G(k, \ell) + G((k, \ell), (m, n)) \leq G(m, n). \tag{1.8}$$

Kingman's subadditive ergodic theorem (Durrett, 2004, Ch. 6) and some estimation implies the existence of a deterministic limit:

$$\gamma(x, y) = \lim_{n \to \infty} n^{-1} G([nx], [ny]) \quad \text{for } (x, y) \in \mathbb{R}_+^2. \tag{1.9}$$

Moment assumptions under which the limit function γ is continuous up to the boundary were investigated by Martin (2004). We turn to the problem of computing γ explicitly, and for this we need very specialized assumptions. Essentially only one distribution can be currently handled: the exponential, and its discrete counterpart, the geometric. Take the $\{Y_{i,j}\}$ to be IID rate 1 exponential random variables. In other words their common density is e^{-y}.

The difficulty with finding the explicit limit has to do with the superadditivity. The limit in Birkhoff's ergodic theorem is simply the expectation of the function averaged over shifts: $n^{-1} \sum_{k=1}^{n} f \circ \theta_k \to \mathbf{E}f$ (Durrett 2004, Ch. 6). But the subadditive ergodic theorem gives only an asymptotic expression for the limit. We need a new ingredient. We shall embed the last-passage model into the totally asymmetric simple exclusion process (TASEP). This has explicitly identifiable invariant distributions ('steady states') with which we can do explicit calculations.

Originally TASEP was introduced as a particle model. We wish to link TASEP with the last-passage model in a way that preserves the original formulation of TASEP, while mapping particle occupation variables into height increments and particle current into column growth. To achieve this we transform the coordinates (i, j) of Fig. 1.1 via the bijection $(i, j) \mapsto (i - j, -j)$. The result is the last-passage model of Fig. 1.3. Weights are relabelled as $X_{i,j} = Y_{i-j,-j}$. The transformation of admissible paths is illustrated by Fig. 1.4. Let the new last-passage times be denoted by $H(k, \ell)$. For $\ell < 0 \wedge k$ the maximizing-path formulation uses now paths of the kind represented in Fig. 1.4:

$$H(k, \ell) = \max_{\sigma \in \Sigma_{k,\ell}} \sum_{(i,j) \in \sigma} X_{i,j} \tag{1.10}$$

where $\Sigma_{k,\ell}$ is the collection of paths σ from $(0, -1)$ to (k, ℓ) that take steps of two types: $(1, 0)$ and $(-1, -1)$. The connection with the previous last-passage

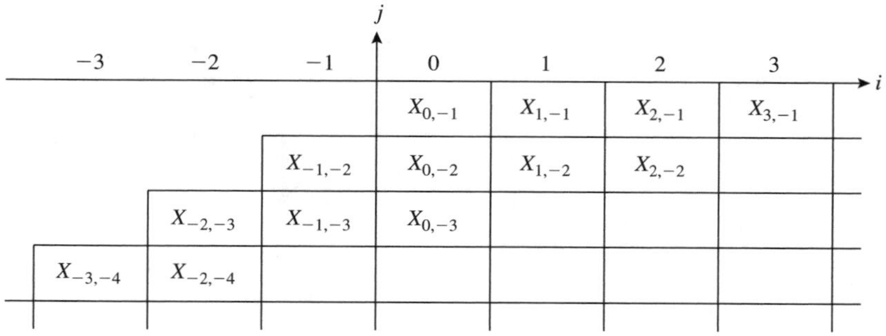

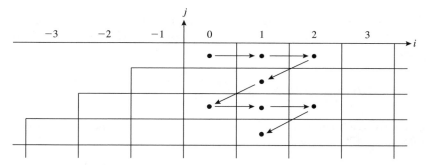

FIG. 1.3: Last-passage model for TASEP. The horizontal i-axis and the vertical j-axis are labeled, and points on the i-axis from -3 to 3 are marked. Weights $X_{i,j}$ are attached to points (i,j) such that $i \in \mathbb{Z}$, $j \in -\mathbb{N}$ and $j < 0 \wedge i$.

FIG. 1.4: The image of the path in Fig. 1.2. Now it goes from $(0,-1)$ to $(1,-4)$.

process is $H(k,\ell) = G(k-\ell,-\ell)$. The process $\{H(k,\ell)\}$ is also defined by the recursion:
$$H(k,\ell) = H(k-1,\ell) \vee H(k+1,\ell+1) + X_{k,\ell} \quad (\ell < 0 \wedge k) \tag{1.11}$$
together with the boundary values $H(k,\ell) = 0$ for $\ell \geq 0 \wedge k$. It satisfies the limit:
$$\lambda(x,y) = \lim_{n \to \infty} n^{-1} H([nx],[ny]) = \gamma(x-y,-y) \quad \text{for } y < 0 \wedge x. \tag{1.12}$$

To establish the TASEP connection we first define a height process $w(t) = (w_i(t) : i \in \mathbb{Z})$ that will turn out to be an alternative description of the last-passage process $\{H(i,j)\}$. Initially at time $t = 0$ the height is given by:
$$w_i(0) = \begin{cases} i, & i \leq -1 \\ 0, & i \geq 0. \end{cases} \tag{1.13}$$
This is the boundary of the region $\{j < 0 \wedge i\}$ filled with $X_{i,j}$'s in Fig. 1.3. This initial shape is a wedge, hence the symbol w.

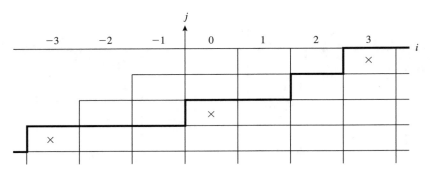

FIG. 1.5: A possible height function $w(t)$ (thickset graph) with column values $w_{-1}(t) = -3$, $w_0(t) = -2$, $w_1(t) = -2$, etc. 8 jumps have taken place during time $(0, t]$. The columns grow downward. ×s mark the allowable jumps from this state.

Give each column i an independent rate 1 Poisson clock N_i. Variable w_i jumps downward according to this rule: if t is a jump time for Poisson process N_i, then:

$$w_i(t) = w_i(t-) - 1 \text{ provided } \begin{cases} w_{i-1}(t-) = w_i(t-) - 1 \text{ and} \\ w_{i+1}(t-) = w_i(t-). \end{cases} \quad (1.14)$$

Equivalently, each column variable w_i jumps down independently at rate 1, as long as the state $w(t)$ remains in the state space:

$$\mathcal{X}_1 = \{h \in \mathbb{Z}^\mathbb{Z} : h_i - h_{i-1} \in \{0, 1\} \text{ for all } i \in \mathbb{Z}\}. \quad (1.15)$$

(See Fig. 1.5 for an example.) The *interaction* between the variables is encoded in rule (1.14). It forces the time-evolution of each variable w_i to depend on the evolution of its neighborus. By contrast, without the interaction the variables w_i would simply march along as Poisson processes independently of each other.

This construction defines the height process $w(t) = (w_i(t) : i \in \mathbb{Z})$ for all times $t \in [0, \infty)$ in terms of the family of Poisson clocks $\{N_i\}$. Given this process $w(t)$ define the stopping times:

$$T(i, j) = \inf\{t \geq 0 : w_i(t) \leq j\} \quad (1.16)$$

that mark the time when column i first reaches level j. *Stopping time* is a technical term for a random time whose arrival can be verified without looking into the future.

Initial condition (1.13) implies $T(i, j) = 0$ for $j \geq i \wedge 0$. Rule (14) tells us that $T(i-1, j) \vee T(i+1, j+1)$ is the stopping time at which the system is ready for w_i to jump from level $j+1$ to j. (Note that w_i must have reached level $j+1$ already earlier because if the rules are followed, $T(i+1, j+1)$ comes after

$T(i, j+1)$.) By the forgetfulness property of the exponential distribution, after the stopping time $T(i-1, j) \vee T(i+1, j+1)$ it takes another independent rate 1 exponential time $\widetilde{X}_{i,j}$ until w_i jumps from $j+1$ to j. Consequently the process $\{T(i,j)\}$ satisfies the recursion:

$$T(i,j) = T(i-1,j) \vee T(i+1, j+1) + \widetilde{X}_{i,j}. \tag{1.17}$$

This is of the same form as the recursion (1.11) satisfied by $\{H(i,j)\}$. From this one can prove that indeed the processes $\{T(i,j)\}$ and $\{H(i,j)\}$ are equal in distribution. Therefore (1.12) gives also $n^{-1}T([nx], [ny]) \to \lambda(x,y)$. This is the precise meaning of the earlier claim that the height process $w(t)$ gives an alternative description of the last-passage process $\{H(i,j)\}$.

Subadditivity and some estimation justifies the existence of a concave function g on $\mathbb{R}$ such that:

$$\lim_{t \to \infty} t^{-1} w_{[xt]}(t) = g(x) \quad \text{a.s. for } x \in \mathbb{R}. \tag{1.18}$$

Since rates are 1, g records only the initial height outside the interval $[-1,1]$, and so:

$$g(x) = 0 \wedge x \quad \text{for } |x| > 1.$$

Since the interface is a level curve of passage times, $\lambda(x, g(x)) = 1$ for $-1 \leq x \leq 1$. The last-passage limits are homogeneous in the sense that $\lambda(cx, cy) = c\lambda(x,y)$ for $c > 0$. Consequently λ and then γ can be obtained from g.

To summarize, thus far we have converted the original task of computing $\gamma(x,y)$ of (9) to finding the function g of (1.18) on the interval $[-1,1]$. Now consider the general height process $h(t)$ with state space $\mathcal{X}_1$ from (1.15) and dynamics defined as for $w(t)$ above: height variable h_i jumps one step down at every jump epoch of the Poisson clock N_i, provided this jump does not take the height function out of $\mathcal{X}_1$. A jump attempt that would violate the state space restriction is simply ignored.

Certain technical issues may trouble the reader. An infinite family of Poisson clocks has infinitely many jumps in any non-empty time interval $(0, \varepsilon)$. So there is no first jump attempt in the system and it is not obvious that the local rule leads to a well-defined global evolution: to determine the evolution of h_i on $[0, t]$ we need to look at the evolution of its neighbours $h_{i\pm 1}$, and then their neighbours, ad infinitum. However, given any $T < \infty$, almost surely there are indices $i_k \searrow -\infty$ and $i'_k \nearrow \infty$ such that N_{i_k} and $N_{i'_k}$ have no jumps during $(0, T]$. Consequently the system decomposes into (random) finite pieces that do not communicate before time T. The evolution can be determined separately in each finite segment which do experience only finitely many jumps up to time T (again almost surely). Another technical point is that the clocks $\{N_i\}$ have no simultaneous jumps (almost surely) so one never needs to consider more than one jump at a time.

Given that the height process $h(t)$ has been constructed, next define the increment process $\eta(t) = (\eta_i(t) : i \in \mathbb{Z})$ by:

$$\eta_i(t) = h_i(t) - h_{i-1}(t). \tag{1.19}$$

Process $\eta(t)$ has compact state space $\{0,1\}^{\mathbb{Z}}$ and its dynamics inherited from $h(t)$ can be succinctly stated as follows: each 10 pair becomes a 01 pair at rate 1, independently of the rest of the system. To see this connection, observe that if Poisson clock N_i jumps at time t, the height process undergoes the transformation $h_i(t) = h_i(t-) - 1$ only if $(\eta_i(t-) = 1, \eta_{i+1}(t-) = 0)$, and then after the jump the situation is $(\eta_i(t) = 0, \eta_{i+1}(t) = 1)$. This is a direct translation of the condition that jumps are executed only if the state h remains in the state space $\mathcal{X}_1$.

It is natural to interpret the 1s as particles and the 0s as holes, or vacant sites. The process $\eta(t)$ is the *totally asymmetric simple exclusion process* (TASEP). In this model the only interaction between the particles is the exclusion rule that stipulates that particles are not allowed to jump onto occupied sites. This property is enforced by the evolution because the definitions made above ensure that a jump in Poisson clock N_i sends a particle from site i to site $i+1$ only if site $i+1$ is vacant. Total asymmetry refers to the property that particles jump only to the right, never left. The definitions also entail this connection between the heights and the particles:

$$h_i(0) - h_i(t) = \text{cumulative particle current across the edge } (i, i+1). \tag{1.20}$$

We need to discuss two more properties of these processes, (i) stationary behaviour and (ii) the envelope property. Then we are ready to compute the function g of (1.18).

(i) Stationary behaviour. For $\rho \in [0,1]$, the *Bernoulli* probability measure ν_ρ on $\{0,1\}^{\mathbb{Z}}$ is defined by the requirement that:

$$\nu_\rho\{\eta : \eta_i = 1 \text{ for } i \in I, \eta_j = 0 \text{ for } j \in J\} = \rho^{|I|}(1-\rho)^{|J|} \tag{1.21}$$

for any disjoint $I, J \subseteq \mathbb{Z}$ with cardinalities $|I|$ and $|J|$. Measure ν_ρ corresponds to putting a particle at each site independently with probability ρ.

It is known that the measures $\{\nu_\rho\}_{\rho \in [0,1]}$ are invariant for the process $\eta(t)$, and in fact they are the extremal members of the compact, convex set of invariant probability measures that are also invariant under spatial shifts. Invariance means that if the process $\eta(t)$ is started with a random ν_ρ-distributed initial state $\eta(0)$, then at each time $t \geq 0$ the state $\eta(t)$ is ν_ρ-distributed, and furthermore, the probability distribution of the entire process $\eta(\cdot) = (\eta(t) : t \geq 0)$ is invariant under time shifts.

If we know the current state $h(t)$, then the probability that h_0 jumps down in a short time interval $(t, t+\varepsilon)$ is $\varepsilon\eta_0(t)(1-\eta_1(t)) + O(\varepsilon^2)$. This follows because a jump can happen only when a 10 pair is present, and from properties of Poisson processes. Estimation of this kind proves that:

$$h_0(t) - h_0(0) = -\int_0^t \eta_0(t)(1 - \eta_1(t))\,ds + M(t) \tag{1.22}$$

where $M(t)$ is a mean-zero martingale. This identity is a stochastic 'fundamental theorem of calculus' of sorts. Since things are random the difference between $h_0(t) - h_0(0)$ and the integral of the infinitesimal rate cannot be identically zero. Instead it is a *martingale*. This is a process whose increments have mean-zero in a very strong sense, namely even when conditioned on the entire past.

Let us average over (1.22) in the stationary situation. Let $\mathbf{E}_{\nu_\rho}$ denote expectation of functions of the stationary process $\eta(\cdot)$ whose state $\eta(t)$ is ν_ρ-distributed at each time t. Normalize the height process $h(\cdot)$ at time zero so that $h_0(0) = 0$. Then $h(\cdot)$ is entirely determined by $\eta(\cdot)$. Since $\eta_0(s)$ and $\eta_1(s)$ are independent at any fixed time s, we get:

$$E_{\nu_\rho}[h_0(t)] = -tf(\rho) \tag{1.23}$$

where the *particle flux* is defined by:

$$f(\rho) = \rho(1 - \rho). \tag{1.24}$$

(ii) Envelope property. Even though the flux f is nonlinear and therefore, as we see later, TASEP is governed by a nonlinear PDE, the height process has a valuable additivity property. Suppose a given initial height function $h(0) \in \mathcal{X}_1$ is the envelope of a countable collection $\{z^{(k)}(0)\}_{k \in \mathcal{K}}$ of height functions in the sense that

$$h_i(0) = \sup_{k \in \mathcal{K}} z_i^{(k)}(0) \quad \text{for each site } i \in \mathbb{Z}. \tag{1.25}$$

Take a single collection $\{N_i\}$ of Poisson clocks, and let all processes $h(t)$, $z^{(k)}(t)$ evolve from their initial height functions by following the same clocks $\{N_i\}$. This kind of simultaneous construction of many random objects for the purpose of comparison is called a *coupling*. By induction on jumps one can prove that this coupling preserves the envelope property for all time:

Lemma 1.1 $h_i(t) = \sup_{k \in \mathcal{K}} z_i^{(k)}(t)$ *for all $i \in \mathbb{Z}$ and all $t \geq 0$, almost surely.*

We take as auxiliary processes $z^{(k)}(t)$ suitable translations of the basic wedge process defined in (1.13)–(1.14). For $k \in \mathbb{Z}$ set:

$$w_i^{(k)}(0) = \begin{cases} i - k, & i < k \\ 0, & i \geq k \end{cases} \quad \text{and} \quad z_i^{(k)}(t) = h_k(0) + w_i^{(k)}(t). \tag{1.26}$$

The apex of the wedge $z^{(k)}(0)$ is at the point $(k, h_k(0))$, and then the definition of the wedge ensures that $h(0) \geq z^{(k)}(0)$. Hypothesis (1.25) holds and Lemma 1.1

gives this variational equality:

$$h_i(t) = \sup_{k \in \mathbb{Z}} \{h_k(0) + w_i^{(k)}(t)\}. \tag{1.27}$$

Now we extract two results from the assembled ingredients: the function g of (1.18), and a general 'hydrodynamic limit' that describes the large scale evolution of the process.

First specialize (1.27) to the stationary situation where $h_0(0) = 0$ and the increments are ν_ρ-distributed, and write (1.27) in the form:

$$t^{-1}h_0(t) = \sup_{y \in \mathbb{R}} \{t^{-1}h_{[ty]}(0) + t^{-1}w_0^{([ty])}(t)\}. \tag{1.28}$$

Let $t \to \infty$. Inside the braces on the right $t^{-1}h_{[ty]}(0) \to \rho y$ a.s. by the law of large numbers. $t^{-1}w_0^{([ty])}(t) \to g(-y)$ by a translation of the limit (1.18). With some work take the limit outside the supremum. Then we know $t^{-1}h_0(t)$ converges. By supplying some moment bounds we can take expectations over the limits, and with (1.23) arrive at:

$$-f(\rho) = \sup_{y \in \mathbb{R}} \{\rho y + g(-y)\}. \tag{1.29}$$

This is a convex duality (Rockafellar 1970). From the explicit invariant distributions (1.21) we obtained f in (1.24), and then we can solve (1.29) for g. (Without the invariant distributions we can carry out part of this reasoning but we cannot find f and g explicitly.) Let us record the results.

Theorem 1.1 *For the limit* (1.18) $g(x) = -\frac{1}{4}(1-x)^2$ *for* $-1 \leq x \leq 1$. *For the limit* (1.9) $\gamma(x,y) = (\sqrt{x} + \sqrt{y})^2$ *for* $x, y \geq 0$.

We turn to the hydrodynamic limit. Assume given a function u_0 on $\mathbb{R}$ and a sequence of random initial height functions $h^n(0) \in \mathcal{X}_1$ ($n \in \mathbb{N}$) such that:

$$n^{-1}h^n_{[nx]}(0) \to u_0(x) \quad \text{a.s. as } n \to \infty \text{ for each } x \in \mathbb{R}. \tag{1.30}$$

For this to be possible u_0 has to be Lipschitz with $0 \leq u_0'(x) \leq 1$ Lebesgue-a.e.

Theorem 1.2 *For* $x \in \mathbb{R}$ *and* $t > 0$ *we have the limit:*

$$n^{-1}h^n_{[nx]}(nt) \to u(t,x) \quad \text{a.s. as } n \to \infty \tag{1.31}$$

where

$$u(t,x) = \sup_{y \in \mathbb{R}} \{u_0(y) + tg\left(\frac{x-y}{t}\right)\}. \tag{1.32}$$

Equation (1.32) is a Hopf–Lax formula (Evans 1998) and it says that u is the entropy solution of the Hamilton–Jacobi equation:

$$u_t + f(u_x) = 0, \quad u|_{t=0} = u_0. \tag{1.33}$$

In other words this equation governs the macroscopic evolution of the height process. Theorem 1.2 is proved by showing that, as $n \to \infty$, variational formula (1.27) for $n^{-1} h^n_{[nx]}(nt)$ turns into (1.32). Details can be found in (Seppäläinen 1999).

Further remarks. The function g in Theorem 1.1 was first calculated by Rost (1981) in one of the seminal papers of hydrodynamic limits, but without the last-passage representation and with a different approach than the one presented here.

Let us discuss various avenues of generalization. We immediately encounter difficult open problems.

(i) Generalizations that retain the envelope property. The argument sketched above that combines the envelope property with the duality of the flux and the wedge shape to derive hydrodynamic limits was introduced in (Seppäläinen 1998a,c; 1999). An earlier instance of the variational connection appeared in Aldous and Diaconis (1995) for Hammersley's process. This work itself was based on the classic paper (Hammersley 1972); see Section 1.3.2 below. Also, in queueing literature, similar variational expressions arise (Szczotka and Kelly 1990).

To define the K-exclusion process we replace the state space $\mathcal{X}_1$ of (1.15) with:

$$\mathcal{X}_K = \{h \in \mathbb{Z}^{\mathbb{Z}} : 0 \leq h_i - h_{i-1} \leq K \text{ for all } i \in \mathbb{Z}\} \tag{1.34}$$

for some $2 \leq K < \infty$. Otherwise keep the model the same: rate 1 Poisson clocks $\{N_i\}$ govern the jumps of height variables h_i, and jumps that take the state h outside the space $\mathcal{X}_K$ are prohibited. The increment process is now called totally asymmetric K-exclusion (some authors use 'generalized exclusion'). The variational coupling (Lemma 1.1) works as before. But invariant distributions are unknown, and there is even no proof of existence of an extremal invariant distribution for each density value $\rho \in [0, K]$. No alternative way to compute f and g has been found. Theorem 1.2 is valid, but the most that can be said about f and g is that they exist as concave functions.

Interestingly, the situation becomes again explicitly analysable for $K = \infty$ where the only constraint on h is $h_i \leq h_{i+1}$. The increment process is a special case of a *zero range process*. Its state space is $(\mathbb{Z}_+)^{\mathbb{Z}}$ and i.i.d. geometric distributions are invariant (Liggett 1973; Andjel 1982). As a final step of generalization, away from monotone height functions, let us mention *bricklayer processes* (Balázs 2003; Balázs et al. 2007) whose increments $\eta_i = h_i - h_{i-1}$ can be positive or negative.

The variational coupling of Lemma 1.1 works equally well for certain multidimensional height processes $h(t) = (h_i(t) : i \in \mathbb{Z}^d)$ of the type discussed here. Examples appear in (Rezakhanlou 2002b; Seppäläinen 2000, 2007). No explicit invariant distributions are known for multidimensional height models. The variational scheme proves that scaled height processes converge to solutions of Hamilton–Jacobi equations as in Theorem 1.2. But again one can only assert the existence of f and g instead of giving them explicitly.

Another direction of generalization is to let the weights $\{Y_{i,j}\}$ have distributions other than exponential or geometric. The height process $h(t)$ ceases to be Markovian but the last-passage model of Figs. 1.1 and 1.2 makes sense. As mentioned, the limit $\gamma(x,y)$ in (1.9) is explicitly known only for the exponential and geometric cases. A distribution as simple as Bernoulli ($Y_{i,j}$ takes only two values) cannot be handled. However, if the paths are altered to require that one or both coordinates increase strictly, then the variational approach does find the explicit shape for the Bernoulli case (Seppäläinen 1997, 1998b).

Thus the present situation is that an explicit limit shape can be found only for some fortuitous combinations of path geometries and weight distributions.

(ii) Partially asymmetric models. Let us next address the case where the column variables h_i are allowed to jump both up and down. Fix two parameters $0 < q < p$ such that $p + q = 1$ (convenient normalization). Give each column i two independent Poisson clocks, $N_i^{(-)}$ with rate p and $N_i^{(+)}$ with rate q. At jump times of $N_i^{(\pm)}$ variable h_i attempts to jump to $h_i \pm 1$, and as before, a jump is completed if its execution does not take the state out of the state space $\mathcal{X}_1$. For the increment process this means that a 10 pair becomes a 01 pair at rate p, and the opposite move happens at rate q. Bernoulli distributions (1.21) are still invariant. This increment process is the *asymmetric simple exclusion process* (ASEP). In the same vein one can allow K particles per site and talk about asymmetric K-exclusion.

The envelope property of Lemma 1.1 is now lost. An alternative approach from (Rezakhanlou 2001) utilizes compactness of the random semigroups of the height process. Limit points are characterized as Hamilton–Jacobi semigroups via the Lions–Nisio theorem (Lions and Nisio 1982). Thereby Theorem 1.2 is derived for one-dimensional asymmetric K-exclusion. For $K = 1$ the flux f in (1.33) must be replaced by $f(\rho) = (p - q)\rho(1 - \rho)$, while for $2 \leq K < \infty$ the flux is unknown. In the multidimensional case it is not known if the resulting equation itself is random or not.

1.3 Fluctuations

A simple way to create initial height functions $h^n(0)$ that satisfy assumption (1.30) is to take independent increments with distributions:

$$\mathbf{P}[\eta_i^n(0) = 1] = n\big(u_0(\tfrac{i}{n}) - u_0(\tfrac{i-1}{n})\big),$$

and at the origin assign the deterministic value $h_0^n(0) = [nu_0(0)]$. The stationary situation is of this type with $u_0(x) = \rho x$. Then initial fluctuations:

$$n^{-1/2}\{h_{[nx]}^n(0) - nu_0(x)\}$$

are Gaussian in the limit $n \to \infty$. This makes it natural to look for a distributional limit at later times $t > 0$ on the central limit scale $n^{1/2}$:

$$n^{-1/2}\{h_{[nx]}^n(nt) - nu(x,t)\} \longrightarrow \zeta(t,x) \quad \text{as } n \to \infty, \tag{1.35}$$

for some limit process $\zeta(t,x)$. Such limits can be proved, but process $\{\zeta(t,x)\}$ turns out to be a deterministic function of the initial fluctuations $\{\zeta(0,x)\}$. Consequently limit (1.35) does not record any fluctuations created by the dynamics. Theorem 1.3 below gives a precise statement of this type.

In asymmetric systems the fluctuations created by the dynamics occur on a scale smaller than $n^{1/2}$. Two types of such phenomena have been found. Processes related to the last-passage model and exclusion process discussed in Section 1.2 have order $n^{1/3}$ fluctuations whose limits are distributions from random matrix theory. A class of linear processes has order $n^{1/4}$ fluctuations and Gaussian limits related to fractional Brownian motion with Hurst parameter $H = 1/4$. To see these lower order fluctuations one can start the system with a deterministic initial state, or one can start the system in the stationary distribution or some other random state, but then follow the evolution along characteristic curves of the macroscopic PDE. The fluctuation situation is very different for symmetric systems; the reader can consult (Kipnis and Landim 1999, Ch. 11).

1.3.1 Exclusion process

Probability distributions from random matrix theory were discovered as limit laws for last-passage growth models almost a decade ago.

Theorem 1.3 (Johansson 2000). *For the corner growth model:*

$$\mathbf{P}\left[\frac{G([xn],[ny]) - n\gamma(x,y)}{c(x,y)n^{1/3}} \leq s\right] \longrightarrow F(s) \quad \text{as } n \to \infty, \tag{1.36}$$

where F is the Tracy–Widom GUE distribution.

The distribution F first appeared as the limit distribution of the scaled largest eigenvalue of a random Hermitian matrix from the GUE (Tracy and Widom 1994). GUE is short for *Gaussian Unitary Ensemble*. This means that a random Hermitian matrix is constructed by putting IID complex-valued Gaussian random variables above the diagonal, IID real-valued Gaussian random variables on the diagonal, and letting the Hermitian property determine the entries below the diagonal. Then as the matrix grows in size, the variances of the entries are scaled appropriately to obtain limits. The standard reference is (Mehta 2004).

Theorem 1.3 and related results initially arose entirely outside probability theory (except for the statements themselves), involving the RSK correspondence and Gessel's identity from combinatorics and techniques from integrable systems to analyse the asymptotics of the resulting determinants. The *RSK correspondence*, named after Robinson, Schensted, and Knuth, is a bijective mapping between certain arrays of integers or integer matrices (in this case the matrix in Fig. 1.1 if the $Y_{i,j}$'s are integers) and pairs of Young tableaux. These latter objects are ubiquitous in combinatorics. Standard references are (Fulton 1997; Sagan 2001). More recently determinantal point processes have appeared as the link between the growth processes and random matrix theory. We shall not pursue these topics further for many excellent reviews are available: Baik (2005), Deift (2000), Johansson (2002), König (2005), and Spohn (2006).

Precise limits such as (1.36) have so far been restricted to totally asymmetric systems. Next we discuss ideas that fall short of exact limits but do give the correct order of the variance of the height for partially asymmetric systems.

Consider the height process $h(t)$ whose increments $\eta_i(t) = h_i(t) - h_{i-1}(t)$ form the asymmetric simple exclusion process (ASEP). This process was introduced in the remarks at the end of Section 1.2. Each height variable h_i attempts downward jumps with rate p and upward jumps with rate q, and $p > q$. A jump is suppressed if it would lead to a violation of the restrictions $h_{i-1} \leq h_i$ or $h_i \geq h_{i+1} - 1$ encoded in the state space $\mathcal{X}_1$ of (1.15). In the increment process each 10 pair becomes a 01 pair at rate p and each 01 pair becomes 10 pair at rate q.

On large space and time scales the height process obeys the Hamilton–Jacobi equation:
$$u_t + f(u_x) = 0 \text{ with } f(\rho) = (p-q)\rho(1-\rho).$$
This PDE carries information along the curves $\dot{x} = f'(u_x(t,x))$, in the sense that the slope u_x is constant along these curves as long as it is continuous. At constant slope $u_x = \rho$ the characteristic speed is $V^\rho = f'(\rho) = (p-q)(1-2\rho)$.

Consider the stationary process: $0 < \rho < 1$ is fixed, and at each time $t \geq 0$ the increments $\{\eta_i(t)\}_{i \in \mathbb{Z}}$ have Bernoulli ν_ρ-distribution from (1.21). Normalize the heights by setting initially $h_0(0) = 0$. We determine the order of magnitude of the variance of the height as seen by an observer traveling at speed V^ρ.

Theorem 1.4 (Balázs and Seppäläinen 2007b) *Height fluctuations along the characteristic satisfy:*

$$0 < \liminf_{t \to \infty} t^{-2/3} \mathbf{Var}\{h_{[V^\rho t]}(t)\} \leq \limsup_{t \to \infty} t^{-2/3} \mathbf{Var}\{h_{[V^\rho t]}(t)\} < \infty.$$

If the observer choses any other speed $v \neq V^\rho$, only a translation of initial Gaussian fluctuations would be observed. Take $v > V^\rho$ to be specific. Due to the normalization $h_0(0) = 0$ we can write:

$$h_{[vt]}(t) = \left(h_{[vt]}(t) - h_{[(v-V^\rho)t]}(0)\right) + \sum_{i=1}^{[(v-V^\rho)t]} \eta_i(0). \tag{1.37}$$

On the right the first expression in parentheses is a height increment along a characteristic and so by the theorem has fluctuations of order $t^{1/3}$. The last sum of initial increments has Gaussian fluctuations of order $t^{1/2}$ and consequently drowns out the first term.

The proof of Theorem 1.4 is entirely different from the proofs of Theorem 1.3. As it involves an important probabilistic idea let us discuss it briefly.

Couplings enable us to study the evolution of discrepancies between processes. In exclusion processes these discrepancies are called second class particles. Consider two initial ASEP configurations $\eta(0), \zeta(0) \in \{0,1\}^{\mathbb{Z}}$. The configurations differ at the origin: $\zeta(0)$ has a particle at the origin ($\zeta_0(0) = 1$) but $\eta(0)$ does not ($\eta_0(0) = 0$). At all other sites $i \neq 0$ we give the configurations a common but random value $\eta_i(0) = \zeta_i(0)$ according to the mean ρ Bernoulli distribution. Let the joint process $(\eta(t), \zeta(t) : t \geq 0)$ evolve together governed by the *same* Poisson clocks. The effect of this coupling is that there is always exactly one site $Q(t)$ such that $\zeta_{Q(t)}(t) = 1$, $\eta_{Q(t)}(t) = 0$, and $\eta_i(t) = \zeta_i(t)$ for all $i \neq Q(t)$.

$Q(t)$ is the location of a *second class particle* relative to the process $\eta(t)$. (Relative to $\zeta(t)$ one should say 'second class *anti*particle'.) In addition to ordinary exclusion jumps, Q yields to η-particles: if an η-particle at $Q + 1$ jumps left (rate q) then Q jumps right to switch places with the η-particle. Similarly an η-particle at $Q - 1$ switches places with Q at rate p. These special jumps follow from considering the effects of clocks $N^{(\mp)}_{Q\pm 1}$ on the discrepancy between η and ζ.

Proof of Theorem 1.4 utilizes couplings of several processes with different initial conditions. Evolution of second class particles is directly related to differences in particle current (height) between processes. On the other hand Q and the height variance are related through this identity:

$$\mathbf{Var}\{h_{[vt]}(t)\} = \rho(1-\rho)\mathbf{E}\{\,|Q(t) - [vt]|\,\} \quad \text{for any } v. \tag{1.38}$$

The right-hand side can be expected to have order smaller than t precisely when $v = V^\rho$ on account of this second identity:

$$\mathbf{E}Q(t) = tV^\rho. \tag{1.39}$$

From these ingredients the bounds in Theorem 1.4 arise.

Further remarks. As already suggested at the end of Section 1.2, a major problem for growth models is to find robust techniques that are not dependent on particular choices of probability distributions or path geometries. Progress on fluctuations of the corner growth model beyond the exponential case has come in situations that are in some sense extreme: for distributions with heavy tails (Hambly and Martin 2007) or for points close to the boundary of the quadrant (Baik and Suidan 2005; Bodineau and Martin 2005). See review by Martin (2006).

The second class particle appears in many places in interacting particle systems. In the hydrodynamic limit picture second class particles converge to characteristics and shocks of the macroscopic PDE (Ferrari and Fontes 1994b; Rezakhanlou 1995; Seppäläinen 2001). Versions of identities (1.38) and (1.39) are valid for zero range and bricklayer processes (Balázs and Seppäläinen 2007a). Equation (1.39) is surprising because the process as seen by the second class particle is *not* stationary.

In general the view of the process from the second class particle is complicated. Studies of invariant distributions seen by second class particles appear in (Derrida et al. 1993; Ferrari et al. 1994; Ferrari and Martin 2007). There are special cases of parameter values for certain processes where unexpected simplification takes place and the process seen from the second class particle has a product-form invariant distribution (Derrida et al. 1997; Balázs 2001).

1.3.2 Hammersley process

We began this paper with the exclusion process because this process is by far the most studied among its kind. It behooves us to introduce also Hammersley's process for which several important results were proved first. It has an elegant graphical construction that is related to a classical combinatorial question, namely the maximal length of an increasing subsequence of a random permutation. This goes back to (Hammersley 1972); see also (Aldous and Diaconis 1995, 1999).

We begin with the growth model. Put a homogeneous rate 1 Poisson point process on the plane. This is a random discrete subset of the plane characterized by the following property: the number of points in a Borel set B is Poisson distributed with mean given by the area of B and independent of the points outside B. Call a sequence $(x_1, t_1), (x_2, t_2), \ldots, (x_k, t_k)$ of these Poisson points *increasing* if $x_1 < x_2 < \cdots < x_k$ and $t_1 < t_2 < \cdots < t_k$. Let $L((a,s),(b,t))$ be the maximal number of points on an increasing sequence in the rectangle $(a,b] \times (s,t]$ (Fig. 1.6). The random permutation comes from mapping the ordered x-coordinates to ordered t-coordinates in the rectangle, and $L((a,s),(b,t))$ is precisely the maximal length of an increasing subsequence of this permutation.

The limit:

$$\lim_{n \to \infty} n^{-1} L((0,0),(nx,nt)) = 2\sqrt{xt} \quad \text{a.s.} \tag{1.40}$$

holds for $x, t > 0$. The limit exists by superadditivity exactly as for (1.9). The functional form $c\sqrt{xt}$ follows from scaling properties of the Poisson process. The value $c = 2$ was first derived by Veršik and Kerov (1977), while Logan and Shepp (1977) independently proved $c \geq 2$. The fluctuation result for L is analogous to (1.36), with normalization $n^{1/3}$ and the Tracy–Widom limit (Baik et al. 1999).

We embed the increasing sequences in the graphical construction of the *Hammersley process*. This process consists of point particles that move on $\mathbb{R}$ by jumping. Put a rate 1 Poisson point process on the space–time plane and

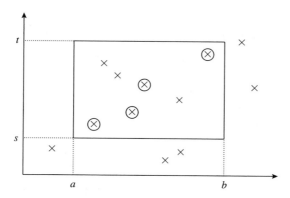

FIG. 1.6: Increasing sequences among planar Poisson points marked by ×s. $L((a,s),(b,t)) = 4$ as shown by the circled Poisson points that form an increasing sequence.

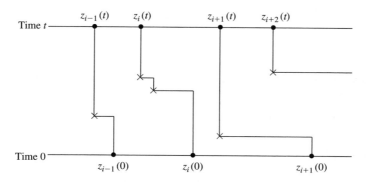

FIG. 1.7: Portion of the graphical construction of Hammersley's process. ×'s mark space–time Poisson points. •'s mark particle locations at time 0 and at a later time $t > 0$. Space–time trajectories of particles are shown. The horizontal segments are traversed instantaneously and the vertical segments at constant speed 1.

place the particles initially on the real axis. Move the real axis up at constant speed 1. Each Poisson point (x,t) instantaneously pulls to x the next particle to the right of x. We label the particles from left to right: $z_i(t) \in \mathbb{R}$ is the position of particle i at time t. We could regard the variables z_i as heights again, but the particle picture seems more compelling. This construction is illustrated by Fig. 1.7. In terms of infinitesimal rates, the construction realizes this rule: independently of other particles, at rate $z_i - z_{i-1}$ variable z_i jumps to a uniformly chosen location in the interval (z_{i-1}, z_i).

As in Section 1.2, there is a variational characterization for this construction. Define an inverse for the maximal path variable $L((a,s),(b,t))$ by:

$$\Gamma((a,s),t,w) = \inf\{h \geq 0 : L((a,s),(a+h,t)) \geq w\}.$$

Take an initial particle configuration $\{z_i(0)\} \in \mathbb{R}^{\mathbb{Z}}$ that satisfies $z_{i-1}(0) \leq z_i(0)$ and $i^{-2}z_i(0) \to 0$ as $i \to -\infty$. Then the graphical construction leads to a well-defined evolution $\{z_i(t)\}$ that satisfies:

$$z_i(t) = \inf_{k:k\leq i}\{z_k(0) + \Gamma((z_k(0),0),t,i-k)\}. \tag{1.41}$$

Here is the hydrodynamic limit. Consider a sequence of processes $z^n(t)$ that satisfies $n^{-1}z_{[ny]}^n(0) \to u_0(y)$ for each $y \in \mathbb{R}$, say in probability. The initial function u_0 is nondecreasing, locally Lipschitz, and satisfies:

$$y^{-2}u_0(y) \to 0 \quad \text{as } y \to -\infty. \tag{1.42}$$

Define:

$$u(t,x) = \inf_{y:y\leq x}\left\{u_0(y) + \frac{(x-y)^2}{4t}\right\}, \quad (t,x) \in (0,\infty) \times \mathbb{R}. \tag{1.43}$$

Since rates are unbounded, we need to assume a left tail bound to prevent the particles from disappearing to $-\infty$: given $\varepsilon > 0$ there exist $0 < q, n_0 < \infty$ such that:

$$\mathbf{P}\{z_i^n(0) < -\varepsilon i^2/n \text{ for some } i \leq -qn\} \leq \varepsilon \quad \text{for } n \geq n_0.$$

Under these assumptions:

$$n^{-1}z_{[nx]}^n(nt) \to u(t,x) \quad \text{in probability} \tag{1.44}$$

(Seppäläinen 1996). The function defined by the Hopf–Lax formula (1.43) solves the Hamilton–Jacobi equation $u_t + (u_x)^2 = 0$.

Let us state a precise result about the central limit scale fluctuations (1.35) that covers also shocks. For $(t,x) \in (0,\infty) \times \mathbb{R}$ let:

$$I(t,x) = \left\{y \in (-\infty,x] : u(t,x) = u_0(y) + \frac{(x-y)^2}{4t}\right\} \tag{1.45}$$

be the set of minimizers in (1.43), guaranteed nonempty and compact by hypothesis (1.42). Then (t,x) is a *shock* if $I(t,x)$ is not a singleton. This is equivalent to the nonexistence of the x-derivative $u_x(t,x)$.

Fluctuations on the scale $n^{1/2}$ from the limit (1.44) are described by the process:

$$\zeta_n(t,x) = n^{-1/2}\{z_{[nx]}^n(nt) - nu(x,t)\}.$$

Assume the existence of a continuous random function ζ_0 on $\mathbb{R}$ such that the convergence in distribution $\zeta_n(0,\,\cdot\,) \to \zeta_0$ holds in the topology of uniform convergence on compact sets. Define the process ζ by:

$$\zeta(t,x) = \inf_{y \in I(t,x)} \zeta_0(y)$$

where $I(t,x)$ is the (deterministic) set defined in (1.45).

Theorem 1.5 *For each (t,x), $\zeta_n(t,x) \to \zeta(t,x)$ in distribution.*

As stated in the beginning of Section 1.3, this distributional limit reflects no contribution from dynamical fluctuations as the process ζ is a deterministic transformation of ζ_0. The underlying reason is that the dynamical fluctuations of order $n^{1/3}$ are not visible on the $n^{1/2}$ scale. The dynamical fluctuations are the universal ones described by the Tracy–Widom laws. See again the discussion and references that follow Theorem 1.3.

Further remarks. The *polynuclear growth model* (PNG) is another related (1+1)-dimensional growth model used by several authors for studies of Tracy–Widom fluctuations and the Airy process in the KPZ scaling picture (Baik and Rains 2000; Ferrari 2004; Johansson 2003; Prähofer and Spohn 2002, 2004). Like the Hammersley process, the graphical construction of the PNG utilizes a planar Poisson process, and in fact the same underlying last-passage model of increasing paths. This time the Poisson points mark space–time nucleation events from which new layers grow laterally at a fixed speed. Roughly speaking, this corresponds to putting the time axis at a 45 degree angle in Figs. 1.6 and 1.7.

More about the phenomena related to Theorem 1.5 can be found in article (Seppäläinen 2002). A similar theorem for TASEP appears in (Rezakhanlou 2002a). Earlier work on the diffusive fluctuations of ASEP was done by Ferrari and Fontes (1994a,b).

1.3.3 Linear models

We turn to systems macroscopically governed by linear first order equations $u_t + b u_x = 0$. Fluctuations across the characteristic occur now on the scale $n^{1/4}$ and converge to a Gaussian process related to fractional Brownian motion.

The *random average process* (RAP) was first studied by Ferrari and Fontes (1998). The state of the process is a height function $\sigma : \mathbb{Z} \to \mathbb{R}$ with $\sigma_i \in \mathbb{R}$ denoting the height over site i. (More generally the domain can be $\mathbb{Z}^d$.) The basic step of the evolution is that a value σ_i is replaced by a weighted average of values in a neighbourhood, and the randomness comes in the weights. This time we consider a discrete time process. The basic step is carried out simultaneously at all sites i.

Now for precise formulations. Let $\{u(k,\tau) : k \in \mathbb{Z},\ \tau \in \mathbb{N}\}$ be an IID collection of random probability vectors indexed by space–time $\mathbb{Z} \times \mathbb{N}$. In terms of coordinates $u(k,\tau) = (u_j(k,\tau) : -M \leq j \leq M)$. We assume the system has finite range defined by the fixed parameter M. We impose a minimal assumption that guarantees that the weight vectors are not entirely degenerate:

$$\mathbf{P}\{\max_j u_j(0,0) < 1\} > 0. \tag{1.46}$$

For technical convenience we also assume that the process is 'on the correct lattice': there does not exist an integer $h \geq 2$ such that for some $b \in \mathbb{Z}$ the mean weights $p(j) = \mathbf{E}u_j(0,0)$ satisfy $\sum_{j \in b+h\mathbb{Z}} p(j) = 1$.

To start the dynamics let $\sigma(0)$ be a given random or deterministic initial height function. The process $\sigma(\tau)$, $\tau = 0, 1, 2, \ldots$, is defined iteratively by:

$$\sigma_i(\tau) = \sum_j u_j(i,\tau)\sigma_{i+j}(\tau-1), \quad \tau \geq 1, i \in \mathbb{Z}. \tag{1.47}$$

As before, we can define the process of increments $\eta_i(\tau) = \sigma_i(\tau) - \sigma_{i-1}(\tau)$. The increments also evolve via random linear mappings and are conserved like particles in exclusion processes.

As in Section 1.2 we create suitable initial conditions for a hydrodynamic limit. Consider a sequence of processes $\sigma^n(\tau)$ indexed by $n \in \mathbb{N}$, initially normalized by $\sigma_0^n(0) = 0$. For each n assume independent initial increments $\{\eta_i^n(0) : i \in \mathbb{Z}\}$ with:

$$\mathbf{E}[\eta_i^n(0)] = \rho(i/n) \quad \text{and} \quad \mathbf{Var}[\eta_i^n(0)] = v(i/n) \tag{1.48}$$

for given Hölder $1/2 + \varepsilon$ functions ρ and v. Assume a uniform moment bound: $\sup_{n,i} \mathbf{E}[\,|\eta_i^n(0)|^{2+\delta}\,] < \infty$ for some $\delta > 0$.

The hydrodynamic limit is rather trivial for it consists only of translation. Define a function u on $\mathbb{R}$ by $u(0) = 0$ and $u'(x) = \rho(x)$. The characteristic speed is:

$$b = -\sum_j jp(j).$$

Then for each $(t,x) \in \mathbb{R}_+ \times \mathbb{R}$:

$$n^{-1}\sigma_{[nx]}^n([nt]) \longrightarrow u(x-bt) \quad \text{as } n \to \infty, \text{ in probability.}$$

In other words, the height obeys the linear PDE $u_t + bu_x = 0$.

On the central limit scale $n^{1/2}$ one would also see only translation of initial fluctuations. To see something nontrivial we look at fluctuations around a characteristic line. Fix a point $\bar{y} \in \mathbb{R}$ and consider the characteristic line $t \mapsto \bar{y} + tb$ emanating from $(\bar{y},0)$. Define space–time process:

$$Z_n(t,r) = \sigma_{[n\bar{y}]+[r\sqrt{n}]+[ntb]}^n([nt]) - \sigma_{[n\bar{y}]+[r\sqrt{n}]}^n(0), \quad (t,r) \in \mathbb{R}_+ \times \mathbb{R}.$$

The spatial variable r describes fluctuations around the characteristic on the spatial scale $n^{1/2}$. For the increment process $Z_n(t,0)$ represents net current from right to left across the characteristic.

Theorem 1.6 (Balázs et al. 2006) *The finite-dimensional distributions of the process $n^{-1/4}Z_n$ converge to those of the Gaussian process $\{z(t,r) : t \geq 0,\ r \in \mathbb{R}\}$ described below.*

The statement means that for any finite collection of space–time points $(t_1, r_1), \ldots, (t_k, r_k)$, the $\mathbb{R}^k$-valued random vector $n^{-1/4}(Z_n(t_1, r_1), \ldots, Z_n(t_k, r_k))$ converges in distribution to the vector $(z(t_1, r_1), \ldots, z(t_k, r_k))$. The limiting process z has the following representation in terms of stochastic integrals:

$$z(t,r) = \rho(\bar{y})\sigma_a\sqrt{\kappa} \iint_{[0,t]\times\mathbb{R}} \varphi_{\sigma_a^2(t-s)}(r-z)\, dW(s,z) \qquad (1.49)$$
$$+ \sqrt{v(\bar{y})} \int_{\mathbb{R}} \operatorname{sign}(x-r)\, \Phi_{\sigma_a^2 t}(-|x-r|)\, dB(x).$$

Above W is a two-parameter Brownian motion on $\mathbb{R}_+ \times \mathbb{R}$ and B is a one-parameter Brownian motion on $\mathbb{R}$ independent of W. The first integral represents dynamical noise generated by the random weights, and the second the initial noise propagated by the evolution. The functions in the integrals are Gaussian densities and distribution functions:

$$\varphi_{\sigma^2}(x) = \frac{1}{\sqrt{2\pi\sigma^2}} \exp\left\{-\frac{x^2}{2\sigma^2}\right\} \quad \text{and} \quad \Phi_{\sigma^2}(x) = \int_{-\infty}^{x} \varphi_{\sigma^2}(y)\, dy.$$

The only effects from the initial height are the mean $\rho(\bar{y})$ and variance $v(\bar{y})$ of the increments around the point $n\bar{y}$. The parameter σ_a^2 is the variance of the probabilities $p(j)$, and κ another parameter determined by the distribution of the weights. Process z has a self-similarity property: $\{z(at, a^{1/2}r)\} \stackrel{d}{=} \{a^{1/4}z(t,r)\}$.

In the special case where $v(\bar{y}) = \kappa\rho(\bar{y})^2$ the temporal process $\{z(t,r) : t \in \mathbb{R}_+\}$ (for any fixed r) has covariance:

$$\mathbf{E}z(s,r)z(t,r) = \frac{\sigma_a \kappa \rho^2}{\sqrt{2\pi}}\left(\sqrt{s} + \sqrt{t} - \sqrt{|t-s|}\right).$$

This identifies $z(\cdot, r)$ as *fractional Brownian motion* with Hurst parameter $H = 1/4$. In particular, this limit arises in a stationary case where the averaging involves two points and the weight is beta distributed (Balázs et al. 2006 Example 2.1).

The proof of Theorem 1.6 utilizes a special case of another stochastic model of great contemporary interest, namely random walk in random environment (RWRE). Here is how the RWRE arises. An environment $\omega = \{u(k, \tau)\}$ is determined by the weight vectors. Given ω, define a 'backward' walk $\{X_s^{i,\tau} : s \in \mathbb{Z}_+\}$

on $\mathbb{Z}$ with initial position $X_0^{i,\tau} = i$ and transition probability:

$$P^\omega(X_{s+1}^{i,\tau} = y \mid X_s^{i,\tau} = x) = u_{y-x}(x, \tau - s), \quad s = 0, 1, 2, \ldots$$

The superscript ω on P^ω indicates that it is the *quenched* path measure of $\{X_s^{i,\tau} : s \in \mathbb{Z}_+\}$ under a fixed ω. The basic step (1.47) of RAP evolution can be rewritten so that $\sigma_i(\tau)$ equals the average value of the previous height function $\sigma(\tau - 1)$ seen by a walk started at i after one step:

$$\sigma_i(\tau) = \sum_j u_{j-i}(i,\tau) \sigma_j(\tau - 1) = E^\omega\big[\sigma_{X_1^{i,\tau}}(\tau - 1)\big].$$

This can be iterated all the way down to the initial height function:

$$\sigma_i(\tau) = E^\omega\big[\sigma_{X_\tau^{i,\tau}}(0)\big].$$

Note that the expectation E^ω over paths of the walk $X_s^{i,\tau}$ under fixed weights ω sees the initial height function $\{\sigma_i(0)\}_{i \in \mathbb{Z}}$ as a constant.

We have here a special type of RWRE called 'space–time'. Another term used is 'dynamical environment' because after each step the walk sees a new sample of its environment. Proof of Theorem 1.6 requires limits for the walk itself and its quenched mean process $E^\omega(X_s^{i,\tau})$. These results appear in (Balázs et al. 2006; Rassoul-Agha and Seppäläinen 2005).

Independent walks on $\mathbb{Z}$ display the same behavior as RAP. Let the process $Z_n(t,r)$ be the net particle current across the characteristic $t \mapsto [n\bar{y}] + [r\sqrt{n}] + [ntb]$ where b is the common average speed of the particles. Then under suitable assumptions on the initial particle arrangements and their jump kernel, Z_n satisfies a stronger form of Theorem 1.6 that also contains process-level convergence. One adjustment is necessary: the constants in front of the stochastic integrals in (1.49) are different for the random walk case. The stationary system sees again fractional Brownian motion as the limit of the current. Details for the random walk case appear in (Seppäläinen 2005; Kumar 2007). Earlier related results for a Poisson system of independent Brownian motions appeared in (Dürr et al. 1985).

1.4 Large deviations

We present the large deviation picture for the Hammersley process, so we continue in the setting of Section 1.3.2. Recall the definition of the longest path model among planar Poisson points illustrated in Fig. 1.6. Abbreviate $L_n = L((0,0),(n,n))$. Then the limit (1.40) is $n^{-1}L_n \to 2$. Here is the large deviation theorem for L_n. It was completed shortly before the fluctuation result of Baik et al. (1999), through a combination of several independent papers: Logan and Shepp (1977), Kim (1996), Seppäläinen (1998d), and Deuschel and Zeitouni (1999).

Theorem 1.7 *We have the following upper and lower tail large deviation bounds.*

$$\lim_{n\to\infty} n^{-1} \log \mathbf{P}\{L_n \geq nx\} = -I(x) \quad \text{for } x \geq 2 \tag{1.50}$$

with rate function $I(x) = 2x \cosh^{-1}(x/2) - 2\sqrt{x^2 - 4}$.

$$\lim_{n\to\infty} n^{-2} \log \mathbf{P}\{L_n \leq nx\} = -U(x) \quad \text{for } 0 \leq x \leq 2 \tag{1.51}$$

with rate function $U(x) = \int_x^2 R_2(s)\,ds$ *where* $R_2(s) = s\log(s/2) - s + 2$ *is the rate function for IID mean 2 Poisson random variables.*

To develop this theme further we state a lower tail LDP for the tagged particle in Hammersley's process. An interesting feature is that the large deviation rate functions again obey the Hopf–Lax semigroup formula, as did the limit (1.44). The assumption is that lower tail rate functions exist initially: for all $y, s \in \mathbb{R}$ the limit:

$$J_0(y, s) = -\lim_{n\to\infty} n^{-1} \log \mathbf{P}\{z^n_{[ny]}(0) \leq ns\}$$

exists and is left continuous in y for each fixed s. Define:

$$\Psi(w, r) = -\lim_{n\to\infty} n^{-1} \log \mathbf{P}\{\Gamma((0,0), (n, nw)) \leq nr\}. \tag{1.52}$$

This limit exists by superadditivity. For technical reasons a uniform tail bound is needed for the initial particle locations: there exist constants $0 < C_j < \infty$ such that:

$$\mathbf{P}\{z^n_i(0) \leq -C_1|i|\} \leq e^{-C_2|i|} \quad \text{for } i \leq -C_3 n, \text{ for large enough } n.$$

Theorem 1.8 (Seppäläinen 1998d) *The limit:*

$$J_t(x, r) = -\lim_{n\to\infty} n^{-1} \log \mathbf{P}\{z^n_{[nx]}(nt) \leq nr\}$$

exists for all $x, r \in \mathbb{R}$ *and* $t > 0$, *and is given by:*

$$J_t(x, r) = \inf_{(y,s): y \leq x,\, s \leq r} \left\{ J_0(y, s) + t\Psi\left(\frac{x-y}{t}, \frac{r-s}{t}\right) \right\}. \tag{1.53}$$

The approach of Section 1.2 can be adapted to prove the upper tail LDP (1.50) and Theorem 1.8. The stationary systems make explicit calculation again possible. Presently it is not clear how to include the lower tail LDP (1.51) in the variational framework. Hence this part requires a separate proof. Details appear in Deuschel and Zeitouni (1999) and Seppäläinen (1998d).

Further remarks. Even though Theorem 1.7 has explicit rate functions, this large deviation problem remains unfinished in an important sense. It is not understood how the system behaves to create a deviation, and it is not clear what the rate functions I and U represent. The present proofs are too indirect.

Let us illustrate through the random walk LDP (1.4)–(1.5) how a large deviation problem ideally should be understood. To create a deviation $S_n \approx nu$ with $u > v$, the entire walk behaves as a random walk with mean step u. Namely, it can be proved that the conditioned measure $\mathbf{P}(\,\cdot\,|\,S_n \geq nu)$ converges on the path space to the distribution $\mathbf{Q}^{(u)}$ of a mean u random walk. The value $I(u)$ of the rate function in (1.5) is the entropy of this measure $\mathbf{Q}^{(u)}$ relative to the original $\mathbf{P}$.

Results of the type presented in this section appear for TASEP in (Seppäläinen 1998a). The asymptotic analysis of Baik et al. (1999) and Johansson (2000) gives also LDP's for the growth models. Concentration results for a Brownian last-passage model appear in (Hambly et al. 2002) and a general discussion of deviation inequalities for growth models in the lectures of Ledoux (2007).

Acknowledgement

The author thanks M. Balázs for valuable consultation during the preparation of this article.

References

Aldous, D. and Diaconis, P. (1995). Hammersley's interacting particle process and longest increasing subsequences. *Probab. Theory Related Fields* **103**(2), 199–213.

Aldous, D. and Diaconis, P. (1999). Longest increasing subsequences: from patience sorting to the Baik-Deift-Johansson theorem. *Bull. Amer. Math. Soc. (N.S.)* **36**(4), 413–32.

Andjel, E. D. (1982). Invariant measures for the zero range processes. *Ann. Probab.* **10**(3), 525–47.

Baik, J. (2005). Limiting distribution of last passage percolation models. In *XIVth International Congress on Mathematical Physics*, pp. 339–46. World Sci. Publ., Hackensack, NJ.

Baik, J., Deift, P. and Johansson, K. (1999). On the distribution of the length of the longest increasing subsequence of random permutations. *J. Amer. Math. Soc.* **12**(4), 1119–78.

Baik, J. and Rains, E. M.(2000). Limiting distributions for a polynuclear growth model with external sources. *J. Statist. Phys.* **100**(3–4), 523–41.

Baik, J. and Suidan, T. M. (2005). A GUE central limit theorem and universality of directed first and last passage site percolation. *Int. Math. Res. Not.* (6), 325–37.

Balázs, M. (2001). Microscopic shape of shocks in a domain growth model. *J. Statist. Phys.* **105**(3–4), 511–24.

Balázs, M. (2003). Growth fluctuations in a class of deposition models. *Ann. Inst. H. Poincaré Probab. Statist.* **39**(4), 639–85.

Balázs, M., Rassoul-Agha, F. and Seppäläinen, T. (2006). The random average process and random walk in a space-time random environment in one dimension. *Comm. Math. Phys.* **266**, 499–545.

Balázs, M., Rassoul-Agha, F., Sethuraman, S. and Seppäläinen, T. (2007). Existence of the zero range process and a deposition model with superlinear growth rates. *Ann. Probab.* **35**(4), 1201–49.

Balázs, M. and Seppäläinen, T. (2007a). Exact connections between current fluctuations and the second class particle in a class of deposition models. *J. Statist. Phys.* **127**, 431–55.

Balázs, M. and Seppäläinen, T. (2007b). Order of current variance and diffusivity in the asymmetric simple exclusion process. http://front.math.ucdavis.edu/math.PR/0608400.

Bodineau, T. and Martin, J. (2005). A universality property for last-passage percolation paths close to the axis. *Electron. Comm. Probab.* **10**, 105–12 (electronic).

De Masi, A. and Presutti, E. (1991). *Mathematical methods for hydrodynamic limits*, Volume 1501 of *Lecture Notes in Mathematics*. Springer-Verlag Berlin.

Deift, P. (2000). Integrable systems and combinatorial theory. *Notices Amer. Math. Soc.* **47**(6), 631–40.

Derrida, B., Janowsky, S. A., Lebowitz, J. L. and Speer, E. R. (1993). Exact solution of the totally asymmetric simple exclusion process: shock profiles. *J. Statist. Phys.* **73**(5-6), 813–42.

Derrida, B., Lebowitz, J. L. and Speer, E. R. (1997). Shock profiles for the asymmetric simple exclusion process in one dimension. *J. Statist. Phys.* **89** (1-2), 135–67. Dedicated to Bernard Jancovici.

Deuschel, J.-D. and Zeitouni, O. (1999). On increasing subsequences of I.I.D. samples. *Combin. Probab. Comput.* **8**(3), 247–63.

Dürr, D., Goldstein, S. and Lebowitz, J. (1985). Asymptotics of particle trajectories in infinite one-dimensional systems with collisions. *Comm. Pure Appl. Math.* **38**(5), 573–97.

Durrett, R. (1988). *Lecture Notes on Particle Systems and Percolation*. The Wadsworth & Brooks/Cole Statistics/Probability Series. Wadsworth & Brooks/Cole Advanced Books & Software: Pacific Grove, CA.

Durrett, R. (2004). *Probability: Theory and Examples* (Third ed.). Duxbury Advanced Series. Brooks/Cole–Thomson: Belmont, CA.

Evans, L. C. (1998). *Partial differential equations*, Volume 19 of *Graduate Studies in Mathematics*. American Mathematical Society: Providence, RI.

Ferrari, P. A. and Fontes, L. R. G. (1994a). Current fluctuations for the asymmetric simple exclusion process. *Ann. Probab.* **22**(2), 820–32.

Ferrari, P. A. and Fontes, L. R. G. (1994b). Shock fluctuations in the asymmetric simple exclusion process. *Probab. Theory Related Fields* **99**(2), 305–19.

Ferrari, P. A. and Fontes, L. R. G. (1998). Fluctuations of a surface submitted to a random average process. *Electron. J. Probab.* **3**, no. 6, 34 pp. (electronic).

Ferrari, P. A., Fontes, L. R. G. and Kohayakawa, Y. (1994). Invariant measures for a two-species asymmetric process. *J. Statist. Phys.* **76**(5-6), 1153–77.

Ferrari, P. A. and Martin, J. B. (2007). Stationary distributions of multi-type totally asymmetric exclusion processes. *Ann. Probab.* **35**(3), 807–32.

Ferrari, P. L. (2004). Polynuclear growth on a flat substrate and edge scaling of GOE eigenvalues. *Comm. Math. Phys.* **252**(1–3), 77–109.

Fulton, W. (1997). *Young tableaux*, Volume 35 of London Mathematical Society Student Texts. Cambridge University Press: Cambridge. With applications to representation theory and geometry.

Hambly, B. and Martin, J. B. (2007). Heavy tails in last-passage percolation. *Probab. Theory Related Fields* **137**(1–2), 227–75.

Hambly, B. M., Martin, J. B. and O'Connell, N. (2002). Concentration results for a Brownian directed percolation problem. *Stochastic Process. Appl.* **102**(2), 207–20.

Hammersley, J. M. (1972). A few seedlings of research. In *Proceedings of the Sixth Berkeley Symposium on Mathematical Statistics and Probability (Univ. California, Berkeley, Calif., 1970/1971), Vol. I: Theory of statistics*, pp. 345–94. Berkeley, Calif., Univ. California Press.

Johansson, K. (2000). Shape fluctuations and random matrices. *Comm. Math. Phys.* **209**(2), 437–76.

Johansson, K. (2002). Toeplitz determinants, random growth and determinantal processes. In *Proceedings of the International Congress of Mathematicians, Vol. III (Beijing, 2002)*, pp. 53–62. Beijing, Higher Ed. Press.

Johansson, K. (2003). Discrete polynuclear growth and determinantal processes. *Comm. Math. Phys.* **242**(1–2), 277–329.

Kallenberg, O. (2002). *Foundations of Modern Probability* (Second ed.). Probability and its Applications. Springer-Verlag: New York.

Kim, J. H. (1996). On increasing subsequences of random permutations. *J. Combin. Theory Ser. A* **76**(1), 148–55.

Kipnis, C. and Landim, C. (1999). *Scaling limits of interacting particle systems*, Volume 320 of *Grundlehren der Mathematischen Wissenschaften [Fundamental Principles of Mathematical Sciences]*. Springer-Verlag: Berlin.

König, W. (2005). Orthogonal polynomial ensembles in probability theory. *Probab. Surv.* **2**, 385–447 (electronic).

Krug, J. and Spohn, H. (1992). Kinetic roughening of growing surfaces. In C. Godrèche (Ed.), *Solids Far From Equilibrium*, Collection Aléa-Saclay: Monographs and Texts in Statistical Physics, 1, pp. 117–30. Cambridge University Press: Cambridge.

Kumar, R. (2007). UW-Madison doctoral thesis under preparation.

Ledoux, M. (2007). Deviation inequalities on largest eigenvalues. Lectures for 3rd Cornell Probability Summer School.

Liggett, T. M. (1973). An infinite particle system with zero range interactions. *Ann. Probability* **1**, 240–53.

Liggett, T. M. (1985). *Interacting particle systems*, Volume 276 of *Grundlehren der Mathematischen Wissenschaften [Fundamental Principles of Mathematical Sciences]*. Springer-Verlag: New York.

Liggett, T. M. (1999). *Stochastic interacting systems: contact, voter and exclusion processes*, Volume 324 of *Grundlehren der Mathematischen Wissenschaften [Fundamental Principles of Mathematical Sciences]*. Springer-Verlag: Berlin.

Liggett, T. M. (2004). Interacting particle systems—an introduction. In *School and Conference on Probability Theory*, ICTP Lect. Notes, XVII, pp. 1–56 (electronic). Abdus Salam Int. Cent. Theoret. Phys., Trieste.

Lions, P.-L. and Nisio, M. (1982). A uniqueness result for the semigroup associated with the Hamilton-Jacobi-Bellman operator. *Proc. Japan Acad. Ser. A Math. Sci.* **58**(7), 273–76.

Logan, B. F. and Shepp, L. A. (1977). A variational problem for random Young tableaux. *Advances in Math.* **26**(2), 206–22.

Martin, J. B. (2004). Limiting shape for directed percolation models. *Ann. Probab.* **32**(4), 2908–37.

Martin, J. B. (2006). Last-passage percolation with general weight distribution. *Markov Process. Related Fields* **12**(2), 273–99.

Mehta, M. L. (2004). *Random matrices* (Third ed.), Volume 142 of *Pure and Applied Mathematics (Amsterdam)*. Elsevier/Academic Press, Amsterdam.

Prähofer, M. and Spohn, H. (2002). Scale invariance of the PNG droplet and the Airy process. *J. Statist. Phys.* **108**(5–6), 1071–06. Dedicated to David Ruelle and Yasha Sinai on the occasion of their 65th birthdays.

Prähofer, M. and Spohn, H. (2004). Exact scaling functions for one-dimensional stationary KPZ growth. *J. Statist. Phys.* **115**(1–2), 255–79.

Rassoul-Agha, F. and Seppäläinen, T. (2005). An almost sure invariance principle for random walks in a space-time random environment. *Probab. Theory Related Fields* **133**(3), 299–314.

Rezakhanlou, F. (1995). Microscopic structure of shocks in one conservation laws. *Ann. Inst. H. Poincaré Anal. Non Linéaire* **12**(2), 119–53.

Rezakhanlou, F. (2001). Continuum limit for some growth models. II. *Ann. Probab.* **29**(3), 1329–72.

Rezakhanlou, F. (2002a). A central limit theorem for the asymmetric simple exclusion process. *Ann. Inst. H. Poincaré Probab. Statist.* **38**(4), 437–64.

Rezakhanlou, F. (2002b). Continuum limit for some growth models. *Stochastic Process. Appl.* **101**(1), 1–41.

Rockafellar, R. T. (1970). *Convex Analysis*. Princeton Mathematical Series, No. 28. Princeton University Press: Princeton, N.J..

Rost, H. (1981). Nonequilibrium behaviour of a many particle process: density profile and local equilibria. *Z. Wahrsch. Verw. Gebiete* **58**(1), 41–53.

Sagan, B. E. (2001). *The Symmetric Group* (Second ed.), Volume 203 of *Graduate Texts in Mathematics*. Springer-Verlag: New York. Representations, combinatorial algorithms, and symmetric functions.

Seppäläinen, T. (1996). A microscopic model for the Burgers equation and longest increasing subsequences. *Electron. J. Probab.* **1**, no. 5, approx. 51 pp. (electronic).

Seppäläinen, T. (1997). Increasing sequences of independent points on the planar lattice. *Ann. Appl. Probab.* **7**(4), 886–98.

Seppäläinen, T. (1998a). Coupling the totally asymmetric simple exclusion process with a moving interface. *Markov Process. Related Fields* **4**(4), 593–628. I Brazilian School in Probability (Rio de Janeiro, 1997).

Seppäläinen, T. (1998b). Exact limiting shape for a simplified model of first-passage percolation on the plane. *Ann. Probab.* **26**(3), 1232–50.

Seppäläinen, T. (1998c). Hydrodynamic scaling, convex duality and asymptotic shapes of growth models. *Markov Process. Related Fields* **4**(1), 1–26.

Seppäläinen, T. (1998d). Large deviations for increasing sequences on the plane. *Probab. Theory Related Fields* **112**(2), 221–44.

Seppäläinen, T. (1999). Existence of hydrodynamics for the totally asymmetric simple K-exclusion process. *Ann. Probab.* **27**(1), 361–415.

Seppäläinen, T. (2000). Strong law of large numbers for the interface in ballistic deposition. *Ann. Inst. H. Poincaré Probab. Statist.* **36**(6), 691–736.

Seppäläinen, T. (2001). Second class particles as microscopic characteristics in totally asymmetric nearest-neighbor K-exclusion processes. *Trans. Amer. Math. Soc.* **353**(12), 4801–29 (electronic).

Seppäläinen, T. (2002). Diffusive fluctuations for one-dimensional totally asymmetric interacting random dynamics. *Comm. Math. Phys.* **229**(1), 141–82.

Seppäläinen, T. (2005). Second-order fluctuations and current across characteristic for a one-dimensional growth model of independent random walks. *Ann. Probab.* **33**(2), 759–97.

Seppäläinen, T. (2007). A growth model in multiple dimensions and the height of a random partial order. In *Asymptotics: Particles, Processes and Inverse Problems*, Volume 55 of IMS Lecture Notes–Monograph Series, pp. 204–33. Institute mathematical statistics: Ohio.

Spitzer, F. (1970). Interaction of Markov processes. *Advances in Math.* **5**, 246–290.

Spohn, H. (1991). *Large scale Dynamics of Interacting Particles*. Springer-Verlag: Berlin.

Spohn, H. (2006). Exact solutions for KPZ-type growth processes, random matrices, and equilibrium shapes of crystals. *Phys. A* **369**(1), 71–99.

Szczotka, W. and Kelly, F. P. (1990). Asymptotic stationarity of queues in series and the heavy traffic approximation. *Ann. Probab.* **18**(3), 1232–48.

Tracy, C. A. and Widom, H. (1994). Level-spacing distributions and the Airy kernel. *Comm. Math. Phys.* **159**(1), 151–74.

Varadhan, S. R. S. (2000). Lectures on hydrodynamic scaling. In *Hydrodynamic Limits and Related Topics (Toronto, ON, 1998)*, Volume 27 of *Fields Inst. Commun.*, pp. 3–40. Amer. Math. Soc: Providence, RI.

Veršik, A. M. and Kerov, S. V. (1977). Asymptotic behavior of the Plancherel measure of the symmetric group and the limit form of Young tableaux. *Dokl. Akad. Nauk SSSR* **233**(6), 1024–27.

2

THE PLEASURES AND PAINS OF STUDYING THE TWO-TYPE RICHARDSON MODEL

Maria Deijfen and Olle Häggström

Abstract

This paper provides a survey of known results and open problems for the two-type Richardson model, which is a stochastic model for competition on $\mathbb{Z}^d$. In its simplest formulation, the Richardson model describes the evolution of a single infectious entity on $\mathbb{Z}^d$, but more recently the dynamics have been extended to comprise two competing growing entities. For this version of the model, the main question is whether there is a positive probability for both entities to simultaneously grow to occupy infinite parts of the lattice, the conjecture being that the answer is yes if, and only if, the entities have the same intensity. In this paper attention focuses on the two-type model, but the most important results for the one-type version are also described.

2.1 Introduction

Consider an interacting particle system in which, at any time t, each site $x \in \mathbb{Z}^d$ is in either of two states, denoted by 0 and 1. A site in state 0 flips to a 1 at a rate proportional to the number of nearest neighbours in state 1, while a site in state 1 remains a 1 forever. We may think of sites in state 1 as being occupied by some kind of infectious entity, and the model then describes the propagation of an infection where each infected site tries to infect each of its nearest neighbours on $\mathbb{Z}^d$ at some constant rate $\lambda > 0$. More precisely, if at time t a vertex x is infected, and a neighbouring vertex y is uninfected, then, conditional on the dynamics up to time t, the probability that x infects y during a short time window $(t, t + h)$ is $\lambda h + o(h)$. Here, and in what follows, sites in state 0 and 1 are referred to as uninfected and infected respectively. This is the intuitive description of the model; a formal definition is given in Section 2.2.

The model is a special case of a class of models introduced by Richardson (1973), and is commonly referred to as the Richardson model. It has several cousins among processes from mathematical biology, see e.g., Eden (1961), Williams and Bjerknes (1972), and Bramson and Griffeath (1981). The model is also a special case of so called first-passage percolation, which was introduced in Hammersley and Welsh (1965) as a model for describing the passage of a fluid through a porous medium. In first-passage percolation, each edge of the $\mathbb{Z}^d$-lattice is equipped with a random variable representing the time it takes for the fluid to traverse the edge, and the Richardson model is obtained by letting these passage times be i.i.d. exponential.

Since an infected site stays infected forever, the set of infected sites in the Richardson model increases to cover all of $\mathbb{Z}^d$ as $t \to \infty$, and attention focuses on *how* this set grows. The main result is, roughly, that the infection grows linearly in time in each fixed direction and that, scaled by a factor $1/t$, the set of infected points converges to a non-random asymptotic shape as $t \to \infty$. To prove that the growth is linear in a fixed direction involves Kingman's subadditive ergodic theorem—in fact, the study of first-passage percolation was one of the main motivations for the development of subadditive ergodic theory. That the linear growth is preserved when all directions are considered simultaneously is stated in the celebrated shape theorem (Theorem 2.1 in Section 2.2) which originates from Richardson (1973).

Now consider the following extension of the Richardson model, known as the two-type Richardson model and introduced in Häggström and Pemantle (1998). Instead of two possible states for the sites there are three states, which we denote by 0, 1, and 2. The process then evolves in such a way that, for $i = 1, 2$, a site in state 0 flips to state i at rate λ_i times the number of nearest neighbours in state i and, once in state 1 or 2, a site remains in that state forever. Interpreting states 1 and 2 as two different types of infection and state 0 as absence of infection, this gives rise to a model describing the simultaneous spread of two infections on $\mathbb{Z}^d$. To rigorously define the model requires a bit more work; see Section 2.3. In what follows we will always assume that $d \geq 2$; the model makes sense also for $d = 1$ but the questions considered here become trivial.

A number of similar extensions of (one-type) growth models to (two-type) competition models appear in the literature; see for instance Neuhauser (1992), Durrett and Neuhauser (1997), Kordzakhia and Lalley (2005), and Ferrari et al. (2006). These tend to require somewhat different techniques, and results tend not to be easily translated from these other models to the two-type Richardson model (and vice versa). Closer to the latter are (non-Markovian) competition models based on first-passage percolation models with non-exponential passage time variables—Garet and Marchand (2005), Hoffman (2005:1), Hoffman (2005:2), Garet and Marchand (2006), Gouéré (2007), Pimentel (2007)—and a certain continuum model—Deijfen et al. (2004), Deijfen and Häggström (2004), Gouéré (2007). For ease of exposition, we shall not consider these variations even in cases where results generalize.

The behaviour of the two-type Richardson model depends on the initial configuration of the infection and on the ratio between the intensities λ_1 and λ_2 of the infection types. Assume first, for simplicity, that the model is started at time 0 from two single sites, the origin being type 1 infected and the site

$(1, 0, \ldots, 0)$ next to the origin being type 2 infected. Three different scenarios for the development of the infection are conceivable:

(a) The type 1 infection at some point completely surrounds type 2, thereby preventing type 2 from growing any further.
(b) Type 2 similarly strangles type 1.
(c) Both infections grow to occupy infinitely many sites.

It is not hard to see that, regardless of the intensities of the infections, outcomes (a) and (b) where one of the infection types at some point encloses the other have positive probability regardless of λ_1 and λ_2. This is because each of (a) and (b) can be guaranteed through some finite initial sequence of infections. In contrast, scenario (c)—referred to as infinite coexistence—can never be guaranteed from any finite sequence of infections, and is therefore harder to deal with: the main challenge is to decide whether, for given values of the parameters λ_1 and λ_2, this event (c) has positive probability or not. Intuitively, infinite coexistence represents some kind of power balance between the infections, and it seems reasonable to suspect that such a balance is possible if and only if the infections are equally powerful, that is, when $\lambda_1 = \lambda_2$. This is Conjecture 2.1 in Section 2.3, which goes back to Häggström and Pemantle (1998), and, although a lot of progress have been made, it is not yet fully proved. We describe the state of the art in Sections 2.4 and 2.5.

As mentioned above, apart from the intensities, the development of the infections in the two-type model also depends on the initial state of the model. However, if we are only interested in deciding whether the event of infinite coexistence has positive probability or not, it turns out that, as long as the initial configuration is bounded and one of the sets does not completely surround the other, the precise configuration does not matter, that is, whether infinite coexistence is possible or not is determined only by the relation between the intensities. This is proved in Deijfen and Häggström (2006a); see Theorem 2.2 in Section 2.3 for a precise formulation. Of course one may also consider unbounded initial configurations. Starting with both infection types occupying infinitely many sites means—apart from in very laboured cases—that they will both infect infinitely many sites. A more interesting case is when one of the infection types starts from an infinite set and the other one from a finite set. We may then ask if outcomes where the finite type infects infinitely many sites have positive probability or not. This question is dealt with in Deijfen and Häggström (2007), and we describe the results in Section 2.6.

The dynamics of the two-type Richardson model is deceptively simple, and yet gives rise to intriguing phenomena on a global scale. In this lies a large part of the pleasure indicated in the title. Furthermore, proofs tend to involve elegant probabilistic techniques such as coupling, subadditivity, and stochastic comparisons, adding more pleasure. The pain alluded to (which by the way is not so severe that it should dissuade readers from entering this field) comes from the stubborn resistance that some of the central problems have so far put up against attempts to solve them. A case in point is the 'only if' direction of the aforementioned Conjecture 2.1, saying that infinite coexistence starting from a bounded initial configuration does not occur when $\lambda_1 \neq \lambda_2$.

2.2 The one-type model

As mentioned in the introduction, the one-type Richardson model is equivalent to first-passage percolation with i.i.d. exponential passage times. To make the construction of the model more precise, first define $E_{\mathbb{Z}^d}$ as the edge set for the $\mathbb{Z}^d$ lattice (i.e., each pair of vertices $x, y \in \mathbb{Z}^d$ at Euclidean distance 1 from each other have an edge $e \in E_{\mathbb{Z}^d}$ connecting them). Then attach i.i.d. non-negative random variables $\{\tau(e)\}_{e \in E_{\mathbb{Z}^d}}$ to the edges. We take each $\tau(e)$ to be exponentially distributed with parameter $\lambda > 0$, meaning that:

$$P(\tau(e) > t) = \exp(-\lambda t)$$

for all $t \geq 0$. For $x, y \in \mathbb{Z}^d$, define:

$$T(x, y) = \inf_\Gamma \sum_{e \in \Gamma} \tau(e) \qquad (2.1)$$

where the infimum is over all paths Γ from x to y. The Richardson model with a given set $S_0 \subset \mathbb{Z}^d$ of initially infected sites is now defined by taking the set S_t of sites infected at time t to be:

$$S_t = \{x \in \mathbb{Z}^d : T(y, x) \leq t \text{ for some } y \in S_0\}. \qquad (2.2)$$

It turns out that the infimum in (2.1) is a.s. a minimum and attained by a unique path. That S_t grows in the way described in the introduction is a consequence of the memoryless property of the exponential distribution: for any $s, t > 0$ we have that $P(\tau(e) > s + t \,|\, \tau(e) > s) = \exp(-\lambda t)$.

Note that for any $x, y, z \in \mathbb{Z}^d$ we have $T(x, y) \leq T(x, z) + T(z, y)$. This subadditivity property opens up for the use of subadditive ergodic theory in analysing the model. To formulate the basic result, let $T(x)$ be the time when the point $x \in \mathbb{Z}^d$ is infected when starting from a single infected site at the origin and write $\mathbf{n} = (n, 0, \ldots, 0)$. It then follows from the subadditive ergodic theorem— see e.g. Kingman (1968)—that there is a constant μ_λ such that $T(\mathbf{n})/n \to \mu_\lambda$ almost surely and in L_1 as $n \to \infty$. Furthermore, a simple time scaling argument implies that $\mu_\lambda = \lambda \mu_1$ and hence, writing $\mu_1 = \mu$, we have that:

$$\lim_{n \to \infty} \frac{T(\mathbf{n})}{n} = \lambda \mu \quad \text{a.s. and in } L_1. \qquad (2.3)$$

The constant μ indicates the inverse asymptotic speed of the growth along the axes in a unit rate process and is commonly referred to as the time constant. It turns out that $\mu > 0$, so that indeed the growth is linear in time. Similarly, an analog of (2.3) holds in any direction, that is, for any $x \in \mathbb{Z}^d$, there is a constant

$\mu(x) > 0$ such that $T(nx)/n \to \lambda\mu(x)$. The infection hence grows linearly in time in each fixed direction and the asymptotic speed of the growth in a given direction is an almost sure constant.

We now turn to the shape theorem, which asserts roughly that the linear growth of the infection is preserved also when all directions are considered simultaneously. More precisely, when scaled down by a factor $1/t$ the set S_t converges to a non-random shape A. To formalize this, let $\tilde{S}_t \subset \mathbb{R}^d$ be a continuum version of S_t obtained by replacing each $x \in S_t$ by a unit cube centred at x.

Theorem 2.1 (Shape Theorem) *There is a compact convex set A such that, for any $\varepsilon > 0$, almost surely*

$$(1-\varepsilon)\lambda A \subset \frac{\tilde{S}_t}{t} \subset (1+\varepsilon)\lambda A$$

for large t.

In the above form, the shape theorem was proved in Kesten (1973) as an improvement on the original 'in probability' version, which appears already in Richardson (1973). See also Cox and Durrett (1988) and Boivin (1990) for generalizations to first-passage percolation processes with more general passage times. Results concerning fluctuations around the asymptotic shape can be found, e.g., in Kesten (1993), Alexander (1993), and Newman and Piza (1995), and, for certain other passage time distributions, in Benjamini et al. (2003).

Working out exactly, or even approximately, what the asymptotic shape A is has turned out to be difficult. Obviously the asymptotic shape inherits all symmetries of the $\mathbb{Z}^d$ lattice—invariance under reflection and permutation of coordinate hyperplanes—and it is known to be compact and convex, but, apart from this, not much is known about its qualitative features. These difficulties with characterizing the shape revolve around the fact that $\mathbb{Z}^d$ is not rotationally invariant, which causes the growth to behave differently in different directions. For instance, simulations on $\mathbb{Z}^2$ indicate that the asymptotic growth is slightly faster along the axes as compared to the diagonals. There is, however, no formal proof of this.

Before proceeding with the two-type model, we mention some work concerning properties of the time-minimizing paths in (2.1), also known as geodesics. Starting at time 0 with a single infection at the origin $\mathbf{0}$, we denote by $\Gamma(x)$ the (unique) path Γ for which the infimum $T(\mathbf{0}, x)$ in (2.1) is attained. Define $\Psi = \cup_{x \in \mathbb{Z}^d} \Gamma(x)$, making Ψ a graph specifying which paths the infection actually takes. It is not hard to see that Ψ is a tree spanning all of $\mathbb{Z}^d$ and hence there must be at least one semi-infinite self-avoiding path from the origin (called an end) in Ψ. The issue of whether Ψ has more than one end was noted by Häggström and Pemantle (1998) to be closely related to the issue of infinite coexistence in the two-type Richardson model with $\lambda_1 = \lambda_2$: such infinite coexistence happens with positive probability starting from a finite initial configuration if and only if Ψ has at least two ends with positive probability.

We say that an infinite path $x_1, x_2, \ldots$ has asymptotic direction $\hat{x}$ if $x_k/|x_k| \to \hat{x}$ as $k \to \infty$. In $d = 2$, it has been conjectured that every end in Ψ has an asymptotic direction and that, for every $x \in \mathbb{R}^2$, there is at least one end (but never more than two) in Ψ with asymptotic direction $\hat{x}$. In particular, this would mean that Ψ has uncountably many ends. For results supporting this conjecture, see Newman (1995) and Licea and Newman (1996). In the former of these papers, the conjecture is shown to be true provided an unproven but highly plausible assumption on the asymptotic shape A, saying roughly that the boundary is sufficiently smooth. See also Lalley (2003) for related work.

Results not involving unproven assumptions are comparatively weak: the coexistence result of Häggström and Pemantle (1998) shows for $d = 2$ that Ψ has at least two ends with positive probability. This was later improved to Ψ having almost surely at least $2d$ ends, by Hoffman (2005b) for $d = 2$ and by Gouéré (2007) for higher dimensions.

2.3 Introducing two types

The definition of the two-type Richardson model turns out to be simplest in the symmetric case $\lambda_1 = \lambda_2$, where the same passage time variables $\{\tau(e)\}_{e \in E_{\mathbb{Z}^d}}$ as in the one-type model can be used, with $\lambda = \lambda_1 = \lambda_2$. Suppose we start with an initial configuration (S_0^1, S_0^2) of infected sites, where $S_0^1 \subset \mathbb{Z}^d$ are those initially containing type 1 infection, and $S_0^2 \subset \mathbb{Z}^d$ are those initially containing type 2 infection. We wish to define the sets S_t^1 and S_t^2 of type 1 and type 2 infected sites for all $t > 0$. To this end, set $S_0 = S_0^1 \cup S_0^2$, and take the set $S_t = S_t^1 \cup S_t^2$ of infected sites at time t to be given by precisely the same formula (2.2) as in the one-type model; a vertex $x \in S_t$ is then assigned infection 1 or 2 depending on whether the $y \in S_0$ for which:

$$\inf\{T(y, x) : y \in S_0\}$$

is attained is in S_0^1 or S_0^2.

As in the one-type model, it is a straightforward exercise involving the memoryless property of the exponential distribution to verify that $(S_t^1, S_t^2)_{t \geq 0}$ behaves in terms of infection intensities as described in the introduction.

This construction demonstrates an intimate link between the one-type and the symmetric two-type Richardson model: if we watch the two-type model while wearing a pair of glasses preventing us from distinguishing the two types of infection, what we see behaves exactly as the one-type model. The link between infinite coexistence in the two-type model and the number of ends in the tree of infection Ψ of the one-type model claimed in the previous section is also a consequence of the construction.

In the asymmetric case $\lambda_1 \neq \lambda_2$, the two-type model is somewhat less trivial to define due to the fact that the time it takes for infection to spread along a path depends on the type of infection. There are various ways to deal with this, one being to assign, independently to each $e \in E_{\mathbb{Z}^d}$, two independent random variables $\tau_1(e)$ and $\tau_2(e)$, exponentially distributed with respective parameters

λ_1 and λ_2, representing the time it takes for infections 1 resp. 2 to traverse e. Starting from an initial configuration (S_0^1, S_0^2), we may picture the infections as spreading along the edges, taking time $\tau_1(e)$ or $\tau_2(e)$ to cross e depending on the type of infection, with the extra condition that once a vertex becomes hit by one type of infection it becomes inaccessible for the other type. This is intuitively clear, but readers with a taste for detail may require a more rigorous definition, which however we refrain from here; see Häggström and Pemantle (2000) and Deijfen and Häggström (2006a).

We now move on to describing conjectures and results. Write G_i for the event that type i infects infinitely many sites on $\mathbb{Z}^d$ and define $G = G_1 \cap G_2$. The question at issue is:

$$\text{Does } G \text{ have positive probability?} \tag{2.4}$$

A priori, the answer to this question may depend both on the initial configuration–that is, on the choice of the sets S_0^1 and S_0^2—and on the ratio between the infection intensities λ_1 and λ_2. However, it turns out that, if we are not interested in the actual value of the probability of G, but only in whether it is positive or not, then the initial configuration is basically irrelevant, as long as neither of the initial sets completely surrounds the other. This motivates the following definition.

Definition 2.1 *Let ξ_1 and ξ_2 be two disjoint finite subsets of $\mathbb{Z}^d$. We say that one of the sets (ξ_i) strangles the other (ξ_j) if there exists no infinite self-avoiding path in $\mathbb{Z}^d$ that starts at a vertex in ξ_j and that does not intersect ξ_i. The pair (ξ_1, ξ_2) is said to be* fertile *if neither of the sets strangles the other.*

Now write $P_{\xi_1, \xi_2}^{\lambda_1, \lambda_2}$ for the distribution of a two-type process started from $S_0^1 = \xi_1$ and $S_0^2 = \xi_2$. We then have the following result.

Theorem 2.2 *Let (ξ_1, ξ_2) and (ξ_1', ξ_2') be two fertile pairs of disjoint finite subsets of $\mathbb{Z}^d$, where $d \geq 2$. For all choices of (λ_1, λ_2), we have:*

$$P_{\xi_1, \xi_2}^{\lambda_1, \lambda_2}(G) > 0 \Leftrightarrow P_{\xi_1', \xi_2'}^{\lambda_1, \lambda_2}(G) > 0.$$

For connected initial sets ξ_1 and ξ_2 and $d = 2$, this result is proved in Häggström and Pemantle (1998). The idea of the proof in that case is that, by controlling the passage times of only finitely many edges, two processes started from (ξ_1, ξ_2) and (ξ_1', ξ_2') respectively can be made to evolve to the same total infected set after some finite time, with the same configuration of the infection types on the boundary. Coupling the processes from this time on and observing that the development of the infections depends only on the boundary configuration yields the result. This argument however breaks down when the initial sets are not connected (since it is then not sure that the same boundary configuration can be obtained in the two processes) and it is unclear whether it applies for $d \geq 3$. Theorem 2.2 is proved in full generality in Deijfen and Häggström (2006a), using a more involved coupling construction.

It follows from Theorem 2.2 that the answer to question (2.4) depends only on the value of the intensities λ_1 and λ_2. Hence it is sufficient to consider a process started from $S_0^1 = \mathbf{0}$ and $S_0^2 = \mathbf{1}$ (recall that $\mathbf{n} = (n, 0, \ldots, 0)$), and in this case we drop subscripts and write P^{λ_1, λ_2} for $P_{\mathbf{0},\mathbf{1}}^{\lambda_1,\lambda_2}$. Also, by time-scaling, we may assume that $\lambda_1 = 1$. The following conjecture, where we write $\lambda_2 = \lambda$, goes back to Häggström and Pemantle (1998).

Conjecture 2.1 *In any dimension $d \geq 2$, we have that $P^{1,\lambda}(G) > 0$ if and only if $\lambda = 1$.*

The conjecture is no doubt true, although proving it has turned out to be a difficult task. In fact, the 'only if' direction is not yet fully established. In the following two sections we describe the existing results for $\lambda = 1$ and $\lambda \neq 1$ respectively.

2.4 The case $\lambda = 1$

When $\lambda = 1$, we are dealing with two equally powerful infections and Conjecture 2.1 predicts a positive probability for infinite coexistence. This part of the conjecture has been proved:

Theorem 2.3 *If $\lambda = 1$, we have, for any $d \geq 2$, that $P^{1,\lambda}(G) > 0$.*

This was first proved in the special case $d = 2$ by Häggström and Pemantle (1998). That proof has a very ad hoc flavour, and heavily exploits not only the two-dimensionality but also other specific properties of the square lattice, including a lower bound on the time constant μ in (2.3) that just happens to be good enough. When eventually the result was generalized to higher dimensions, which was done simultaneously and independently by Garet and Marchand (2005) and Hoffman (2005a), much more appealing proofs were obtained. Yet another distinct proof of Theorem 2.3 was given by Deijfen and Häggström (2007). All four proofs are different, though if you inspect them for a smallest common denominator you find that they all make critical use of the fact that the time constant μ is strictly positive. We will give the Garet–Marchand proof below. In Hoffman's proof, ergodic theory is applied to the tree of infection Ψ and a so-called Busemann function which is shown to exhibit contradictory behavior under the assumption that infinite coexistence has probability zero. The Deijfen–Häggström proof proceeds via the two-type Richardson model with certain infinite initial configurations (cf. Section 2.6.)

Proof of Theorem 2.3 The following argument is due to Garet and Marchand (2005), though our presentation follows more closely the proof of an analogous result in a continuum setting in Deijfen and Häggström (2004)—a paper that, despite the publication dates, was preceded by and also heavily influenced by Garet and Marchand (2005).

Fix a small $\varepsilon > 0$. By Theorem 2.2, we are free to choose any finite starting configuration we want, and here it turns out convenient to begin with a single type 1 infection at the origin $\mathbf{0}$, and a single type 2 infection at a vertex

$\mathbf{n} = (n, 0, \ldots, 0)$, where n is large enough so that:
(i) $E[T(\mathbf{0}, \mathbf{n})] \leq (1+\varepsilon)n\mu$, and
(ii) $P(T(\mathbf{0}, \mathbf{n}) < (1-\varepsilon)n\mu) < \varepsilon$;
note that both (i) and (ii) hold for n large enough due to the asymptotic speed result (2.3). The reader may easily check, for later reference, that (i) and (ii) together with the nonnegativity of $T(\mathbf{0}, \mathbf{n})$ imply for any event B with $P(B) = \alpha$ that:

$$E[T(\mathbf{0}, \mathbf{n}) \mid \neg B] \leq \left(1 + \frac{3\varepsilon}{1-\alpha}\right) n\mu. \tag{2.5}$$

Next comes an important telescoping idea: for any positive integer k we have:

$$E[T(\mathbf{0}, k\mathbf{n})] = E[T(\mathbf{0}, \mathbf{n})] + E[T(\mathbf{0}, 2\mathbf{n}) - T(\mathbf{0}, \mathbf{n})] + E[T(\mathbf{0}, 3\mathbf{n}) - T(\mathbf{0}, 2\mathbf{n})]$$
$$+ \cdots + E[T(\mathbf{0}, k\mathbf{n}) - T(\mathbf{0}, (k-1)\mathbf{n})].$$

Since $\lim_{k \to \infty} k^{-1} E[T(\mathbf{0}, k\mathbf{n})] = n\mu$, there must exist arbitrarily large k such that:

$$E[T(\mathbf{0}, (k+1)\mathbf{n}) - T(\mathbf{0}, k\mathbf{n})] \geq (1-\varepsilon)n\mu.$$

By taking $\mathbf{m} = k\mathbf{n}$, and by translation and reflection invariance, we may deduce that:

$$E[T(\mathbf{n}, -\mathbf{m}) - T(\mathbf{0}, -\mathbf{m})] \geq (1-\varepsilon)n\mu \tag{2.6}$$

for some arbitrarily large m. We will pick such an m; how large will soon be specified.

The goal is to show that $P(G) > 0$, so we may assume for contradiction that $P(G) = 0$. By symmetry of the initial configuration, we then have that $P(G_1) = P(G_2) = \frac{1}{2}$. This implies that:

$$\lim_{m \to \infty} P(-\mathbf{m} \text{ gets infected by type } 2) = \lim_{m \to \infty} P(T(\mathbf{n}, -\mathbf{m}) < T(\mathbf{0}, -\mathbf{m})) = \frac{1}{2}$$

so let us pick m in such a way that:

$$P(T(\mathbf{n}, -\mathbf{m}) < T(\mathbf{0}, -\mathbf{m})) \geq \frac{1}{4} \tag{2.7}$$

while also (2.6) holds. Write B for the event in (2.7). The expectation $E[T(\mathbf{n}, -\mathbf{m}) - T(\mathbf{0}, -\mathbf{m})]$ may be decomposed as:

$$\begin{aligned}
E[T(\mathbf{n}, -\mathbf{m}) - T(\mathbf{0}, -\mathbf{m})] &= E[T(\mathbf{n}, -\mathbf{m}) - T(\mathbf{0}, -\mathbf{m}) \mid B] P(B) \\
&\quad + E[T(\mathbf{n}, -\mathbf{m}) - T(\mathbf{0}, -\mathbf{m}) \mid \neg B] P(\neg B) \\
&\leq E[T(\mathbf{n}, -\mathbf{m}) - T(\mathbf{0}, -\mathbf{m}) \mid \neg B] P(\neg B) \\
&\leq \frac{3}{4} E[T(\mathbf{n}, -\mathbf{m}) - T(\mathbf{0}, -\mathbf{m}) \mid \neg B] \\
&\leq \frac{3}{4} E[T(\mathbf{n}, \mathbf{0}) \mid \neg B] \\
&\leq \frac{3}{4}(1 + 4\varepsilon)n\mu
\end{aligned}$$

where the second-to-last inequality is due to the triangle inequality $T(\mathbf{n}, -\mathbf{m}) \leq T(\mathbf{n}, \mathbf{0}) + T(\mathbf{0}, -\mathbf{m})$, and the last one uses (2.5). For small ε, this contradicts (2.6), so the proof is complete. □

2.5 The case $\lambda \neq 1$

Let us move on to the case when $\lambda \neq 1$, that is, when the type 2 infection has a different intensity than type 1. It then seems unlikely that the kind of equilibrium which is necessary for infinite coexistence to occur would persist in the long run. However, this part of Conjecture 2.1 is not proved. The best result to date is the following theorem from Häggström and Pemantle (2000).

Theorem 2.4 *For any $d \geq 2$, we have $P^{1,\lambda}(G) = 0$ for all but at most countably many values of λ.*

We leave it to the reader to decide whether this is a very strong or a very weak result: it is very strong in the sense of showing that infinite coexistence has probability 0 for (Lebesgue)-almost all λ, but very weak in the sense that infinite coexistence is not ruled out for any given λ.

The result may seem a bit peculiar at first sight and we will spend some time explaining where it comes from and where the difficulties arise when one tries to strengthen it. Indeed, as formulated in Conjecture 2.1, the belief is that the set $\{\lambda : P^{1,\lambda}(G) > 0\}$ in fact consists of the single point $\lambda = 1$, but Theorem 2.4 only asserts that the set is countable.

First note that, by time-scaling and symmetry, we have $P^{1,\lambda}(G) = P^{1,1/\lambda}(G)$ and hence it is enough to consider $\lambda \leq 1$. An essential ingredient in the proof of Theorem 2.4 is a coupling of the two-type processes $\{P^{1,\lambda}\}_{\lambda \in (0,1]}$ obtained by associating two independent exponential mean 1 variables $\tau_1(e)$ and $\tau_2'(e)$ to each edge $e \in \mathbb{Z}^d$ and then letting the type 2 passage time at parameter value λ be given by $\tau_2(e) = \lambda^{-1}\tau_2'(e)$ and the type 1 time (for any λ) by $\tau_1(e)$. Write Q for the probability measure underlying this coupling and let G^λ be the event that infinite coexistence occurs at parameter value λ. Theorem 2.4 is obtained by showing that:

$$\text{with } Q\text{-probability 1 the event } G^\lambda \text{ occurs} \atop \text{for at most one value of } \lambda \in (0, 1]. \quad (2.8)$$

Hence, $Q(G^\lambda)$ can be positive for at most countably many λ, and Theorem 2.4 then follows by noting that $P^{1,\lambda}(G) = Q(G^\lambda)$.

But why is (2.8) true? Let G_i^λ be the event that the type i infection grows unboundedly at parameter value λ. Then the coupling defining Q can be shown to be monotone in the sense that G_1^λ is decreasing in λ – that is, if G_1^λ occurs then $G_1^{\lambda'}$ occurs for all $\lambda' < \lambda$ as well – and G_2^λ is increasing in λ. This kind of monotonicity of the coupling is crucial for proving (2.8), as is the following result, which asserts that, on the event that the type 2 infection survives, the total infected set in a two-type process with distribution $P^{1,\lambda}$, where $\lambda < 1$, grows to a first approximation like a one-type process with intensity λ. More precisely, the speed of the growth in the two-type process is determined by the weaker type 2 infection type. We take $\tilde{S}_t^i$ to denote the union of all unit cubes centered at points in S_t^i and A is the limiting shape for a one-type process with rate 1.

Theorem 2.5 *Consider a two-type process with distribution $P^{1,\lambda}$ for some $\lambda \leq 1$. On the event G_2 we have, for any $\varepsilon > 0$, that almost surely:*

$$(1-\varepsilon)\lambda A \subset \frac{\tilde{S}_t^1 \cup \tilde{S}_t^2}{t} \subset (1+\varepsilon)\lambda A$$

for large t.

Theorem 2.4 follows readily from this result and the monotonicity properties of the coupling Q. Indeed, fix $\varepsilon > 0$ and suppose G^λ occurs. Then Theorem 2.5 guarantees that on level λ the type 1 infection is eventually contained in $(1+\varepsilon)\lambda t A$, a conclusion that extends to all $\lambda' > \lambda$, because increasing the type 2 infection rate does not help type 1. On the other hand, for any $\lambda' > \lambda$ we get on level λ' that the union of the two infections will—again by Theorem 2.5— eventually contain $(1-\varepsilon)\lambda' t A$, so by taking ε sufficiently small we see that the type 1 infection is strangled on level λ', implying (2.8), and Theorem 2.4 follows.

We will not prove Theorem 2.5, but mention that the hard work in proving it lies in establishing a certain key result (Proposition 2.2 in Häggström and Pemantle 2000) that asserts that if the strong infection type reaches outside $(1+\varepsilon)\lambda t A$ infinitely often, then the weak type is doomed. The proof of this uses geometrical arguments, the most important ingredient being a certain spiral construction, emanating from the part of the strong of infection reaching beyond $(1+\varepsilon)\lambda t A$, and designed to allow the strong type to completely surround the weak type before the weak type catches up from inside.

How would one go about strengthening Theorem 2.4 and ruling out infinite coexistence for all $\lambda \neq 1$? One possibility would be to try to derive a contradiction with Theorem 2.5 from the assumption that the strong infection type grows unboundedly. For instance, intuitively it seems likely that the strong type occupying a positive fraction of the boundary of the infected set would cause the speed of the growth to exceed the speed prescribed by the weak infection type. This type of argument is indeed used in Garet and Marchand (2007) to show, for $d = 2$, that on the event of infinite coexistence the fraction of infected sites occupied by the strong infection will tend to 0 as $t \to \infty$. This feels like a strong indication that infinite coexistence does not happen.

Another approach to strengthening Theorem 2.4 in order to obtain the only-if direction of Conjecture 2.1 is based on the observation that, since coexistence represents a power balance between the infections, it is reasonable to expect that

$P^{1,\lambda}(G)$ decreases as λ moves away from 1. We may formulate that intuition as a conjecture:

Conjecture 2.2 *For the two-type Richardson model on $\mathbb{Z}^d$ with $d \geq 2$, we have, for $\lambda < \lambda' \in (0,1]$, that $P^{1,\lambda}(G) \leq P^{1,\lambda'}(G)$.*

A confirmation of this conjecture would, in combination with Theorem 2.4, clearly establish the only-if direction of Conjecture 2.1: If $P^{1,\lambda}(G) > 0$ for some $\lambda < 1$, then, according to Conjecture 2.2, we would have $P^{1,\lambda'}(G) > 0$ for all $\lambda' \in (\lambda, 1]$ as well. But the interval $(\lambda, 1]$ is uncountable, yielding a contradiction to Theorem 2.4.

Although Conjecture 2.2 might seem close to obvious, it has turned out to be very difficult to prove. A natural first attempt would be to use coupling. Consider for instance the coupling Q described above. As pointed out, the events G_1^λ and G_2^λ that the individual infections grow unboundedly at parameter value λ are then monotone in λ, but one of them is increasing and the other is decreasing, so monotonicity of their intersection G^λ does not follow. Hence more sophisticated arguments are needed.

Observing how our colleagues react during seminars and corridor chat, we have noted that it is very tempting to go about trying to prove Conjecture 2.2 by abstract and 'easy' arguments, here meaning arguments that do not involve any specifics about the geometry or graph structure of $\mathbb{Z}^d$. To warn against such attempts, Deijfen and Häggström (2006b) constructed graphs on which the two-type Richardson model fails to exhibit the monotonicity behaviour predicted in Conjecture 2.2. Let us briefly explain the results.

The dynamics of the two-type Richardson model can of course be defined on graphs other than the $\mathbb{Z}^d$ lattice. For a graph $\mathcal{G}$, write $\text{Coex}(\mathcal{G})$ for the set of all $\lambda \geq 1$ such that there exists a finite initial configuration (ξ_1, ξ_2) for which the two-type Richardson model with infection intensities 1 and λ started from (ξ_1, ξ_2) yields infinite coexistence with positive probability. Note that, by time-scaling and interchange of the infections, coexistence is possible at parameter value λ if and only if it is possible at λ^{-1}, so no information is lost by restricting to $\lambda \geq 1$. In Deijfen and Häggström (2006b) examples of graphs $\mathcal{G}$ are given that demonstrate that, among others, the following kinds of coexistence sets $\text{Coex}(\mathcal{G})$ are possible:

(i) $\text{Coex}(\mathcal{G})$ may be an interval (a, b) with $1 < a < b$.
(ii) For any positive integer k the set $\text{Coex}(\mathcal{G})$ may consist of exactly k points.
(iii) $\text{Coex}(\mathcal{G})$ may be countably infinite.

All these phenomena show that the monotonicity suggested in Conjecture 2.2 fails for general graphs. However, a reasonable guess is that Conjecture 2.2 is true on transitive graphs. Indeed, all counterexamples provided by Deijfen and Häggström are highly nonsymmetric (one might even say ugly) with certain parts of the graph being designed specifically with propagation of type 1 in mind, while other parts are meant for type 2. We omit the details.

2.6 Unbounded initial configurations

Let us now go back to the $\mathbb{Z}^d$ setting and describe some results from our most recent paper, Deijfen and Häggström (2007), concerning the two-type model with unbounded initial configurations. Roughly, the model will be started from configurations where one of the infections occupies a single site in an infinite 'sea' of the other type. The dynamic is as before and also the question at issue is the same: can both infection types simultaneously infect infinitely many sites? With both types initially occupying infinitely many sites the answer is (apart from in particularly silly cases) obviously yes, so we will focus on configurations where type 1 starts with infinitely many sites and type 2 with finitely many—for simplicity only one. The question then becomes whether type 2 is able to survive.

To describe the configurations in more detail, write $(x_1, \ldots, x_d)$ for the coordinates of a point $x \in \mathbb{Z}^d$, and define $\mathcal{H} = \{x : x_1 = 0\}$ and $\mathcal{L} = \{x : x_1 \leq 0 \text{ and } x_i = 0 \text{ for } i = 2, \ldots, d\}$. We will consider the following starting configurations.

$$\begin{aligned} I(\mathcal{H}) &: \text{all points in } \mathcal{H} \backslash \{\mathbf{0}\} \text{ are type 1 infected and} \\ &\quad \mathbf{0} \text{ is type 2 infected, and} \\ I(\mathcal{L}) &: \text{all points in } \mathcal{L} \backslash \{\mathbf{0}\} \text{ are type 1 infected and} \\ &\quad \mathbf{0} \text{ is type 2 infected.} \end{aligned} \quad (2.9)$$

Interestingly, it turns out that the set of parameter values for which type 2 is able to grow indefinitely is slightly different for these two configurations. First note that, as before, we may restrict to the case $\lambda_1 = 1$. Write $P^{1,\lambda}_{\mathcal{H},\mathbf{0}}$ and $P^{1,\lambda}_{\mathcal{L},\mathbf{0}}$ for the distribution of the process started from $I(\mathcal{H})$ and $I(\mathcal{L})$ respectively and with type 2 intensity λ. The following result, where G_2 denotes the event that type 2 grows unboundedly, is proved in Deijfen and Häggström (2007).

Theorem 2.6 *For the two-type Richardson model in $d \geq 2$ dimensions, we have:*

(a) $P^{1,\lambda}_{\mathcal{H},\mathbf{0}}(G_2) > 0$ *if and only if $\lambda > 1$;*

(b) $P^{1,\lambda}_{\mathcal{L},\mathbf{0}}(G_2) > 0$ *if and only if $\lambda \geq 1$.*

In other words, a strictly stronger type 2 infection will be able to survive in both configurations, but, when the infections have the same intensity, type 2 can survive only in the configuration $I(\mathcal{L})$.

The proof of the if-direction of Theorem 2.6 (a) is based on a lemma stating roughly that the speed of a hampered one-type process, living only inside a tube which is bounded in all directions except one, is close to the speed of an unhampered process when the tube is large. For a two-type process started from $I(\mathcal{H})$, this lemma can be used to show that, if the strong type 2 infection at the origin is successful in the beginning of the time course, it will take off along the x_1-axis and grow faster than the surrounding type 1 infection inside a tube around the x_1-axis, thereby escaping eradication. The same scenario—that the type 2 infection rushes away along the x_1-axis—can, by different means, be proved to have positive probability in a process with $\lambda = 1$ started from $I(\mathcal{L})$.

Infinite growth for type 2 when $\lambda < 1$ is ruled out by the key proposition from Häggström and Pemantle (2000) mentioned in Section 3. Proving that type 2 cannot survive in a process with $\lambda = 1$ started from $I(\mathcal{H})$ is the most tricky part. The idea is basically to divide $\mathbb{Z}^d$ in different levels, the l-th level being all sites with x_1-coordinate l, and then show that the expected number of type 2 infected sites at level l is constant and equal to 1. It then follows from a certain comparison with a one-type process on each level combined with an application of Levy's 0-1 law that the number of type 2 infected sites at the l-th level converges almost surely to 0 as $l \to \infty$.

Finally we mention a question formulated by Itai Benjamini as well as by an anonymous referee of Deijfen and Häggström (2007). We have seen that, when $\lambda = 1$, the type 2 infection at the origin can grow unboundedly from $I(\mathcal{L})$ but not from $I(\mathcal{H})$. It is then natural to ask what happens if we interpolate between these two configurations. More precisely, instead of letting type 1 occupy only the negative x_1-axis (as in $I(\mathcal{L})$), we let it occupy a cone of constant slope around the same axis. The question then is what the critical slope is for this cone such that there is a positive probability for type 2 to grow unboundedly. That type 2 cannot survive when the cone occupies the whole left half-space follows from Theorem 2.6, as this situation is equivalent to starting the process from $I(\mathcal{H})$. It seems likely, as suggested by Itai Benjamini, that this is actually also the critical case, that is, infinite growth for type 2 most likely have positive probability for any smaller type 1 cone. This, however, remains to be proved.

References

Alexander, K. (1993): A note on some rates of convergence in first-passage percolation, *Ann. Appl. Probab.* **3**, 81–90.

Benjamini, I., Kalai, G. and Schramm, O. (2003): First passage percolation has sublinear distance variation, *Ann. Probab.* **31**, 1970–78.

Bramson, M. and Griffeath, D. (1981): On the Williams–Bjerknes tumour growth model I, *Ann. Probab.* **9**, 173–85.

Boivin, D. (1990): First passage percolation: the stationary case. *Probab. Theory Related Fields* **86**(4), 491–99.

Cox, J.T. and Durrett, R. (1981): Some limit theorems for percolation processes with necessary and sufficient conditions, *Ann. Probab.* **9**, 583–603.

Cox, J.T. and Durrett, R. (1988): Limit theorems for the spread of epidemics and forest fires. *Stochastic Process. Appl.* **30**(2), 171–91.

Deijfen, M. and Häggström, O. (2004): Coexistence in a two-type continuum growth model, *Adv. Appl. Probab.* **36**, 973–80.

Deijfen, M. and Häggström, O. (2006a): The initial configuration is irrelevant for the possibility of mutual unbounded growth in the two-type Richardson model, *Comb. Probab. Computing* **15**, 345–53.

Deijfen, M. and Häggström, O. (2006b): Nonmonotonic coexistence regions for the two-type Richardson model on graphs, *Electr. J. Probab.* **11**, 331–44.

Deijfen, M. and Häggström, O. (2007): The two-type Richardson model with unbounded initial configurations, *Ann. Appl. Probab.* **17**, 1639–56

Deijfen, M., Häggström, O. and Bagley, J. (2004): A stochastic model for competing growth on R^d, *Markov Proc. Relat. Fields* **10**, 217–48.

Durrett, R. (1988): *Lecture Notes on Particle Systems and Percolation*, Wadsworth & Brooks/Cole: Pacific Crare C.A.

Durrett, R. and Neuhauser, C. (1997): Coexistence results for some competition models *Ann. Appl. Probab.* **7**, 10–45.

Eden, M. (1961): A two-dimensional growth process, *Proceedings of the 4th Berkeley Symposium on Mathematical Statistics and Probability* vol. **IV**, 223–39, University of California Press: California.

Ferrari, P., Martin, J. and Pimentel, L. (2006): Roughening and inclination of competition interfaces, *Phys Rev E* **73**, 031602 (4 p).

Garet, O. and Marchand, R. (2005): Coexistence in two-type first-passage percolation models, *Ann. Appl. Probab.* **15**, 298–330.

Garet, O. and Marchand, R. (2006): Competition between growths governed by Bernoulli percolation, *Markov Proc. Relat. Fields* **12**, 695–734.

Garet, O. and Marchand, R. (2007): First-passage competition with different speeds: positive density for both species is impossible, preprint, ArXiV math.PR/0608667.

Gouéré, J.-B. (2007): Shape of territories in some competing growth models, *Ann. Appl. Probab.* **17**, 1273–1305.

Häggström, O. and Pemantle, R. (1998): First passage percolation and a model for competing spatial growth, *J. Appl. Probab.* **35**, 683–692.

Häggström, O. and Pemantle, R. (2000): Absence of mutual unbounded growth for almost all parameter values in the two-type Richardson model, *Stoch. Proc. Appl.* **90**, 207–22.

Hammersley, J. and Welsh D. (1965): First passage percolation, subadditive processes, stochastic networks and generalized renewal theory, *1965 Proc. Internat. Res. Semin., Statist. Lab., Univ. California, Berkeley*, 61–110, Springer: New York.

Hoffman, C. (2005a): Coexistence for Richardson type competing spatial growth models, *Ann. Appl. Probab.* **15**, 739–47.

Hoffman, C. (2005b): Geodesics in first passage percolation, preprint, ArXiV math.PR/0508114.

Kesten, H. (1973): Discussion contribution, *Ann. Probab.* **1**, 903.

Kesten, H. (1993): On the speed of convergence in first-passage percolation, *Ann. Appl. Probab.* **3**, 296–338.

Kingman, J.F.C. (1968): The ergodic theory of subadditive stochastic processes, *J. Roy. Statist. Soc. Ser. B* **30**, 499–510.

Kordzakhia, G. and Lalley, S. (2005): A two-species competition model on Z^d, *Stoch. Proc. Appl.* **115**, 781–96.

Lalley, S. (2003): Strict convexity of the limit shape in first-passage percolation, *Electr. Comm. Probab.* **8**, 135–41.

Licea, C. and Newman, C. (1996): Geodesics in two-dimensional first-passage percolation, *Ann. Probab.* **24**, 399–410.

Neuhauser, C. (1992): Ergodic theorems for the multitype contact process, *Probab. Theory Relat. Fields* **91**, 467–506.

Newman, C. (1995): A surface view of first passage percolation, *Proc. Int. Congr. Mathematicians* **1,2** (Zurich 1994), 1017–23.

Newman, C. and Piza, M. (1995): Divergence of shape fluctuations in two dimensions, *Ann. Probab.* **23**, 977–1005.

Pimentel, L. (2007): Multitype shape theorems for first passage percolation models, *Adv. Appl. Probab.* **39**, 53–76.

Richardson, D. (1973): Random growth in a tessellation, *Proc. Cambridge Phil. Soc.* **74**, 515–528.

Williams, T. and Bjerknes R. (1972): Stochastic model for abnormal clone spread through epithelial basal layer, *Nature* **236**, 19–21.

3
BALLISTIC PHASE OF SELF-INTERACTING RANDOM WALKS

Dmitry Ioffe and Yvan Velenik

Abstract

We explain a unified approach to a study of ballistic phase for a large family of self-interacting random walks with a drift and self-interacting polymers with an external stretching force. The approach is based on a recent version of the Ornstein–Zernike theory developed in (Campanino, Ioffe, and Velenik, 2003; Campanino, Ioffe, and Velenik, 2004; Campanino, Ioffe, and Velenik, 2007). It leads to local limit results for various observables (e.g., displacement of the end-point or number of hits of a fixed finite pattern) on paths of n-step walks (polymers) on all possible deviation scales from CLT to LD. The class of models, which display ballistic phase in the 'universality class' discussed in the paper, includes self-avoiding walks, Domb–Joyce model, random walks in an annealed random potential, reinforced polymers, and weakly reinforced random walks.

3.1 Introduction and results

Self-interacting polymers and random walks have received much attention by both physicists and probabilists. As the resulting models are non-Markovian, their analysis requires new techniques, and many basic questions remain open.

Chayes 2007 recently pointed out that the presence of an arbitrary nonzero drift turns a self-avoiding walk on $\mathbb{Z}^d$ into a 'massive' model, and used this observation to prove the existence of a positive speed, using renewal techniques of Ornstein–Zernike-type, in the spirit of (Chayes and Chayes 1986). A similar approach was used by Flury (2007a) in order to study non-directed polymers in a quenched random environment, under the influence of a drift: after suitable reduction to an annealed setting (at the cost of considering two interacting copies of the polymer), he used the same Ornstein–Zernike approach in order to prove equality of quenched and annealed free energies. Notice that introducing a drift in such polymer models is very reasonable from the point of view of physics, as it is equivalent to pulling the polymer with a constant force at one endpoint, the other being pinned.

Ornstein–Zernike renewal techniques have known considerable progress in recent years, allowing nonperturbative control of percolation connectivities (Campanino and Ioffe 2002), Ising correlation functions (Campanino et al. 2003,

2004) and more generally connectivities in random cluster models (Campanino et al. 2007), in the whole subcritical regime.

The purpose of this note is to point out that the approach of (Campanino et al. 2003, 2004, 2007) allows a unified and a detailed treatment of self-interacting polymers or self-interacting walks in ballistic regime. In particular, we shall explain how to use it in order to prove that:

- For self-interacting polymers with a repulsive interaction (see below), a positive drift gives rise to a ballistic phase, in a very strong sense (local limit theorem for the endpoint).

- For self-interacting polymers with an attractive interaction (see below), there is a sharp transition between a confined phase and a ballistic phase as the drift increases; in the latter regime, we establish again a local limit theorem for the endpoint.

- These behaviours are stable under small perturbations (in a sense defined below). In addition to allowing us to consider mixed attractive/repulsive models, this also makes it possible to prove local limit theorems for some models of reinforced random walk with drift, strengthening the recent results obtained by van der Hofstad and Holmes (2007) using the lace expansion.

- In all cases mentioned, the local limit theorem can be readily complemented by a functional CLT for the ballistic path.

- In the ballistic phase, these results can be complemented by sharp local limit theorems for general local observables of the path. This permits for example a strong version of Kesten's pattern theorem (Kesten 1963) in this regime.

We would also like to remark that the approach could be pushed to analyse the interaction of paths, e.g., as considered in (Flury 2007a), in order to analyse random polymers in quenched environment.

For the sake of readability, we make some simplifying assumptions in the sequel (for example, on the nearest neighbour nature of paths or on the form of the local observables), but the technique is flexible enough to be applicable in numerous other situations.

3.1.1 Class of models

For each nearest neighbour path $\gamma = (\gamma(0), \ldots, \gamma(n))$ on $\mathbb{Z}^d$ define:

- Length $|\gamma| = n$ and displacement $D(\gamma) = \gamma(n) - \gamma(0)$.
- Local times: for a given a site $x \in \mathbb{Z}^d$ set:

$$l_x(\gamma) = \sum_{k=0}^{|\gamma|} 1_{\{\gamma(k) = x\}}.$$

Similarly, one defines local times $l_b(\gamma)$ for either un-oriented or oriented nearest neighbour bonds b.

- Potential $\Phi(\gamma)$: in this paper we shall concentrate on potentials Φ which depend only on bond or edge local times of γ. That is:

$$\Phi(\gamma) = \sum_{x \in \mathbb{Z}^d} \phi\left(l_x(\gamma)\right) \quad \text{or} \quad \Phi(\gamma) = \sum_b \phi\left(l_b(\gamma)\right). \tag{3.1}$$

To every $h \in \mathbb{R}^d$ and $\lambda \in \mathbb{R}$, we associate grandcanonical weights defined by:

$$W_{h,\lambda}(\gamma) = e^{-\Phi(\gamma)+(h,D(\gamma))-\lambda|\gamma|}, \tag{3.2}$$

where $(\cdot, \cdot)$ is the usual scalar product on $\mathbb{R}^d$. If h or λ equal to zero, then the corresponding entry is dropped from the notation.

The important assumptions are those imposed on the potential Φ in (3.1). In all the models we are going to consider:

(N) $\phi(0) = 0$ and ϕ is a nonnegative and nondecreasing function on $\mathbb{N}$.

Furthermore, we shall distinguish between the repulsive case:

(R) For all $l, m \in \mathbb{N}$, $\phi(l+m) \geq \phi(l) + \phi(m)$;

and the attractive case:

(A) For all $l, m \in \mathbb{N}$, $\phi(l+m) \leq \phi(l) + \phi(m)$.

In the attractive setup, there is no loss of generality in restricting our attention to sublinear potentials:

(SL) $\lim_{n \to \infty} \phi(n)/n = 0$.

Finally, we shall consider small perturbations of the above two pure (attractive or repulsive) cases.

Whichever model we are going to consider, the main object of our study is the canonical path measure:

$$\mathbb{P}_n^h(\cdot) = \frac{W_h(\cdot)}{Z_n^h},$$

where:

$$Z_n^h = \sum_{\gamma(0)=0, |\gamma|=n} W_h(\gamma).$$

Note that in the attractive case condition **(SL)** simply boils down to a choice of normalization for the partition functions Z_n^h.

3.1.2 Collection of examples

Two main examples of repulsive interactions are:

1. Site or bond self-avoiding walks with $\phi(l) = \infty \cdot \delta\{l > 1\}$.
2. Domb–Joyce model with $\phi(l) = \beta l(l-1)/2$.

Two examples of attractive interactions are:
1. Annealed random walks in random potential,
$$\phi(l) = -\log \mathbb{E}e^{lV},$$
where V is a *nonpositive* random variable.
2. Edge or site reinforced polymers with
$$\phi(l) = \sum_{1}^{l} \beta_k,$$
where $\{\beta_k\}$ is a *nonnegative* and *nonincreasing* sequence.

3.1.3 *Connectivity constants, Lyapunov exponents and Wulff shapes*
The connectivity constant:
$$\lambda_0 \triangleq \lim_{n\to\infty} \frac{\log Z_n}{n}$$
is well defined and finite in both repulsive and attractive cases. Indeed, since ϕ is nonnegative and nondecreasing:
$$e^{-\phi(1)n} \leq Z_n \leq (2d)^n.$$
On the other hand, $Z_{n+m} \leq Z_n Z_m$ in the case of repulsion, whereas $Z_{n+m} \geq Z_n Z_m$ in the attractive case.

As we shall check in the Appendix it is always the situation in the attractive case (under the normalization (**SL**)) that:
$$\lambda_0 = \log(2d). \tag{3.3}$$
There are two slightly different approaches to Lyapunov exponents which we are going to employ: for every $x \in \mathbb{Z}^d$, let $\mathcal{D}_x$ be the family of nearest neighbour paths from 0 to x. The family $\mathcal{H}_x$ comprises those paths from $\mathcal{D}_x$ which hit x for the first time. Formally:
$$\mathcal{H}_x = \{\gamma \in \mathcal{D}_x : l_x(\gamma) = 1\}.$$
Define:
$$\mathbb{D}_\lambda(x) = \sum_{\gamma \in \mathcal{D}_x} W_\lambda(\gamma) \quad \text{and} \quad \mathbb{H}_\lambda(x) = \sum_{\gamma \in \mathcal{H}_x} W_\lambda(\gamma), \tag{3.4}$$
where, as before, $W_\lambda(\gamma) = e^{-\Phi(\gamma) - \lambda|\gamma|}$. Notice that:
$$\mathbb{H}_\lambda(x) \leq \mathbb{D}_\lambda(x) \leq \sum_{n \geq \|x\|_1} Z_n e^{-\lambda n}, \tag{3.5}$$
which is converging for every $\lambda > \lambda_0$.

For $\lambda > \lambda_0$ define now:
$$\xi_\lambda(x) = -\lim_{M\to\infty} \frac{1}{M} \log \mathbb{H}_\lambda(\lfloor Mx \rfloor) = -\lim_{M\to\infty} \frac{1}{M} \log \mathbb{D}_\lambda(\lfloor Mx \rfloor). \tag{3.6}$$

As we shall recall in the Appendix:

Proposition 3.1 *For every $\lambda > \lambda_0$ the function ξ_λ in (3.6) is well-defined. Furthermore, ξ_λ is an equivalent norm on $\mathbb{R}^d$, and*

$$\max_{x \in \mathbb{S}^{d-1}} \xi_\lambda(x) \leq c(d) \min_{x \in \mathbb{S}^{d-1}} \xi_\lambda(x). \tag{3.7}$$

Finally, $\xi_{\lambda_0} \triangleq \lim_{\lambda \downarrow \lambda_0} \xi_\lambda$ is identically zero in the repulsive case (**R**), *whereas it is strictly positive in the attractive case* (**A**) + (**SL**).

For every $\lambda \geq \lambda_0$ define the Wulff shape:

$$\mathbf{K}_\lambda = \left\{ h \in \mathbb{R}^d : (h, x) \leq \xi_\lambda(x) \,\forall\, x \in \mathbb{R}^d \right\}. \tag{3.8}$$

The name Wulff shape is inherited from continuum mechanics, where $\mathbf{K}_\lambda$ is the equilibrium crystal shape once ξ_λ is interpreted to be a surface tension. Alternatively, one can describe $\mathbf{K}_\lambda$ in terms of polar norms, as was done, e.g., in (Flury 2007b): introducing the polar norm:

$$\xi_\lambda^*(h) = \max_{x \neq 0} \frac{(h, x)}{\xi_\lambda(x)} = \max_{\xi_\lambda(x)=1} (h, x).$$

we see that $\mathbf{K}_\lambda$ can be identified with the corresponding unit ball:

$$\mathbf{K}_\lambda = \{ h : \xi_\lambda^*(h) \leq 1 \}.$$

This way or another, in view of Proposition 3.1, the limiting shape $\mathbf{K}_{\lambda_0}$ has non-empty interior in the attractive case, whereas $\mathbf{K}_{\lambda_0} = \{0\}$ in the repulsive case.

3.1.4 *Main result*

Let $h \in \mathbb{R}^d \setminus \{0\}$. Assume that the potential Φ in (3.1) satisfies condition (**N**). In all the cases under consideration the displacement per step $D(\gamma)/n$ satisfies under $\mathbb{P}_n^h$ a large deviation (LD) principle with a convex rate function J_h. This, of course, does not say much. However:

1) For repulsive potentials (**R**), it is always the case that the probability measures $\mathbb{P}_n^h$ are asymptotically concentrated on ballistic trajectories: For every $h \neq 0$ there exist $\bar{v} = \bar{v}_h \in \mathbb{R}^d \setminus \{0\}$, a constant $\kappa > 0$ and a small ϵ ($\epsilon < \|\bar{v}_h\|$), such that:

$$\mathbb{P}_n^h \left(\frac{D(\gamma)}{n} \notin B_\epsilon(\bar{v}) \right) \leq e^{-\kappa n}, \tag{3.9}$$

where $B_\epsilon(\bar{v}) = \{u : |u - \bar{v}| < \epsilon\}$. Furthermore, the end-point $\gamma(n)$ complies with the following strong local limit type description: the rate function J_h is real analytic and strictly convex on $B_\epsilon(\bar{v}_h)$ with a non-degenerate quadratic minimum at $\bar{v}_h$. Moreover, there exists a strictly positive real analytic function G on $B_\epsilon(\bar{v}_h)$ such that:

$$\mathbb{P}_n^h\left(\frac{D(\gamma)}{n} = u\right) = \frac{G(u)}{\sqrt{(n)^d}} e^{-nJ_h(u)}\left(1 + \mathrm{o}(1)\right), \tag{3.10}$$

uniformly in $u \in B_\epsilon(\bar{v}_h) \cap \mathbb{Z}^d/n$. In particular, the displacement $D(\gamma)$ obeys under $\mathbb{P}_n^h$ a local CLT and local moderate deviations on all possible scales.

2) In the (normalized) attractive case **(A)** + **(SL)**:

- If $h \in \mathrm{int}\,(\mathbf{K}_{\lambda_0})$, then $D(\gamma)/n$ behaves sub-ballistically: Namely, the rate function J_h is bounded below in a neighbourhood of the origin as $J_h(u) \geq \alpha\|u\|$ for some $\alpha = \alpha(h) > 0$.
- If, on the other hand, $h \notin \mathbf{K}_{\lambda_0}$, then both (3.9) and the local limit description (3.10) hold.

Remark 3.1 *It remains an open question to determine what happens when $h \in \partial \mathbf{K}_{\lambda_0}$. Nevertheless, we shall show that in this case the rate function satisfies $J_h(0) = 0$, and thus no ballistic behaviour in the sense of (3.9) is to be expected. It still remains to understand whether the sub-ballistic to ballistic transition is of first order. One way to formulate this question is: assume that $h \in \partial \mathbf{K}_{\lambda_0}$; does $\liminf_{\epsilon \to 0} \|\bar{v}_{(1+\epsilon)h}\| > 0$ hold or not? In Mehra and Grassberger 2002 it is claimed on the basis of simulations that the transition is of first order in dimensions $d \geq 2$ whereas it is of second order in $d = 1$.*

In the sequel, values of *positive* constants $\delta, \epsilon, \nu, c, C_1, c_2, \ldots$ may vary from section to section.

3.2 Coarse graining

Let $\lambda > \lambda_0$. In this section we are going to develop a rough description of typical paths which contribute to $\mathbb{H}_\lambda(x)$. Their contribution is going to be measured in terms of the probability distribution:

$$\mathbb{P}_\lambda^x(\gamma) = \frac{W_\lambda(\gamma)}{\mathbb{H}_\lambda(x)}$$

on $\mathcal{H}_x$. Our results will be particularly relevant in the asymptotic regime $\|x\| \to \infty$.

The key idea behind the renormalization is that although on the microscopic scale paths γ are entitled to wiggle as much as they wish, on large but yet finite scales they exibit a much more rigid behaviour and, in a sense, go ballistically towards x.

The skeleton scale K is going to be chosen the same for all x. Given such K and a path $\gamma \in \mathcal{H}_x$ we shall use $\hat{\gamma}_K$ for the K-skeleton of γ. The construction of

FIG. 3.1: The construction of the K-skeleton in the repulsive case.

$\hat{\gamma}_K$ will be different for repulsive and attractive interactions, but in both cases $\hat{\gamma}_K$ will enjoy the following two properties:

(P1) γ passes through *all* the vertices of $\hat{\gamma}_K$.
(P2) If η is a portion of γ which connects two vertices u and v of $\hat{\gamma}_K$ and η does not pass through any other vertex of $\hat{\gamma}_K$, then:

$$\eta \subseteq K\mathbf{U}_\lambda(u) \cup K\mathbf{U}_\lambda(v), \qquad (3.11)$$

where $\mathbf{U}_\lambda$ is the unit ball in the ξ_λ norm.

Below we shall rely on the following inequality which holds for all the models we consider: Let $\lambda > \lambda_0$ be fixed. There exists a sequence $\epsilon_K \to 0$, such that:

$$\mathbb{H}_\lambda(u) \leq e^{-\xi_\lambda(u)(1-\epsilon_K)}, \qquad (3.12)$$

uniformly in K and in $u \notin K\mathbf{U}_\lambda$.

Indeed, in the attractive case (3.12) follows from super-multiplicativity of $\mathbb{H}_\lambda(x)$ (with $\epsilon_K \equiv 0$). In the repulsive case (3.12) follows from the estimate (A.2) of the Appendix.

3.2.1 Skeletons in the repulsive case

Let us fix a scale K. The K-skeleton $\hat{\gamma}_K$ of $\gamma = (\gamma(0), \ldots, \gamma(n)) \in \mathcal{H}_x$ is constructed as follows (see Fig. 3.1):

STEP 0. Set $u_0 = 0$, $\tau_0 = 0$ and $\hat{\gamma}_K = \{u_0\}$. Go to STEP 1.
STEP (l+1). If $(\gamma(i_l), \ldots, \gamma(n)) \subseteq K\mathbf{U}_\lambda(u_l)$ then stop. Otherwise set

$$\tau_{l+1} = \min\{j > \tau_l : \gamma(j) \notin K\mathbf{U}_\lambda(u_l)\}.$$

Define $u_{l+1} = \gamma(\tau_{l+1})$, update $\hat{\gamma}_K = \hat{\gamma}_K \cup \{u_{l+1}\}$ and go to STEP (l+2).

Given a skeleton $\hat{\gamma}_K = (0, u_1, \ldots, u_m)$ we use $\#(\hat{\gamma}_K) = m$ for the number of *full* K-steps. If γ is compatible with $\hat{\gamma}_K$, then it is decomposable as the concatenation:

$$\gamma = \gamma_1 \cup \gamma_2 \cup \cdots \cup \gamma_m \cup \gamma_{m+1},$$

where $\gamma_l : u_{l-1} \mapsto u_l$ for $l = 1, \ldots, m$ and $\gamma_{m+1} : u_{m+1} \mapsto x$. By construction, $\gamma_l \in \mathcal{H}_{u_l - u_{l-1}}$. Since we are in the repulsive case:

$$W_\lambda(\gamma) \leq \prod_1^{m+1} W_\lambda(\gamma_l) \tag{3.13}$$

Therefore, the probability of appearence of $\hat{\gamma}_K$ is bounded above as (see (3.12) for the second inequality):

$$\mathbb{P}_\lambda^x(\hat{\gamma}_K) \leq \frac{\prod_1^m \mathbb{H}_\lambda(u_l - u_{l-1})}{\mathbb{H}_\lambda(x)} \leq \frac{e^{-mK(1-\epsilon_K)}}{\mathbb{H}_\lambda(x)}. \tag{3.14}$$

This formula is a key to a study of the geometry of typical skeletons and hence of typical paths. Its implications for path decomposition are discussed in detail in Section 3.3. Notice, however, that one immediate consequence is the following uniform exponential bound on the number of steps in typical skeletons: there exist $\nu = \nu(\lambda, d)$ and $C = C(\lambda, d)$ and a large finite scale $K_0 = K_0(\lambda, d)$, such that:

$$\mathbb{P}_\lambda^x\left(\#(\hat{\gamma}_K) > C\frac{\|x\|}{K}\right) \leq e^{-\nu\|x\|}, \tag{3.15}$$

uniformly in $K \geq K_0$ and $x \in \mathbb{Z}^d$. Indeed, the total number of m-step K-skeletons is bounded above as $\exp\{c_1(d, \lambda) \log K\}$, which, on large K scales, is suppressed by the extra $e^{-K(1-\epsilon_K)}$ per step price in (3.14).

3.2.2 Skeletons in the attractive case

Since in the attractive case (3.13) does not hold we should proceed with more care. Accordingly our construction relies on the following fact: assume that the path γ can be represented as a concatenation:

$$\gamma = \gamma_1 \cup \eta_1 \cup \gamma_2 \cup \cdots \cup \gamma_m \cup \eta_m, \tag{3.16}$$

such that $\gamma_1, \ldots, \gamma_m$ share at most one end-point. Then:

$$W_\lambda(\gamma) \leq \prod_{l=1}^m W_\lambda(\gamma_l) e^{\phi(1)} \cdot \prod_{k=1}^m e^{-\lambda|\eta_k|}. \tag{3.17}$$

For every $\lambda > \lambda_0$, (3.3) enables a comparison with a killed simple random walk. Consequently, since ξ_λ is an equivalent norm, there exists $\delta = \delta(\lambda, d) > 0$, such that:

$$\sum_{\eta \in \mathcal{H}_u} e^{-\lambda|\eta|} \leq e^{-\delta K}, \tag{3.18}$$

uniformly in K and in $u \notin K\mathbf{U}_\lambda$.

In view of (3.17) and (3.18) it happens to be natural to construct skeletons $\hat{\gamma}_K$ as a union $\hat{\gamma}_K = \mathfrak{t}_K \cup \mathfrak{h}_K$, where $\mathfrak{t}_K$ is the trunk and $\mathfrak{h}_K$ is the set of hairs of $\hat{\gamma}_K$ (see Fig. 3.2). Let $\gamma \in \mathcal{H}_x$ and choose a scale K.

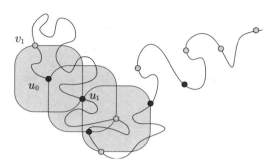

FIG. 3.2: The first stage of the construction of the K-skeleton in the attractive case (notice that $u_5 = v_5$ and $u_6 = v_6$ in this picture). In a second stage, one constructs skeletons for the (reversed) paths connecting u_k to v_k, using the algorithm for the repulsive case.

STEP 0. Set $u_0 = 0$, $\tau_0 = 0$ and $\mathfrak{t}_0 = \{u_0\}$. Go to STEP 1.

STEP $(l+1)$ If $(\gamma(\tau_l), \ldots \gamma(n)) \subseteq K\mathbf{U}_\lambda(u_l)$ then set $\sigma_{l+1} = n$ and stop. Otherwise, define

$$\sigma_{l+1} = \min\{i > \tau_l \ : \ \gamma(i) \notin K\mathbf{U}_\lambda(u_l)\}$$

and

$$\tau_{l+1} = 1 + \max\{i > \tau_l \ : \ \gamma(i) \in K\mathbf{U}_\lambda(u_l)\}.$$

Set $v_{l+1} = \gamma(\sigma_{l+1})$ and $u_{l+1} = \gamma(\tau_{l+1})$. Update $\mathfrak{t}_K = \mathfrak{t}_K \cup \{u_{l+1}\}$ and go to STEP $(l+2)$

Apart from producing $\mathfrak{t}_K$ the above algorithm leads to a decomposition of γ as in (3.16) with:

$$\gamma_l = (\gamma(\tau_l), \ldots, \gamma(\sigma_{l+1})) \quad \text{and} \quad \eta_l = (\gamma(\sigma_l), \ldots, \gamma(\tau_l)).$$

The hairs $\mathfrak{h}_K$ of $\hat{\gamma}_K$ take into account those η_l-s which are long on K-th scale. Recall that $\eta_l : v_l \mapsto u_l$. It is equivalent, but more convenient, to think about η_l as of a reversed path from u_l to v_l. Then the l-th hair $\mathfrak{h}_K^l$ of γ is constructed as follows: if $\eta_l \subseteq K\mathbf{U}_\lambda(u_l)$ then $\mathfrak{h}_K^l = \emptyset$. Otherwise, construct $\mathfrak{h}_K^l$ of η_l following exactly the same rules as in the construction of K-skeletons in the repulsive case.

Putting everything together, using (3.17) and (3.18), we arrive to the following upper bound on the probability of a K-skeleton $\hat{\gamma}_K = \mathfrak{t}_K \cup \mathfrak{h}_K$:

$$\mathbb{P}_\lambda^x (\mathfrak{t}_K \cup \mathfrak{h}_K) \leq \frac{e^{-\#(\mathfrak{t}_K)K(1-\phi(1)/K) - \delta K \#(\mathfrak{h}_K)}}{\mathbb{H}_\lambda(x)}, \quad (3.19)$$

where $\#(\mathfrak{h}_K) \stackrel{\Delta}{=} \sum_1^m \#(\mathfrak{h}_K^l)$ is the total K-length of hairs attached to the trunk $\mathfrak{t}_K$.

As in the repulsive case, (3.19) leads to an exponential upper bound on the number of K-steps in the trunk $\mathfrak{t}_K$: there exists a large finite scale $K_0 = K_0(\lambda, d)$

such that:
$$\mathbb{P}_\lambda^x(\mathfrak{t}_K) \leq \frac{e^{-\#(\mathfrak{t}_K)K(1-\epsilon_K)}}{\mathbb{H}_\lambda(x)}, \quad (3.20)$$

for all $K \geq K_0$ and uniformly in x and $\mathfrak{t}_K$. Consequently, there exist $\nu = \nu(\lambda, d)$ and $C = C(\lambda, d)$ such that:
$$\mathbb{P}_\lambda^x\left(\#(\mathfrak{t}_K) > C\frac{\|x\|}{K}\right) \leq e^{-\nu\|x\|}, \quad (3.21)$$

uniformly in $K \geq K_0$ and $x \in \mathbb{Z}^d$. Furthermore, since the number of different ways to attach hairs $\mathfrak{h}_K$ with $\#(\mathfrak{h}_K) = r$ to vertices of a trunk $\mathfrak{t}_K$ of cardinality $\#(\mathfrak{t}_K) = m$ is bounded above as:
$$e^{c_1 r \log K} \cdot \frac{1}{\max_p p^r(1-p)^m} \leq e^{c_2 r \log K + m/K},$$

formula (3.19) implies that:
$$\mathbb{P}_\lambda^x\left(\#(\mathfrak{h}_K) = r; \#(\mathfrak{t}_K) \leq C\frac{\|x\|}{K}\right) \leq e^{-\delta r K + c_3 \|x\| \log K/K} \max_{K\#(\mathfrak{t}_K) \leq C\|x\|} \frac{e^{-K\#(\mathfrak{t}_K)}}{\mathbb{H}_\lambda(x)}.$$

As we shall point out in the beginning of Section 3.3:
$$\limsup_{K\to\infty} \limsup_{\|x\|\to\infty} \frac{1}{\|x\|} \log \max_{K\#(\mathfrak{t}_K) \leq C\|x\|} \frac{e^{-K\#(\mathfrak{t}_K)}}{\mathbb{H}_\lambda(x)} = 0. \quad (3.22)$$

Consequently, for each $\epsilon > 0$ there exists a finite scale $K_0 = K_0(\epsilon, \lambda, d)$, such that:
$$\mathbb{P}_\lambda^x\left(\#(\mathfrak{h}_K) > \epsilon\frac{\|x\|}{K}\right) \leq e^{-\epsilon\delta\|x\|/2}, \quad (3.23)$$

uniformly in $K \geq K_0$ and $x \in \mathbb{Z}^d$.

3.3 Irreducible decomposition of ballistic paths

In this section we derive an irreducible representation of typical (under $\mathbb{P}_\lambda^x$) paths $\gamma \in \mathcal{H}_x$. This irreducible representation has an effective 1D structure and it enables a local limit treatment of various observables over paths such as, e.g., the displacement $D(\gamma)$ along γ, or the number of steps $|\gamma|$ in γ. It should be kept in mind that in the framework of the theory we develop there are *many alternative* ways to define irreducible paths and, accordingly, to study statistics of other local patterns over γ.

In the sequel, given $x \in \mathbb{Z}^d$, we say that $h \in \partial \mathbf{K}_\lambda$ is dual to x if $(h, x) = \xi_\lambda(x)$. Of course, if h is dual to x, then h is dual to αx for any $\alpha > 0$. Thus it makes sense to talk about $\|x\| \to \infty$ for a fixed dual h.

3.3.1 Surcharge function and surcharge inequality

We shall now analyse the geometry of typical K-skeletons. In order to unify as much as possible our treatment of the repulsive and attractive cases, let us define the trunk $\mathfrak{t}_K$ of the skeleton $\hat{\gamma}_K$ in the repulsive case as being the skeleton itself, $\mathfrak{t}_K = \hat{\gamma}_K$.

The basic quantity in the following analysis is the *surcharge function* of a trunk $\mathfrak{t}_K = (u_0, \ldots, u_m)$ defined by:

$$\mathfrak{s}_h(\mathfrak{t}_K) = \sum_{i=1}^m \mathfrak{s}_h(u_i - u_{i-1}),$$

where the function $\mathfrak{s}_h : \mathbb{R}^d \to \mathbb{R}_+$ is given by $\mathfrak{s}_h(y) = \xi_\lambda(y) - (h, y)$. Note that $\mathfrak{s}_h(y) = 0$ if and only if $h \in \partial \mathbf{K}_\lambda$ and y are dual directions. The surcharge measure $\mathfrak{s}_h$ enters our skeleton calculus in the following fashion: by construction, $\xi_\lambda(u_i - u_{i-1}) \le K + c_1(\lambda)$ for all K-steps of $\mathfrak{t}_K$. As a result:

$$K \#(\mathfrak{t}_K) \ge \sum (\xi_\lambda(u_i - u_{i-1}) - c_1(\lambda)) \ge \xi_\lambda(x) + \mathfrak{s}_h(\mathfrak{t}_K) - c_1 \#(\mathfrak{t}_K).$$

Since we can restrict attention to $\#(\mathfrak{t}_K) \le C\|x\|/K$, (3.22) is an immediate corollary. Moreover, arguing similarly as in (3.14) and (3.20), we obtain that, for any $\epsilon > 0$ fixed:

$$\mathbb{P}_\lambda^x(\mathfrak{t}_K) \le e^{-\mathfrak{s}_h(\mathfrak{t}_K)(1-o_K(1))},$$

uniformly in $\|x\|$ and in $\mathfrak{s}_h(\mathfrak{t}_K) \ge \epsilon\|x\|$, with $\lim_{K \to \infty} o_K(1) = 0$. The following *surcharge inequality* is at the core of our method, allowing us to reduce the characterization of typical trunks to geometrical considerations.

Lemma 3.1 *For every small $\epsilon > 0$ there exists $K_0(d, \lambda, \epsilon)$ such that*

$$\mathbb{P}_\lambda^x(\mathfrak{s}_h(\mathfrak{t}_K) > 2\epsilon\|x\|) \le e^{-\epsilon\|x\|},$$

uniformly in $x \in \mathbb{Z}^d$, $h \in \partial \mathbf{K}_\lambda$ dual to x, and scales $K > K_0$.

Proof: By (3.15) and (3.21), we can assume that the trunk is admissible, that is $\#(\mathfrak{t}_K) \le C\|x\|/K$. Since the number of such trunks being bounded by:

$$\exp\left(c_2(d, \lambda) \frac{\log K}{K} \|x\|\right),$$

we infer that:

$$\mathbb{P}_\lambda^x(\mathfrak{s}_h(\mathfrak{t}_K) > 2\epsilon\|x\|) \le e^{-\nu\|x\|} + \exp\left(c_2(d, \lambda) \frac{\log K}{K} \|x\| - \mathfrak{s}_h(\mathfrak{t}_K)(1 - o_K(1))\right)$$
$$\le e^{-\epsilon\|x\|},$$

as soon as K is chosen large enough. □

Thanks to the surcharge inequality, we can exclude whole families of trunks simply by establishing a lower bound as above on their surcharge function.

3.3.2 Cone points of trunks

Let us fix $\delta \in (0, \frac{1}{3})$. For any $h \in \partial \mathbf{K}_\lambda$, we define the *forward cone* by:

$$Y_\delta^>(h) = \{y \in \mathbb{Z}^d : \mathfrak{s}_h(x) < \delta \xi_\lambda(x)\},$$

and the *backward cone* by $Y_\delta^<(h) = -Y_\delta^>(h)$.

Given a trunk $\mathfrak{t}_K = (u_0, \ldots, u_m)$, we say that the point u_k is an (h, δ)-*forward cone point* if:

$$\{u_{k+1}, \ldots, u_m\} \subset u_k + Y_\delta^>(h).$$

Similarly, u_k is an (h, δ)-*backward cone point* if:

$$\{u_0, \ldots, u_{k-1}\} \subset u_k + Y_\delta^<(h).$$

Finally, u_k is an (h, δ)-*cone point* if it is both an (h, δ)-forward and an (h, δ)-backward cone point.

Proceeding now as in Section 2.6 of (Campanino et al. 2007), we prove that most vertices of typical trunks are (h, δ)-cone points. Let us denote by $\#_{h,\delta}^{\mathrm{non-cone}}(\mathfrak{t}_K)$ the number of those vertices in the trunk $\mathfrak{t}_K$ that are not (h, δ)-cone points.

Lemma 3.2 *Let $\delta \in (0, \frac{1}{3})$ be fixed. Then*

$$\mathfrak{s}_h(\mathfrak{t}_K) \geq c_3 \delta K \#_{h,\delta}^{\mathrm{non-cone}}(\mathfrak{t}_K),$$

uniformly in $x \in \mathbb{Z}^d$, $h \in \partial \mathbf{K}_\lambda$ dual to x, and K large enough. In particular, the estimate

$$\mathbb{P}_\lambda^x \left(\#_{h,\delta}^{\mathrm{non-cone}}(\mathfrak{t}_K) \geq \epsilon \#(\mathfrak{t}_K) \right) \leq e^{-c_4 \epsilon \|x\|},$$

holds uniformly in $x \in \mathbb{Z}^d$, $h \in \partial \mathbf{K}_\lambda$ dual to x, and K large enough.

3.3.3 Cone points of skeletons

In the repulsive case, the previous lemma provides all the control we need. In the attractive case, to which we restrict ourselves temporarily, it is also necessary to control the hairs. Let us start by extending the notion of (h, δ)-cone points from trunks to full skeletons: a point u_k of a trunk $\mathfrak{t}_K = (u_0, \ldots, u_m)$ is an (h, δ)-forward cone point of the skeleton $\hat{\gamma}_K = (\mathfrak{t}_K \cup \mathfrak{h}_K)$ if:

$$\hat{\gamma}_K \subset \left(u_k + Y_{2\delta}^>(h)\right) \cup \left(u_k + Y_{2\delta}^<(h)\right).$$

(Notice that we increased the aperture of the cone from δ to 2δ.) It readily follows (3.23) that most vertices of the trunk have no hair attached to them. Therefore, the only way an (h, δ)-cone point of $\mathfrak{t}_K$ may fail to be an (h, δ)-cone point of $\hat{\gamma}_K$ is when another vertex of the trunk has such a long hair attached to it that the latter exits the (enlarged) cone; in such a case, we say that the (h, δ)-cone point is *blocked*. Lemma 2.5 of (Campanino et al. 2007) implies that most vertices of the trunk are (h, δ)-cone points of $\hat{\gamma}_K$.

Lemma 3.3 *Let $\#_{h,\delta}^{\text{blocked}}(\hat\gamma_K)$ be the number of vertices of the trunk of $\hat\gamma_K$ that are not (h,δ)-cone points of $\hat\gamma_K$. Suppose that t_K is an admissible trunk with $\#_{h,\delta}^{\text{non-cone}}(\mathsf{t}_K) < \epsilon \#(\mathsf{t}_K)$. Then there exists $c_5(d,\delta,\lambda) > 0$ such that*

$$\mathbb{P}_\lambda^x(\#_{h,\delta}^{\text{blocked}}(\hat\gamma_K) \geq \epsilon \#(\mathsf{t}_K) \,|\, \mathsf{t}_K) \leq e^{-c_5 \epsilon \|x\|},$$

uniformly in $x \in \mathbb{Z}^d$, $h \in \partial \mathbf{K}_\lambda$ dual to x, and K large enough.

3.3.4 Cone points of paths

We return now to the general case of attractive or repulsive potentials; in the latter case we identify the notions of cone points of trunks and skeletons, as these two notions coincide.

Now that the skeletons are under control, we can turn to the microscopic path itself. Let $\gamma = (\gamma(0),\ldots,\gamma(n)) \in \mathcal{H}_x$. We say that $\gamma(k)$ is an (h,δ)-cone point of γ if:

$$\gamma \subset \left(\gamma(k) + Y_{3\delta}^>(h)\right) \cup \left(\gamma(k) + Y_{3\delta}^<(h)\right).$$

Notice that we increased again the aperture of the cone to 3δ. Notice also that we tacitly assume that $Y_{3\delta}^>(h)$ contains a lattice direction.

Proceeding as in Section 2.7 of (Campanino et al. 2003), we can show that, up to exponentially small $\mathbb{P}_\lambda^x$-probabilities, a uniformly strictly positive fractions of the (h,δ)-cone points of the trunk of $\hat\gamma_K$ are actually (h,δ)-cone points of the path.

Theorem 3.1 *Let $\#_{h,\delta}^{\text{cone}}(\gamma)$ be the number of (h,δ)-cone points of γ. There exist $\delta \in (0,\frac{1}{3})$ and two positive numbers c and ν, depending only on d, δ and λ, such that*

$$\mathbb{P}_\lambda^x(\#_{h,\delta}^{\text{cone}}(\gamma) < c\|x\|) \leq e^{-\nu\|x\|},$$

uniformly in $x \in \mathbb{Z}^d$, $h \in \partial \mathbf{K}_\lambda$ dual to x, and K large enough.

Remark 3.2 *Observe that the above estimate still holds true (possibly for a smaller constant ν) when x is replaced by an arbitrary site $y \in Y_\delta^>(h)$.*

3.3.5 Decomposition of paths into irreducible pieces

With the help of the Theorem 3.1, we can finally construct the desired decomposition of typical ballistic paths into irreducible pieces.

A path $\gamma = (\gamma(0),\ldots,\gamma(n))$ is said to be (h,δ)-*backward irreducible* if $\gamma(n)$ is the only (h,δ)-cone point of γ. Similarly, γ is said to be (h,δ)-*forward irreducible* if $\gamma(0)$ is the only (h,δ)-cone point of γ. Finally, γ is said to be *irreducible* if $\gamma(0)$ and $\gamma(n)$ are the only (h,δ)-cone points of γ.

Let Ω_L, Ω and Ω_R be the corresponding sets of such irreducible paths.

In view of Theorem 3.1, we can restrict our attention to paths possessing at least $c\|x\|$ (h,δ)-cone points, at least when $\|x\|$ is sufficiently large. We can then

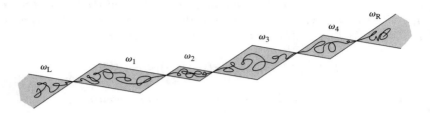

FIG. 3.3: The decomposition of a path into irreducible components.

unambiguously decompose λ into irreducible sub-paths (see Fig. 3.3):

$$\gamma = \omega_L \cup \omega_1 \cup \cdots \cup \omega_m \cup \omega_R. \tag{3.24}$$

We thus have the following expression:

$$e^{(h,x)}\mathbb{H}_\lambda(x) = O\left(e^{-\nu\|x\|}\right) + \sum_{m \geq c\|x\|} \sum_{\omega_L \in \Omega_L} \sum_{\omega_1,\ldots\omega_m \in \Omega} \sum_{\omega_R \in \Omega_R} W_{h,\lambda}(\gamma)\mathbb{1}_{\{D(\gamma)=x\}}. \tag{3.25}$$

In fact, since $e^{(h,y)}\mathbb{H}_\lambda(y) \asymp e^{-s_h(y)}$, we, in view of Remark 3.2, infer that, perhaps for a smaller choice of $\nu > 0$

$$e^{(h,y)}\mathbb{H}_\lambda(y) = O\left(e^{-\nu\|y\|}\right) + \sum_{m \geq c\|y\|} \sum_{\omega_L \in \Omega_L} \sum_{\omega_1,\ldots\omega_m \in \Omega} \sum_{\omega_R \in \Omega_R} W_{h,\lambda}(\gamma)\mathbb{1}_{\{D(\gamma)=y\}}. \tag{3.26}$$

uniformly in $y \in \mathbb{Z}^d$.

3.3.6 Probabilistic structure of the irreducible decomposition

We shall treat Ω, Ω_L and Ω_R as probability spaces. In this way various path observables such as, e.g., $D(\omega)$ or $|\omega|$ will be naturally interpreted as random variables. First of all let us try to rewrite the weights of concatenated paths in the product form:

$$W_{h,\lambda}(\gamma) = W_{h,\lambda}(\omega_L \cup \omega_1 \cup \cdots \cup \omega_m \cup \omega_R) = \hat{W}_{h,\lambda}(\omega_L)\hat{W}_{h,\lambda}(\omega_R) \prod_l \hat{W}_{h,\lambda}(\omega_l) \tag{3.27}$$

If the potential Φ depends on bond local times only, then $\hat{W}_{h,\lambda} = W_{h,\lambda}$ qualifies. However, when Φ depends on site local times, such a choice would imply overcounting of local times at end-points. In that case, we thus set $\hat{W}_{h,\lambda}(\omega) = e^{\phi(1)}W_{h,\lambda}(\omega)$ for $\omega \in \Omega_L$ and $\omega \in \Omega$, and $\hat{W}_{h,\lambda} = W_{h,\lambda}$ on Ω_R. Evidently (3.27) is satisfied. Accordingly, we can rewrite (3.26) as:

$$e^{(h,y)}\mathbb{H}_\lambda(y) = O\left(e^{-\nu\|y\|}\right) + \sum_{m \geq c\|y\|} \sum_{\omega_L \in \Omega_L} \sum_{\omega_1,\ldots\omega_m \in \Omega} \sum_{\omega_R \in \Omega_R} \hat{W}_{h,\lambda}(\omega_L)\hat{W}_{h,\lambda}(\omega_R) \times \prod_1^m \hat{W}_{h,\lambda}(\omega_l)\mathbb{1}_{\{D(\gamma)=y\}}. \tag{3.28}$$

uniformly in $y \in \mathbb{Z}^d$. Since the sums over ω_L and ω_R converge (by Theorem 3.1), and:

$$\mathbf{K}_\lambda = \overline{\{h \in \mathbb{R}^d \,:\, \sum_{y \in \mathbb{Z}^d} e^{(h,y)} \mathbb{H}_\lambda(y) < \infty\}},$$

it follows that $h \in \partial \mathbf{K}_\lambda$ if and only if $\sum_{\omega \in \Omega} \hat{W}_{h,\lambda}(\omega) = 1$, and thus $\hat{W}_{h,\lambda}$ is a probability measure on Ω. Let us use the notation $\mathbb{Q}^{h,\lambda} = \hat{W}_{h,\lambda}$ to stipulate this fact. Similarly we shall use the notation $\mathbb{Q}_L^{h,\lambda}$ and $\mathbb{Q}_R^{h,\lambda}$ for the values of $\hat{W}_{h,\lambda}$ on respectively Ω_L and Ω_R.

In general $\mathbb{Q}_L^{h,\lambda}$ and $\mathbb{Q}_R^{h,\lambda}$ are not probability measures. However, together with $\mathbb{Q}^{h,\lambda}$, they display exponential tails both in the displacement variable $D(\omega)$ and in the number of steps $|\omega|$ (as well as in many other path observables of interest). Indeed, as follows from Theorem 3.1 and in view of the $Y_{3\delta}(h)$-confinement properties of irreducible paths, all three measures in question already display exponential tails in the displacement variable $D(\omega)$. On the other hand, the weights $\hat{W}_{h,\lambda}$ are bounded above as:

$$\hat{W}_{h,\lambda}(\omega) \leq e^{\|h\| \|D(\omega)\| - \lambda |\omega| + \phi(1)}.$$

Consequently, there exists $c_6 = c_6(\lambda, d)$, such that:

$$\hat{W}_{h,\lambda}\left(\|D(\omega)\| = l, |\omega| > 2(l\|h\| + \phi(1))\right) \leq c_6 e^{-l\|h\|(\lambda - \lambda_0)}.$$

Together with the already established exponential tails of $D(\omega)$, this readily implies exponential tails for the variable $|\omega|$ as well. Exactly the same line of reasoning applies to $\mathbb{Q}_L^{h,\lambda}$ and $\mathbb{Q}_R^{h,\lambda}$.

We use $\mathbb{Q}_m^{h,\lambda}$ for the product probability measure on $\times_1^m \Omega$. Then (3.28) in its final form looks like:

$$e^{(h,y)} \mathbb{H}_\lambda(y) = O\left(e^{-\nu \|y\|}\right) + \sum_{m \geq c\|y\|} \mathbb{Q}_L^{h,\lambda} \star \mathbb{Q}_R^{h,\lambda} \star \mathbb{Q}_m^{h,\lambda} \left(D(\omega_L) + D(\omega_R) + \sum_1^m D(\omega_l) = y\right). \tag{3.29}$$

Similarly, let F be some functional on paths, e.g., $F(\gamma) = |\gamma|$ or, for a change, $F(\gamma) = \sum_x \mathbb{I}_{\{l_x(\gamma) > 1\}}$. We then have the following expression for the restricted partition functions:

$$e^{(h,y)} \mathbb{H}_\lambda(y; F(\gamma) = f) = O\left(e^{-\nu \|y\|}\right) + \sum_{m \geq c\|y\|} \mathbb{Q}_L^{h,\lambda} \star \mathbb{Q}_R^{h,\lambda} \star \mathbb{Q}_m^{h,\lambda}\left(D(\gamma) = y; F(\gamma) = f\right). \tag{3.30}$$

The point is that, for good local functionals F such as in the two instances above, sharp asymptotics simply follow from local limit properties of the product measure $\mathbb{Q}_m^{h,\lambda}$.

Note that a very similar analysis applies for partition functions $\mathbb{D}_\lambda(.)$. In particular (3.30) holds for $\mathbb{D}_\lambda(y; F(\lambda) = f)$ with the very same measures $\mathbb{Q}_L^{h,\lambda}, \mathbb{Q}^{h,\lambda}$ and the appropriate modification of $\mathbb{Q}_R^{h,\lambda}$.

3.4 Proof of the main result

3.4.1 *Large deviation rate function*

Since:
$$\Lambda(h) \triangleq \lim_{n\to\infty} \frac{1}{n} \log Z_n^h$$
is well defined (either by sub- or by superadditivity) and the distribution of the displacement per step $D(\gamma)/n$ under $\mathbb{P}_n^h$ is certainly exponentially tight, the random variables $D(\gamma)/n$ satisfy a LD principle with rate function:
$$J_h(u) = \sup_g \{(g, u) - \Lambda_h(g)\},$$
where $\Lambda_h(g) = \Lambda(h + g) - \Lambda(h)$.

We claim that:
$$\Lambda \equiv \lambda_0 \quad \text{on } \mathbf{K}_{\lambda_0}. \tag{3.31}$$
Consequently, if $h \in \text{int}(\mathbf{K}_{\lambda_0})$, then $\Lambda_h \equiv 0$ in a neighbourhood of the origin and hence there exists $\alpha = \alpha(h)$ such that $J_h(u) \geq \alpha \|u\|$, which is the sub-ballistic part of our Main Result.

Furthermore, we claim that if $h \notin \mathbf{K}_{\lambda_0}$, then Λ is real analytic and strictly convex in a neighbourhood of h. In addition:
$$\bar{v}_h = \nabla \Lambda(h) \neq 0 \quad \text{and} \quad \mathrm{d}^2 \Lambda(h) \text{ is non-degenerate.} \tag{3.32}$$
In fact, for $h \notin \mathbf{K}_{\lambda_0}$, we shall see that the log-moment generating function satisfies $\Lambda(h) = \lambda(h)$ with $\lambda = \lambda(h)$ being recovered from $h \in \partial \mathbf{K}_\lambda$. Since $\mathbf{K}_{\lambda_0} = \cap_{\lambda > \lambda_0} \mathbf{K}_\lambda$, (3.31) is an immediate consequence: indeed, since Λ is convex and, by Jensen inequality, $Z_n^h \geq Z_n \asymp e^{\lambda_0 n}$, it satisfies $\Lambda \geq \lambda_0$.

For the rest of this section we shall, therefore, focus on the case $h \notin \mathbf{K}_{\lambda_0}$.

3.4.2 *Surface* $\lambda = \lambda(h) = \Lambda(h)$

Our next task is to explain (3.32). Let $\lambda > \lambda_0$ and $h \in \partial \mathbf{K}_\lambda$. As we shall see below, there exists $\Psi(h) > 0$, such that:
$$e^{-\lambda n} Z_n^h = \sum_\gamma W_{h,\lambda}(\gamma) = \Psi(h)(1 + o(1)). \tag{3.33}$$

This immediately implies that $\Lambda(h) = \lambda$. Let us for the moment accept (3.33) as an apriori lower bound on $e^{-\lambda n} Z_n^h$ (see the paragraph after (3.34) below). Then we are able to rule out $\mathbb{P}_n^h$-negligible x-s as follows:

1) Let $0 < l < \lambda - \lambda_0$. Then:
$$e^{-\lambda n} Z_n^h \left((h, \gamma(n)) \leq ln \right) \leq e^{-(\lambda - l)n} Z_n.$$

Since $Z_n \asymp e^{\lambda_0 n}$, the latter expression is exponentially small and we can ignore x-s such that $(h, x) \leq ln$.

2) Let $(h, x) > ln$, but $x \notin Y_\delta^>(h)$ (see Section 3.3). Then $Z_{n,x}^{h,\lambda} \leq Z_x^{h,\lambda} \leq e^{-\nu \|x\|}$, and, consequently, we can ignore such x-s as well.

It remains to consider $x \in Y_\delta^>(h)$ with $\|x\| > ln/\|h\| \triangleq c_1 n$. For such x-s, (3.30), or rather its modification in the case of $\mathbb{D}_\lambda$-partition functions, implies that:

$$Z_{n,x}^{h,\lambda} = \sum_{\substack{\gamma \in \mathcal{D}_x \\ |\gamma|=n}} W_{h,\lambda}(\gamma)$$

$$= O\left(e^{-c_1 \nu n}\right) + \sum_{m \geq c\|x\|} \mathbb{Q}_L^{h,\lambda} \star \mathbb{Q}_R^{h,\lambda} \star \mathbb{Q}_m^{h,\lambda}(D(\gamma) = x, |\gamma| = n)$$

$$= C'(h) \sum_{m \geq c\|x\|} \mathbb{Q}_m^{h,\lambda} \left(\sum_{i=1}^m (D(\omega_i), |\omega_i|) = (x, n) \right), \quad (3.34)$$

where in the last line, apart from ignoring the $O\left(e^{-c_1 \nu n}\right)$ term, we have (already anticipating local limit behaviour under $\mathbb{Q}_m^{h,\lambda}$) summed out the terms involving $\mathbb{Q}_L^{h,\lambda}$ and $\mathbb{Q}_R^{h,\lambda}$. And indeed, since the random vector $V(\omega) = (D(\omega), |\omega|)$ has exponential moments under $\mathbb{Q}^{h,\lambda}$, we can apply the classical local CLT to (3.34), which, after summing up with respect to x, yields the lower bound in (3.33), thereby justifying the conclusions 1) and 2) above.

Let now $g \in \partial \mathbf{K}_\mu$ with (g, μ) being sufficiently close to (h, λ), say:

$$|\lambda - \mu| + \|g - h\| < \frac{1}{2} \min\{\nu, \lambda - l\}.$$

Then 1) and 2) above still describe $\mathbb{P}_n^g$-negligible x-s. Moreover, the (g, μ)-modification of (3.34) still holds: for $x \in Y_\delta^>(h)$ with $\|x\| \geq c_1 n$:

$$Z_{n,x}^{g,\mu}(1 + o(1)) = C'(g) \sum_{m \geq c_1 \|x\|} \mathbb{Q}_m^{h,\lambda} \left(\sum_{i=1}^m (D(\omega_i), |\omega_i|) = (x, n) \right) e^{(g-h,x)-(\mu-\lambda)n}.$$

(The prefactor $C'(g)$ coming as before from the summation over ω_L and ω_R.) Since we have employed the very same sets of irreducible paths, the positive function $C'(g)$ is real analytic. Now, using an argument similar to the one in Subsection 3.3.6, together with the fact that the lower bound in (3.33) holds for (g, μ) as well, we deduce that:

$$\mathbb{Q}^{h,\lambda}\left(e^{(g-h,D(\omega))-(\mu-\lambda)|\omega|}\right) = 1.$$

In other words, define:

$$F(g, \mu) = \log \sum_{\omega \in \Omega} e^{(g,D(\omega))-\mu|\omega|} \widehat{W}(\omega).$$

We have proved:

Lemma 3.4 *Let* $\lambda > \lambda_0$ *and* $h \in \partial \mathbf{K}_\lambda$. *Then there exists* $\rho > 0$ *such that the graph of the function* $\mu = \Lambda(g)$ *over* $B_\rho(h)$ *is implicitly given by*

$$F(g, \mu) = 0.$$

Since the distribution of the random vector V under $\mathbb{Q}^{h,\lambda}$ is obviously non-degenerate and has finite exponential moments in a neighbourhood of the origin, (3.32) follows. Furthermore, there exists $\epsilon > 0$ such that:

$$B_\epsilon(\bar{v}_h) \subset \cup_{g \in B_\rho(h)} \nabla \Lambda(g). \tag{3.35}$$

3.4.3 Local limit result in the ballistic regime

We already know that under $\mathbb{P}_n^h$ the average displacement $D(\gamma)/n$ satisfies a LD principle with strictly convex rate function J_h. Thereby, (3.9) is justified. Let us restrict our attention to $x \in B_{\epsilon n}(n \bar{v}_h)$ and go back to (3.34). Obviously, $(\bar{v}_h, 1)$ is parallel to $\mathbf{z} = (\bar{w}, \bar{t}) \stackrel{\Delta}{=} \mathbb{Q}^{h,\lambda}((D(\omega), |\omega|))$. Set $x = \lfloor n \bar{v}_h \rfloor$. By the local CLT and Gaussian summation formula applied to (3.34):

$$Z_{n,x}^{h,\lambda} = \frac{C(h)}{\sqrt{n^d}} (1 + o(1)). \tag{3.36}$$

Let $y \in \mathbf{B}_{n\epsilon}(n \bar{v}_h) \cap \mathbb{Z}^d$ and $u \stackrel{\Delta}{=} y/n \in B_\epsilon(\bar{v}_h)$. By (3.35), we can find $g \in B_\rho(h)$ such that $u = \nabla \Lambda(g)$. Set $\mu = \Lambda(g)$ and $y = \lfloor nu \rfloor$. As in (3.36):

$$Z_{n,y}^{g,\mu} = \frac{C(g)}{\sqrt{n^d}} (1 + o(1)), \tag{3.37}$$

where C is a positive real analytic function on $B_\rho(h)$. On the other hand:

$$Z_{n,y}^{h,\lambda} = e^{-n((g-h,u) + \Lambda(h) - \Lambda(g))} Z_{n,y}^{g,\mu}.$$

It remains to notice that:

$$(g - h, u) + \Lambda(h) - \Lambda(g) = J_h(u).$$

Since C in (3.37) is analytic (continuous would be enough) and J_h has quadratic minimum at $\bar{v}_h$, the partition function asymptotics (3.33) follows from Gaussian summation formula. (3.10) is proved.

3.5 Perturbations by small potentials and statistics of patterns

Let Φ be either an attractive or a repulsive potential of the type considered above, $\lambda > \lambda_0$ and $h \in \partial \mathbf{K}_\lambda$. Let $\mathcal{U}_h \subset \mathbb{R}^d$ be a neighbourhood of h.

We would like to consider perturbations of Φ of the form:

$$\widetilde{\Phi}(\gamma) = \Phi(\gamma) + R(\gamma, h). \tag{3.38}$$

We shall assume that for each fixed γ the function $g \mapsto R(\gamma, g)$ is analytic on $\mathcal{U}_h$ and that it is appropriately negligible: for some ϵ sufficiently small:

$$\sup_{g \in \mathcal{U}_h} |R(\gamma, g)| \leq \epsilon |\gamma|, \tag{3.39}$$

simultaneously for all γ.

Furthermore, we shall assume that the perturbation R is in some sense local. This could be quantified on various levels of generality and, in order to fix ideas, we shall restrict our attention to the following case: For every $g \in \mathcal{U}_h$:

$$R(\gamma_1 \cup \cdots \cup \gamma_m, g) = \sum_{1}^{m} R(\gamma_i, g), \tag{3.40}$$

whenever $\gamma_1, \ldots, \gamma_m$ are edge disjoint.

3.5.1 An example

An immediate example is a random walk with small edge reinforcement. Set $F \equiv 0$. For a path $\gamma = (x_0, x_1, x_2, \ldots)$ define the running local times on *unoriented* bonds:

$$l_b^t(\gamma) = \sum_{j=0}^{t-1} \mathbf{1}_{\{b=(x_i, x_{i+1})\}}.$$

Let $\beta : \mathbb{N} \mapsto \mathbb{R}_+$ be a nondecreasing bounded concave function with $\beta(\infty) = \epsilon$. Set:

$$R(\gamma, h) = -\sum_{t=0}^{|\gamma|-1} \log \mathbb{E} \exp\left\{ \beta(l_{x_i, x_i + X_h}^t(\gamma)) - \beta(l_{(x_i, x_{i+1})}^t(\gamma)) \right\},$$

where the random variable X_h is distributed as the step of a simple random walk with drift h.

3.5.2 Ballistic behaviour and local limit theory

Let $\widetilde{Z}_n^h$ be the partition function which corresponds to the perturbed interaction (3.38). We claim that the following generalization of (3.33) holds:

Theorem 3.2 *Let Φ, $\lambda > \lambda_0$ and $h \in \partial \mathbf{K}_\lambda$ be fixed. Then one can choose a number $\epsilon_0 = \epsilon_0(h, \Phi) > 0$, a continuous function $\rho : [0, \epsilon_0] \mapsto \mathbb{R}_+$ with $\rho(0) = 0$*

and a neighbourhood $\mathcal{U}_h$ of h, so that:

For any $\epsilon \leq \epsilon_0$ and for any (analytic) perturbation $\widetilde{\Phi} = \Phi + R$ of Φ satisfying (3.40) and (3.39) above, there exist an analytic function f with:

$$\sup_{g \in \mathcal{U}_h} |f(g)| \leq \rho(\epsilon), \tag{3.41}$$

and a positive analytic function $\widetilde{\Psi}$ on $\mathcal{U}_h$, such that the following asymptotics holds:

$$e^{-\lambda(g)n} \widetilde{Z}_n^g = \widetilde{\Psi}(g) e^{nf(g)} (1 + o(1)), \tag{3.42}$$

uniformly in $g \in \mathcal{U}_h$. The function f can be recovered from the following implicit relation:

$$\log \mathbb{Q}^{g,\lambda(g)} \left(e^{f(g)|\omega| - R(\omega, g)} \right) = 0. \tag{3.43}$$

The proof and the implications of the above theorem essentially boil down to a rerun of the arguments employed in the pure (repulsive or attractive) cases: the crux of the matter is that under assumption (3.39) the random variable $R(\omega, h)$ on the probability space $(\Omega, \mathbb{Q}^{h,\lambda})$ has exponential tails, whereas under assumption (3.40) all the relevant asymptotics are settled via local limit theorems for independent random variables. Let us briefly reiterate the principal steps involved:

STEP 1 Assume (3.42) as an apriori lower bound on $e^{-\lambda(g)n} \widetilde{Z}_n^g$ and use it to rule out trajectories $\gamma = (\gamma(0), \ldots, \gamma(n))$ which fail to comply with:

1) $\min \{(\gamma(n), g), \|\gamma(n)\|\} \geq c_1 n$ and $\gamma_n \in Y_\delta^>(h)$.

2) γ has at least $m \geq c_2 n$ irreducible pieces:

$$\gamma = \omega_L \cup \omega_1 \cup \cdots \cup \omega_m \cup \omega_R.$$

STEP 2 By Assumption (3.39), the random variables $D(\omega)$ and $|\omega|$ have exponential tails under the modified measures $\widetilde{\mathbb{Q}}^{g,\lambda(g)}$:

$$\widetilde{\mathbb{Q}}^{g,\lambda(g)}(\omega) = \frac{\mathbb{Q}^{g,\lambda(g)}(\omega) e^{-R(\omega, g)}}{\mathbb{Q}^{g,\lambda(g)} \left(e^{-R(\cdot, g)} \right)}.$$

Consequently, both the lower bound in (3.42) and then (3.42) itself follow from the local limit analysis of i.i.d. sums $\sum_1^m V_i$, where $V(\omega) = (D(\omega), |\omega|)$. In particular, the limiting log-moment generating functions:

$$\widetilde{\Lambda}(g) = \lim_{n \to \infty} \frac{\log \widetilde{Z}_n^g}{n} \triangleq \lambda(g) + f(g),$$

are analytic on $\mathcal{U}_h$, whereas the asymptotic speed $\tilde{v}_n = \nabla \widetilde{\Lambda}(h)$ has a positive projection on h by STEP 1. The local limit description of the distribution of the end-point $\gamma(n)$ under the perturbed measure $\widetilde{\mathbb{P}}_n^h$ follows exactly as in the pure (repulsive or attractive) case. In particular, we arrive to the following local limit description of ballistic behaviour under $\widetilde{\mathbb{P}}_n^h$: There exists $\epsilon' > 0$, such that:

- Outside $B_{\epsilon'}(\tilde{v}_h)$:

$$\widetilde{\mathbb{P}}_n^h\left(\frac{D(\gamma)}{n} \notin B_{\epsilon'}(\tilde{v}_h)\right) \leq e^{-c_3 n}.$$

- For $nu \in B_{n\epsilon'}(n\tilde{v}_h) \cap \mathbb{Z}^d$:

$$\widetilde{\mathbb{P}}_n^h\left(D(\gamma) = nu\right) = \frac{\tilde{G}(u)}{\sqrt{n^d}} e^{-n\tilde{J}_h(u)} (1 + o(1)),$$

where the rate function $\tilde{J}_h$ on $B_{\epsilon'}(\tilde{v}_h)$ is quadratic at its unique minimum $\tilde{v}_h$.

3.5.3 Statistics of finite patterns and local observables

A finite pattern is a fixed nearest neighbour (and self-avoiding in case of SAW-s) path:

$$\eta = (u_0, \ldots, u_p).$$

For example, we may take η as an elementary loop:

$$\eta = (0, e_1, e_1 + e_2, e_2). \tag{3.44}$$

Given a trajectory γ define $N_\eta(\gamma)$ as the number of times η is re-incarnated in γ:

$$N_\eta(\gamma) = \sum_{l=0}^{|\gamma|-p} \mathbb{1}_{\{(\gamma(l),\ldots,\gamma(l+p)) \stackrel{m}{=} \eta\}}$$

where $\stackrel{m}{=}$ means matching up to a shift. Obviously, $N(\gamma) \leq |\gamma|$. Also, in the case of (3.44):

$$N_\eta(\gamma) = N_\eta(\omega_{\mathrm{L}}) + N_\eta(\omega_1) + \cdots + N_\eta(\omega_m) + N_\eta(\omega_{\mathrm{R}}), \tag{3.45}$$

whenever γ is given in its irreducible representation (3.24).

For general patterns η the relation (3.45) does not necessarily hold for the particular irreducible decomposition which was constructed in Section 3.3. We could not have worried less: first of all given *any* finite pattern η we could have adjusted the notion of irreducible decomposition in such a way that (3.45) would become true. Furthermore, the import of (3.45) is a possibility to work with independent random variables. In the situation we consider here, sums of finitely dependent variables have qualitatively the same local asymptotics as sums of i.i.d.s. So, for the sake of exposition, let us assume that η satisfies (3.45).

But then, given some small $\delta > 0$, $R(\gamma) = \delta N_\eta(\gamma)$ qualifies as a small perturbation of the type considered in the preceding section. Accordingly, we are in a position to develop a standard local limit analysis of restricted partition functions of the type:
$$Z_n^h\left(N_\eta(\gamma) = \lfloor nx \rfloor\right).$$

Theorem 3.3 *Let η be a finite pattern, $\lambda > \lambda_0$ and $h \in \partial \mathbf{K}_\lambda$. Then there exist $x_\eta \in (0,1)$, $\epsilon > 0$, $\nu > 0$ and a rate function J_h^η on $(x_\eta - \epsilon, x_\eta + \epsilon)$ with quadratic minimum at x_η, such that:*
$$\mathbb{P}_n^h\left(\left|\frac{N_\eta(\gamma)}{n} - x_\eta\right| \geq \epsilon\right) \leq e^{-\nu n},$$

and, for $x \in (x_\eta - \epsilon, x_\eta + \epsilon)$:
$$\mathbb{P}_n^h\left(N_\eta(\gamma) = \lfloor nx \rfloor\right) = \frac{G_\eta(x)}{\sqrt{n}} e^{-n J_h^\eta(x)} \left(1 + \mathrm{o}(1)\right),$$

where, of course, G_η is a positive real analytic function on $[x_\eta - \epsilon, x_\eta + \epsilon]$

Note that we have used only $|N_\eta(\gamma)| \leq |\gamma|$ and (3.45). Therefore, the above theorem holds for a wider class of path observables, as for example:
$$N(\gamma) = \sum_b \mathbf{1}_{\{l_b(\gamma) > 1\}}.$$

3.A Existence of Lyapunov exponents

3.A.1 Attractive case

In the attractive case both $\mathbb{D}_\lambda$ and $\mathbb{H}_\lambda$ are super-multiplicative:
$$\mathbb{D}_\lambda(x+y) \geq \mathbb{D}_\lambda(x)\mathbb{D}_\lambda(y) \quad \text{and} \quad \mathbb{H}_\lambda(x+y) \geq \mathbb{H}_\lambda(x)\mathbb{H}_\lambda(y).$$

Hence both limits in (3.6) exist.

Furthermore, since the potential Φ is monotone in γ, $0 \leq \Phi(\gamma) \leq \Phi(\gamma \cup \eta)$:
$$\mathbb{D}_\lambda(x) \leq \mathbb{H}_\lambda(x) \cdot \sum_{\eta: 0 \mapsto 0} e^{-\lambda |\eta|}. \tag{A.1}$$

The relation (3.3) immediately implies that the right-most term in (A.1) is convergent for every $\lambda > \lambda_0$, and, consequently, that both limits in (3.6) are equal.

In order to check (3.3) note first of all that since Φ is nonnegative, the partition function Z_n is bounded above by the total number of all n-step nearest neighbour trajectories; $Z_n \leq (2d)^n$. For the reverse direction, proceeding as in (Flury 2007b), let us fix a number R and note that in view of the monotonicity of ϕ:
$$Z_n \geq \frac{Z_n(\gamma \subset B_R)}{(2d)^n} e^{n \log(2d) - |B_R \cap \mathbb{Z}^d| \phi(n)} = \mathbb{P}_{\mathrm{SRW}}(\gamma \subset B_R) e^{n \log(2d) - |B_R \cap \mathbb{Z}^d| \phi(n)},$$

where $\mathbb{P}_{SRW}$ is the distribution of simple random walk on $\mathbb{Z}^d$. Since (Hryniv and Velenik 2004)
$$\mathbb{P}_{SRW}\left(\gamma \subset B_R\right) \geq \exp\left\{-c(d)\frac{n}{R^2}\right\},$$
we conclude that for all R and n:
$$\frac{1}{n}\log Z_n \geq \log(2d) - \frac{c(d)}{R^2} - |B_R \cap \mathbb{Z}^d|\frac{\phi(n)}{n}.$$
Thus, (3.3) indeed follows from **(SL)**.

Inequality (3.7) is a trivial consequence of convexity and lattice symmetries. It remains to check that in the attractive case ξ_{λ_0} is strictly positive. However, again as in (Flury 2007b):
$$\mathbb{H}_\lambda(x) = \sum_{n=\|x\|_1}^{\infty} \mathbb{E}_{SRW} e^{-\Phi(\gamma)-(\lambda-\lambda_0)n} \mathbf{1}_{\tau_x=n},$$
where τ_x is the first hitting time of x. As a result, $\xi_\lambda(x) \geq \phi(1)\|x\|_1$ for all $\lambda \geq \lambda_0$.

3.A.2 Repulsive case

First of all:
$$\mathbb{D}_\lambda(x) \leq \mathbb{H}_\lambda(x)\mathbb{D}_\lambda(0).$$
Thus, in view of (3.5), for every $\lambda > \lambda_0$ both exponents in (3.6) are automatically equal once it is proved that at least one of them is defined. We shall show that the first limit in (3.6) exists, using a familiar bubble diagram method. Consider:
$$\mathbb{H}_\lambda(x)\mathbb{H}_\lambda(y) = \sum_{\gamma \in \mathcal{H}_x} \sum_{\eta \in x+\mathcal{H}_y} e^{-\Phi(\gamma)-\Phi(\eta)}.$$
Given a couple of trajectories $\gamma \in \mathcal{H}_x$ and $\eta \in x + \mathcal{H}_y$ define:
$$\underline{n} = \min\{l \ : \ \gamma(l) \in \eta\} \quad \text{and} \quad \overline{n} = \max\{k \ : \ \eta(k) = \gamma(\underline{n})\}.$$
The concatenated path:
$$\tilde{\gamma} \stackrel{\Delta}{=} (\gamma(0),\ldots,\gamma(\underline{n}),\eta(\overline{n}+1),\ldots,\eta(|\eta|)) \stackrel{\Delta}{=} \gamma_1 \cup \eta_1 \in \mathcal{H}_{x+y}.$$
Define also:
$$\gamma_2 = (\gamma(\underline{n}),\ldots,\gamma(|\gamma|)) \quad \text{and} \quad \eta_2 = (\eta(0),\ldots,\eta(\overline{n})).$$
Evidently, one can recover γ and η from $\tilde{\gamma}$, γ_2 and η_2 up to, perhaps, an interchange of γ_2 and η_2. Now, since γ_1 and η_1 intersect only at the end-point:
$$\Phi(\gamma) + \Phi(\eta) \geq \Phi(\gamma_1) + \Phi(\gamma_2) + \Phi(\eta_1) + \Phi(\eta_2)$$
$$= \Phi(\tilde{\gamma}) - \phi(1) + \Phi(\gamma_2) + \Phi(\eta_2).$$

As a result:
$$\mathbb{H}_\lambda(x)\mathbb{H}_\lambda(y) \leq 2e^{\phi(1)}\left(\sum_z \mathbb{D}_\lambda(z)^2\right)\mathbb{H}_\lambda(x+y). \qquad (A.2)$$

But $\sum_z \mathbb{D}_\lambda(z)^2 < \infty$ for every $\lambda > \lambda_0$, and the first limit in (3.6) is indeed well defined.

It remains to show that in the repulsive case $\xi_{\lambda_0} \equiv 0$. If this is not the case, then, as can be easily deduced from (A.2), there exists some finite constant c such that:
$$\mathbb{D}_\lambda(x) \leq ce^{-\xi_\lambda(x)} \leq ce^{-\xi_{\lambda_0}(x)},$$
uniformly in $\lambda > \lambda_0$ and $x \in \mathbb{Z}^d$. By Fatou this would mean that $\sum_x \mathbb{D}_{\lambda_0}(x)$ converges. However, since we are in the repulsive case, $Z_n \geq e^{\lambda_0 n}$, and consequently:
$$\sum_x \mathbb{D}_{\lambda_0}(x) = \sum_n Z_n e^{-\lambda_0 n} = \infty.$$

Acknowledgments

This work was partly supported by the Swiss NSF grant #200020-113256.

References

Campanino, M. and Ioffe, D. (2002). Ornstein-Zernike theory for the Bernoulli bond percolation on $\mathbb{Z}^d$. *Ann. Probab.* **30**(2), 652–82.

Campanino, M., Ioffe, D. and Velenik, Y. (2003). Ornstein-Zernike theory for finite range Ising models above T_c. *Probab. Theory Related Fields* **125**(3), 305–49.

Campanino, M., Ioffe, D. and Velenik, Y. (2004). Random path representation and sharp correlations asymptotics at high-temperatures. In *Stochastic Analysis on Large Scale Interacting Systems*, Volume 39 of *Adv. Stud. Pure Math.*, pp. 29–52. Math. Soc. Japan: Tokyo.

Campanino, M., Ioffe, D. and Velenik, Y. (2007). Fluctuation theory of connectivities for subcritical random cluster models. *Preprint; to appear in Annals of Probability*.

Chayes, J. T. and Chayes, L. (1986). Ornstein-Zernike behavior for self-avoiding walks at all noncritical temperatures. *Comm. Math. Phys.* **105**(2), 221–38.

Chayes, L. (2007). Ballistic behavior for biased self-avoiding walks. *Preprint*.

Flury, M. (2007a). Coincidence of lyapunov exponents for random walks in weak random potentials. *Preprint, arXiv:math/0608357*.

Flury, M. (2007b). Large deviations and phase transition for random walks in random nonnegative potentials. *Stochastic Process. Appl.* **117**(5), 596–612.

Hryniv, O. and Velenik, Y. (2004). Universality of critical behaviour in a class of recurrent random walks. *Probab. Theory Related Fields* **130**(2), 222–58.

Kesten, H. (1963). On the number of self-avoiding walks. *J. Mathematical Phys.* **4**, 960–69.

Mehra, V. and Grassberger, P. (2002). Transition to localization of biased walkers in a randomly absorbing environment. *Physica D: Nonlinear Phenomena* **168–169**, 244–57.

van der Hofstad, R. and Holmes, M. (2007). An expansion for self-interacting random walks. *Preprint, arXiv:0706.0614*.

B

MICROSCOPIC TO MACROSCOPIC TRANSITION

4

STOCHASTIC HOMOGENIZATION AND ENERGY OF INFINITE SETS OF POINTS

Xavier Blanc

Abstract

This article presents in a synthetic way a series of joint works with C. Le Bris and P.-L. Lions (Blanc et al. 2003, 2006, 2007b,a). They are devoted on the one side to the definition of average energies for nonperiodic infinite set of points, and on the other hand to a setting for stochastic homogenization. We will present both aspects, and highlight the link between them.

4.1 Introduction

The aim of this article is to give an overview on the series of papers of Blanc et al. (2003, 2007a) on the one hand, and the papers of Blanc et al. (2006, 2007b) on the other hand. The first ones deal with the thermodynamic limit for some infinite non periodic set of points, and the others with a setting for stochastic homogenization.

This setting for stochastic homogenization of elliptic operators, presented in Section 4.4 below, consists in defining the coefficients of the elliptic PDE as the deformation of periodic ones by a random diffeomorphism, namely:

$$A(y,\omega) = A_{per}\left(\Phi^{-1}(y,\omega)\right),$$

where A_{per} is a periodic square matrix, and Φ is a random diffeomorphism having a stationary gradient (all this is made precise in Section 4.4). We then prove that the problem:

$$-\operatorname{div}\left[A\left(\frac{x}{\varepsilon},\omega\right)\nabla u\right] = f, \qquad (4.1)$$

with Dirichlet boundary condition has a limit as $\varepsilon \to 0$ in the sense of homogenization, and that the homogenized matrix A^* (which is deterministic and constant in space) may be computed using corrector problems.

The definition of average energy for infinite set of points is exposed in Section 4.5. The idea there is to define geometric assumptions on a deterministic set of points $\{X_i\}_{i \in \mathbb{Z}^d}$ of $\mathbb{R}^d$ which allow us to define the average energy of $\{X_i\}_{i \in \mathbb{Z}^d}$. In the special case of a two-body interaction W, this corresponds to the existence of the limit:

$$\lim_{R \to +\infty} \frac{1}{\#\left(\{X_i\}_{i \in \mathbb{Z}^d} \cap B_R\right)} \sum_{X_i \neq X_j \in B_R} W(X_i - X_j). \qquad (4.2)$$

In Blanc et al. (2003), a set of hypotheses was derived which imply that (4.2) exists, even for more sophisticated models than two-body interaction. They are called (H1)–(H2)–(H3) below (see Definition 4.1). Hypotheses (H1) and (H2) are of Delaunay type, meaning that there is no big hole or cluster in the set of points $\{X_i\}_{i\in\mathbb{Z}^d}$, while (H3) is a hypothesis on correlations (in a spatial sense) on $\{X_i\}_{i\in\mathbb{Z}^d}$.

The link between the two aforementioned theories (homogenization and thermodynamic limit) is the following: if the set $\{X_i\}_{i\in\mathbb{Z}^d}$ is random and satisfies:

$$X_i = \Phi(i), \tag{4.3}$$

where Φ is a diffeomorphism such that $\nabla\Phi$ is stationary, then $\{X_i\}_{i\in\mathbb{Z}^d}$ satisfies (H1)–(H2)–(H3) almost surely. But the link between these two settings is even deeper: if $\{X_i\}_{i\in\mathbb{Z}^d}$ is defined by (4.3), then it is possible to define the algebra of functions generated by functions of the form:

$$f(x) = \sum_{i\in\mathbb{Z}^d} \varphi(x - X_i),$$

which we will denote by $\mathcal{A}$. Any $f \in \mathcal{A}$ then satisfies:

$$f(x) = g_{per}\left(\Phi^{-1}(x)\right),$$

where g_{per} is periodic. Hence, if $A \in \mathcal{A}$ is matrix-valued, then the homogenization of the problem (4.1) is exactly the one treated in Section 4.3.

Section 4.2 is devoted to some notations which are useful in the next sections. It can therefore be skipped by readers who are familiar with these notations. Then, Section 4.3 is a (very) short review of homogenization theory, outlining the main ingredients. It is then used as a guideline for the homogenization theory exposed in Section 4.4. Finally, Section 4.5 deals with the above-mentioned notion of average energy and its links with the theory of Section 4.4.

4.2 Definitions and notation

Throughout the article, $\mathcal{D}$ is an open bounded smooth subset of $\mathbb{R}^d$, where d is a positive integer. By 'smooth' we mean that the boundary $\partial\mathcal{D}$ of $\mathcal{D}$ may be locally defined as the nullset of a smooth (i.e, C^∞) function (see Gilbarg and Trudinger 2001 for the details). All the definitions below are valid for $\mathcal{D} = \mathbb{R}^d$.

The set of smooth compactly supported functions is defined as:

$$C_c^\infty(\mathcal{D}) = \{f : \mathcal{D} \to \mathbb{R}, \quad f \text{ is infinitely differentiable,}$$
$$\operatorname{supp}(f) \text{ is a compact subset of } \mathcal{D}\}. \tag{4.4}$$

The space $(C_c^\infty)'(\mathcal{D})$ is the space of distributions on $\mathcal{D}$, i.e., the topological dual of $C_c^\infty(\mathcal{D})$. We will use the standard duality bracket notation:

$$\forall g \in (C_c^\infty)'(\mathcal{D}), \quad \forall f \in C_c^\infty(\mathcal{D}), \quad \langle g, f \rangle := g(f).$$

For any $p \geq 1$, we define $L^p(\mathcal{D})$ as the set of measurable functions $f : \mathcal{D} \to \mathbb{R}$ such that $|f|^p$ is integrable with respect to the Lebesgue measure. If $p = \infty$,

$L^\infty(\mathcal{D})$ is the set of essentially bounded functions. Then, we define the Sobolev spaces:

$$W^{k,p}(\mathcal{D}) = \left\{ f \in L^p(\mathcal{D}), \int_\mathcal{D} |D^k f|^p < +\infty \right\},$$

where k is any nonnegative integer, and $D^k f$ is that k^{th} derivative of f (note that $Df = \nabla f$).[1] In the special case $p = 2$, we will use the standard notation:

$$H^k(\mathcal{D}) = W^{k,2}(\mathcal{D}).$$

The associated norms are:

$$\|f\|_{W^{k,p}(\mathcal{D})} = \sum_{j=0}^{k} \left(\int_\mathcal{D} |D^j f|^p \right)^{1/p}, \quad \|f\|_{H^k(\mathcal{D})} = \sum_{j=0}^{k} \left(\int_\mathcal{D} |D^j f|^2 \right)^{1/2}.$$

This makes $W^{k,p}(\mathcal{D})$ a Banach space. In addition, $H^k(\mathcal{D})$ is a Hilbert space. It is possible to define the trace $f_{|\partial\mathcal{D}}$ of any $f \in H^k(\mathcal{D})$ on the boundary of $\mathcal{D}$. Note that $f_{|\partial\mathcal{D}} \in H^{k-1}(\partial\mathcal{D})$.[2] This allows us to define:

$$H_0^k(\mathcal{D}) = \left\{ f \in H^k(\mathcal{D}), \quad f_{|\partial\mathcal{D}} = 0 \right\}.$$

In other words, $H_0^k(\mathcal{D})$ is the closure of $C_c^\infty(\mathcal{D})$ in $H^k(\mathcal{D})$. Then we define the following.

Definition 4.1 (The spaces $H^{-k}(\mathcal{D})$) *For any smooth open domain $\mathcal{D}$ of $\mathbb{R}^d$ and any $k \in \mathbb{N}$, the space $H^{-k}(\mathcal{D})$ is the dual space (for the L^2 scalar product) of the space $H_0^k(\mathcal{D})$. In other words:*

$$H^{-k}(\mathcal{D}) = \left\{ g \in (C_c^\infty)'(\mathcal{D}), \quad \exists M > 0, \quad \forall f \in H_0^k(\mathcal{D}), \quad |\langle g, f \rangle| \leq M \|f\|_{H^k(\mathcal{D})} \right\}.$$

We define the spaces $W_{\text{loc}}^{k,p}(\mathbb{R}^d)$, or simply $W_{\text{loc}}^{k,p}$, as:

$$W_{\text{loc}}^{k,p}(\mathbb{R}^d) = \left\{ f : \mathbb{R}^d \to \mathbb{R}, \quad f \text{ measurable}, \quad \forall x \in \mathbb{R}^d, \quad f \in W^{k,p}(B_1 + x) \right\},$$

where $B_1 + x$ is the unit ball centred at x. This allows us to define:

Definition 4.2 *For $k \in \mathbb{N}$ and $p \in [1, \infty]$:*

$$W_{\text{unif}}^{k,p}(\mathbb{R}^d) = \left\{ f \in W_{\text{loc}}^{k,p}(\mathbb{R}^d), \quad \sup_{x \in \mathbb{R}^d} \|f\|_{W^{k,p}(B_1+x)} < +\infty \right\},$$

$$\text{and} \|f\|_{W_{\text{unif}}^{k,p}} = \sup_{x \in \mathbb{R}^d} \|f\|_{W^{k,p}(B_1+x)}. \tag{4.5}$$

---

[1] Note that $D^k f$ is a dk-linear form, or equivalently a dk-tensor. Thus, $|D^k f|$ stands for the euclidean norm of $D^k f$ in $\mathbb{R}^{dk}$.

[2] Actually, $f_{|\partial\mathcal{D}} \in H^{k-1/2}(\partial\mathcal{D})$, but we do not want to enter the details of defining H^k for $k \notin \mathbb{N}$.

In the case $k = 0$, we set $L^p_{\text{unif}}(\mathbb{R}^d) = W^{0,p}_{\text{unif}}(\mathbb{R}^d)$, and in the case $p = 2$, we set $H^k_{\text{unif}}(\mathbb{R}^d) = W^{k,2}_{\text{unif}}(\mathbb{R}^d)$.

We now define the notion of weak convergence. For the sake of simplicity, we restrict our attention to the special cases of L^p and H^1. However, these definitions are easily adapted to any Banach space.

Definition 4.3 (Weak convergence in $L^p(\mathcal{D})$) *Let $p \in [1, +\infty)$. Define $q \in (1, +\infty]$ by $\frac{1}{p} + \frac{1}{q} = 1$. We say that a sequence $(f_n)_{n \in \mathbb{N}}$ in $L^p(\mathcal{D})$ converges weakly to $f \in L^p(\mathcal{D})$ if:*

$$\forall g \in L^q(\mathcal{D}), \quad \lim_{n \to \infty} \int_\mathcal{D} f_n g = \int_\mathcal{D} fg.$$

We note this convergence as follows:

$$f_n \rightharpoonup f \quad \text{in} \quad L^p(\mathcal{D}).$$

If $p = +\infty$, we say that the sequence $(f_n)_{n \in \mathbb{N}}$ in $L^\infty(\mathcal{D})$ converges weakly-$*$ to $f \in L^\infty(\mathcal{D})$ if:

$$\forall g \in L^1(\mathcal{D}), \quad \lim_{n \to \infty} \int_\mathcal{D} f_n g = \int_\mathcal{D} fg.$$

We note this convergence as follows:

$$f_n \stackrel{*}{\rightharpoonup} f \quad \text{in} \quad L^\infty(\mathcal{D}).$$

The difference in the notation for $p = +\infty$ is due to the fact that the dual of $L^\infty(\mathcal{D})$ is not $L^1(\mathcal{D})$. We refer to Adams (1975) or Brezis (1983) for the details. Finally, we define the weak convergence in H^1 as follows:

Definition 4.4 (Weak convergence in $H^1(\mathcal{D})$) *A sequence $(f_n)_{n \in \mathbb{N}}$ in $H^1(\mathcal{D})$ converges weakly to $f \in H^1(\mathcal{D})$ if*

$$\forall g \in H^1(\mathcal{D}), \quad \lim_{n \to \infty} \left(\int_\mathcal{D} f_n g + \int_\mathcal{D} \nabla f_n \cdot \nabla g \right) = \int_\mathcal{D} fg + \int_\mathcal{D} \nabla f \cdot \nabla g.$$

We note this convergence as follows:

$$f_n \rightharpoonup f \quad \text{in} \quad H^1(\mathcal{D}).$$

Note that if f_n converges in L^p (resp. H^1), then it converges weakly in L^p (resp. H^1), but the reverse implication is not true. However, any bounded sequence has a subsequence which converges weakly.

We will also need the notion of normalized integral:

Definition 4.5 *For any open subset $\mathcal{D} \subset \mathbb{R}^d$ of finite measure, for any $f \in L^1(\mathcal{D})$, we define the normalized integral of f by:*

$$\fint_\mathcal{D} f := \frac{1}{|\mathcal{D}|} \int_\mathcal{D} f(x) dx. \tag{4.6}$$

4.3 Homogenization of elliptic PDEs

This section is a (very) short review of some aspects of (stochastic) homogenization of linear elliptic PDEs. Subsection 4.3.1 reports on a general theory of homogenization, while subsections 4.3.2 and 4.3.3 indicate how one can go further in the analysis with more specific hypotheses, namely periodicity in Subsection 4.3.2 and stationarity in Subsection 4.3.3.

4.3.1 General theory of homogenization

Let us consider the following model problem:

$$\begin{cases} -\operatorname{div}(A_\varepsilon \nabla u) = f & \text{in } \mathcal{D} \\ u = 0 & \text{on } \partial \mathcal{D}, \end{cases} \tag{4.7}$$

where ε is a small parameter, $f \in L^2(\mathcal{D})$ is independent of ε, and $\mathcal{D}$ is a fixed bounded connected open subset of $\mathbb{R}^d$. The family of matrices $A_\varepsilon = A_\varepsilon(x)$ is assumed to be bounded and uniformly elliptic:

$$\exists \gamma > 0 \quad / \quad \forall \varepsilon > 0, \quad \forall \xi \in \mathbb{R}^d, \quad \xi^T A_\varepsilon(x) \xi \geq \gamma |\xi|^2, \tag{4.8}$$

almost everywhere in $x \in \mathcal{D}$, and:

$$\exists M > 0 \quad / \quad \|A_\varepsilon\|_{L^\infty(\mathcal{D})} \leq M. \tag{4.9}$$

Under these hypotheses, Murat and Tartar (see Murat 1978; Tartar 1979; Murat and Tartar 1997) have proved the following Theorem:

Theorem 4.1 (Murat and Tartar 1997) *Let $\mathcal{D}$ be an open bounded subset of $\mathbb{R}^d$, where d is a positive integer. Consider a set of matrices A_ε satisfying hypotheses (4.8) and (4.9). Let u_ε be the unique solution of (4.7). Then, there exists a sequence $\varepsilon_n \longrightarrow 0$, a matrix A^* satisfying (4.8) and (4.9), and $u^* \in H^1(\mathcal{D})$ such that:*

$$\begin{cases} -\operatorname{div}(A^* \nabla u^*) = f & \text{in } \mathcal{D} \\ u = 0 & \text{on } \partial \mathcal{D}, \end{cases} \tag{4.10}$$

and:

$$u_{\varepsilon_n} \rightharpoonup u^* \text{ in } H^1(\mathcal{D}), \quad A_{\varepsilon_n} \nabla u_{\varepsilon_n} \rightharpoonup A^* \nabla u^* \text{ in } L^2(\mathcal{D}). \tag{4.11}$$

In other words, problem (4.7) has a limit problem, up to extraction of a subsequence, which has the same form, whose coefficients do not depend on f. The main tool for the proof of Theorem 4.1 is the celebrated compensated compactness method (see Murat 1978; Tartar 1979), and more precisely the *div-curl lemma*, which may be stated as follows:

Lemma 4.1 (Murat and Tartar 1997) *Let $\mathcal{D}$ be a bounded open subset of $\mathbb{R}^d$, and $\xi_n \in L^2(\mathcal{D})^d$, $v_n \in L^2(\mathcal{D})$ two sequences such that:*

$$\begin{cases} \xi_n \rightharpoonup \xi \text{ in } L^2(\mathcal{D})^d, \\ \operatorname{div}(\xi_n) \longrightarrow \operatorname{div} \xi \text{ in } H^{-1}(\mathcal{D}), \end{cases}$$

and:
$$\begin{cases} v_n \in H^1(\mathcal{D}), \\ v_n \rightharpoonup v \text{ in } H^1(\mathcal{D}), \end{cases}$$

Then:
$$\xi_n \cdot \nabla v_n \stackrel{*}{\rightharpoonup} \xi \cdot \nabla v \quad \text{in } L^\infty(\mathcal{D}).$$

This Lemma allows us to pass to the limit in product of weakly converging seqences, as for instance $A_\varepsilon \nabla u_\varepsilon$, provided they satisfy some PDE in the limit.

Theorem 4.1 is very powerful in the sense that it provides the existence of a limit problem with very few hypotheses: the assumptions (4.8) and (4.9) on matrices A_ε are very general. However, as one may expect, the lack of information on A_ε implies a lack of information on A^*, which is unknown. The question of giving more explicit formulas, or at least estimates on the coefficients of A^*, is a very difficult question (see for instance Tartar 1997, Jikov et al. 1994 and the references therein). In order to have a more explicit way of computing A^*, one needs more specific assumptions on A_ε. This is the case, for instance, in the two following subsections.

4.3.2 The periodic case

In this subsection, we deal with a special case of matrix A_ε in (4.7), namely:
$$A_\varepsilon(x) = A_{\text{per}}\left(\frac{x}{\varepsilon}\right),$$

where A_{per} is a periodic matrix-valued function. Without loss of generality, we may assume that the periodic cell of A_{per} is the unit cube Q.

We thus consider, in an open bounded domain $\mathcal{D}$ in $\mathbb{R}^d$, the problem:
$$\begin{cases} -\operatorname{div}\left[A_{\text{per}}\left(\frac{x}{\varepsilon}\right)\nabla u_\varepsilon\right] = f & \text{in } \mathcal{D}, \\ u_\varepsilon = 0 & \text{on } \partial \mathcal{D}, \end{cases} \tag{4.12}$$

where the matrix A_{per} is $\mathbb{Z}^d$-periodic.

In order to compute the corresponding homogenized matrix A^*, we define the corrector problem associated with (4.12), which reads, for p fixed in $\mathbb{R}^d$:
$$\begin{cases} -\operatorname{div}\left(A_{per}(y)(p + \nabla w_p)\right) = 0, \\ w_p \text{ is } \mathbb{Z}^d\text{-periodic}. \end{cases} \tag{4.13}$$

It has a unique solution up to the addition of a constant (see Bensoussan et al. 1978). Then, the homogenized coefficients read:
$$\begin{aligned} A^*_{ij} &= \int_Q (e_i + \nabla w_{e_i}(y))^T A_{per}(y)(e_j + \nabla w_{e_j}(y))\, dy \\ &= \int_Q (e_i + \nabla w_{e_i}(y))^T A_{per}(y) e_j\, dy, \end{aligned} \tag{4.14}$$

where Q is the unit cube. As ε goes to zero, the solution u_ε to (4.12) converges to u^*, the solution to:

$$\begin{cases} -\operatorname{div}[A^*\nabla u^*] = f & \text{in } \mathcal{D}, \\ u^* = 0 & \text{on } \partial\mathcal{D}, \end{cases} \quad (4.15)$$

The convergence holds in $L^2(\mathcal{D})$, and weakly in $H_0^1(\mathcal{D})$. The correctors w_{e_i} (for e_i the canonical vectors of $\mathbb{R}^d$) may then also be used to 'correct' u^* in order to identify the behaviour of u_ε in the strong topology $H_0^1(\mathcal{D})$. The proof here again uses the div-curl lemma, and the (trivial) fact that if $f \in L^\infty(\mathbb{R}^d)$ is $\mathbb{Z}^d$-periodic, then:

$$f\left(\frac{x}{\varepsilon}\right) \overset{*}{\rightharpoonup} \int_Q f, \quad (4.16)$$

in $L^\infty(\mathbb{R}^d)$.

This periodic homogenization theory is exposed in many textbooks. See, for instance, Bensoussan et al. 1978; Babuška 1976; Cioranescu and Donato 1999; Persson et al. 1993; and Jikov et al. 1994.

Remark 4.1 *Of course, the above approach using the div-curl lemma is not the only one for the homogenization of linear elliptic problems. Let us mention for instance the two-scale convergence, introduced in Nguetseng (1989) and further developed in Allaire (1992) and Cioranescu et al. (2002) for the periodic setting. It was then adapted to more general settings (Lukkassen et al. 2002; Nguetseng 2003, 2004b; and Visintin 2007).*

4.3.3 The stationary ergodic case

A natural question is, do the explicit formulas of the preceding subsection carry through to some nonperiodic setting? The answer is yes for the special case of stationary ergodic setting, which is a natural extension of the periodic case. In order to introduce it, we first set the notation.

In what follows, $(\Omega, \mathcal{F}, \mathbb{P})$ denotes a probability space, and $\mathbb{E}$ is the expectation value associated with $\mathbb{P}$. We fix $d \in \mathbb{N}^*$, and assume that the group $(\mathbb{Z}^d, +)$ acts on Ω. We denote by $(\tau_k)_{k \in \mathbb{Z}^d}$ this action, and assume that it preserves the measure $\mathbb{P}$, i.e:

$$\forall k \in \mathbb{Z}^d, \quad \forall A \in \mathcal{F}, \quad \mathbb{P}(\tau_k A) = \mathbb{P}(A). \quad (4.17)$$

We assume that τ is *ergodic*, that is:

$$\forall A \in \mathcal{F}, \quad \left(\forall k \in \mathbb{Z}^d, \quad \tau_k A = A\right) \Rightarrow (\mathbb{P}(A) = 0 \text{ or } 1). \quad (4.18)$$

In addition, we define the following notion of stationarity: $F \in L^1_{\text{loc}}\left(\mathbb{R}^d, L^1(\Omega)\right)$ is said to be *stationary* if:

$$\forall k \in \mathbb{Z}^d, \quad F(x+k, \omega) = F(x, \tau_k \omega), \quad (4.19)$$

almost everywhere in x, almost surely.

Note that the case of a periodic function F is a particular case of (4.19), when F is deterministic.

Remark 4.2 *This stationary setting is not the standard one used, for instance in Jikov et al. (1994), which uses $(\mathbb{R}^d, +)$ as the group acting on Ω. However, all the properties we are about to give here extend to this case.*

We thus consider problem (4.7), where:

$$A_\varepsilon(x,\omega) = A\left(\frac{x}{\varepsilon}, \omega\right), \qquad (4.20)$$

with A a stationary ergodic matrix (in the sense of (4.19)). In such a setting, the corrector problem reads, for any $p \in \mathbb{R}^d$:

$$\begin{cases} -\operatorname{div}\left[A(y,\omega)(p + \nabla w_p)\right] = 0, \\ \nabla w_p(y,\omega) \text{ is stationary in the sense of (4.19)}, \\ \int_Q \mathbb{E}(\nabla w_p) = 0. \end{cases} \qquad (4.21)$$

Here again, this system has a unique solution (the proof is easily adapted from Jikov et al. 1994), up to the addition of a (random) constant. Using the solution of the corrector problem, one can compute the matrix A^* with the formula:

$$A^*_{ij} = \int_Q \mathbb{E}\left[(e_i + \nabla w_{e_i})^T A (e_j + \nabla w_{e_j})\right] = \int_Q \mathbb{E}\left[(e_i + \nabla w_{e_i})^T A e_j\right]. \quad (4.22)$$

In this setting again, the proof of convergence to the homogenized problem:

$$\begin{cases} -\operatorname{div}\left[A^* \nabla u^*\right] = f & \text{in } \mathcal{D}, \\ u^* = 0 & \text{on } \partial \mathcal{D}, \end{cases}$$

relies on the div-curl lemma. In addition, one also needs the ergodic theorem (see Kallenberg 2002; Krengel 1985, or Shiryaev 1996, for instance), which in this setting can be stated as follows:

Theorem 4.2 (Ergodic theorem, Kallenberg 2002) *Let $F \in L^\infty(\mathbb{R}^d, L^1(\Omega))$ be a stationary random variable in the sense of (4.19). For $k = (k_1, k_2, \ldots k_d) \in \mathbb{R}^d$, we set $|k|_\infty = \sup_{1 \leq i \leq d} |k_i|$. Then:*

$$\frac{1}{(2N+1)^d} \sum_{|k|_\infty \leq N} F(x, \tau_k \omega) \xrightarrow[N \to \infty]{} \mathbb{E}(F(x, \cdot)) \quad \text{in } L^\infty(\mathbb{R}^d), \text{ almost surely.} \qquad (4.23)$$

This implies that (here, Q is the unit cube):

$$F\left(\frac{x}{\varepsilon}, \omega\right) \xrightharpoonup[\varepsilon \to 0]{*} \mathbb{E}\left(\int_Q F(x, \cdot) dx\right) \quad \text{in } L^\infty(\mathbb{R}^d), \text{ almost surely.} \qquad (4.24)$$

As can be seen in (4.24), this theorem replaces the convergence property (4.16) of the periodic setting. Roughly speaking, the above considerations indicate that the periodic case extends to the stationary one, with the integrals over the cube replaced by expectation values of integrals over the unit cube in all the formulas.

Let us finally point out once again that the present *discrete* setting has a *continuous* counterpart, where the group $(\mathbb{Z}^d, +)$ is replaced by $(\mathbb{R}^d, +)$. In such a case, the operator $\int_Q \mathbb{E}$ is simply replaced by $\mathbb{E}$ in all the above formulas. However, we stick here to the discrete case because it is a more natural setting for the following section.

4.4 Random deformations of periodic problems

We next introduce a different stationary setting: the idea is to use a periodic geometry as a reference, and deform it randomly, with some kind of stationarity. Note that this is not a particular case of the standard (discrete or continuous) stationary setting.

The probability space $(\Omega, \mathcal{F}, \mathbb{P})$ enjoys the same properties as in Subsection 4.3.3, and the action τ of $(\mathbb{Z}^d, +)$ on Ω is ergodic. We consider the model problem (4.7), with:

$$A_\varepsilon(x, \omega) = A_{per}\left(\Phi^{-1}\left(\frac{x}{\varepsilon}, \omega\right)\right), \tag{4.25}$$

that is, we consider the following problem:

$$\begin{cases} -\operatorname{div}\left[A_{per}\left(\Phi^{-1}\left(\frac{x}{\varepsilon}, \omega\right)\right) \nabla u\right] = f & \text{in } \mathcal{D}, \\ u = 0 & \text{on } \partial \mathcal{D}, \end{cases} \tag{4.26}$$

where the function $\Phi(\cdot, \omega)$ is assumed to be a diffeomorphism from $\mathbb{R}^d$ to $\mathbb{R}^d$ for $\mathbb{P}$-almost every ω. The diffeomorphism is assumed to additionally satisfy:

$$\operatorname*{EssInf}_{\omega \in \Omega,\, x \in \mathbb{R}^d} [\det(\nabla \Phi(x, \omega))] = \nu > 0, \tag{4.27}$$

$$\operatorname*{EssSup}_{\omega \in \Omega,\, x \in \mathbb{R}^d} (|\nabla \Phi(x, \omega)|) = M < \infty, \tag{4.28}$$

$$\nabla \Phi(x, \omega) \quad \text{is stationary in the sense of (4.19)}. \tag{4.29}$$

Such a Φ will be called a *random stationary diffeomorphism*.

The following results are proved in Blanc et al. (2006) and Blanc et al. (2007b) (Recall that L^2_{unif} is defined in (4.5)):

Theorem 4.3 (Blanc et al. 2006) *Let A_{per} be a square matrix which is $\mathbb{Z}^d$-periodic and satisfies (4.8)–(4.9) and Φ a random stationary diffeomorphism* satisfying hypotheses (4.27)–(4.28)–(4.29). Then for any $p \in \mathbb{R}^d$, the system:

$$\begin{cases} -\operatorname{div}\left[A_{per}\left(\Phi^{-1}(y,\omega)\right)(p+\nabla w_p)\right] = 0, \\ w_p(y,\omega) = \tilde{w}_p\left(\Phi^{-1}(y,\omega),\omega\right), \quad \nabla \tilde{w}_p \text{ is stationary}, \\ \mathbb{E}\left(\int_{\Phi(Q,\cdot)} \nabla w_p(y,\cdot)dy\right) = 0, \end{cases} \quad (4.30)$$

has a solution in $\{w \in L^2_{\text{loc}}(\mathbb{R}^d, L^2(\Omega)), \ \nabla w \in L^2_{\text{unif}}(\mathbb{R}^d, L^2(\Omega))\}$. Moreover, this solution is unique up to the addition of a (random) constant.

Theorem 4.4 (Blanc et al. 2006) *Let $\mathcal{D}$ be a bounded smooth open subset of $\mathbb{R}^d$, and let $f \in H^{-1}(\mathcal{D})$. Let A_{per} and Φ satisfy the hypotheses of Theorem 4.3. Then the solution $u_\varepsilon(x,\omega)$ of (4.26) satisfies the following properties:*

(i) $u_\varepsilon(x,\omega)$ converges to some $u_0(x)$ strongly in $L^2(\mathcal{D})$ and weakly in $H^1(\mathcal{D})$, almost surely;

(ii) the function u_0 is the solution to the homogenized problem:

$$\begin{cases} -\operatorname{div}(A^* \nabla u) = f \quad \text{in} \quad \mathcal{D}, \\ u = 0 \quad \text{on} \quad \partial \mathcal{D}. \end{cases} \quad (4.31)$$

In (4.31), the homogenized matrix A^* is defined by:

$$A^*_{ij} = \det\left(\mathbb{E}\left(\int_Q \nabla \Phi(z,\cdot)dz\right)\right)^{-1}$$
$$\times \mathbb{E}\left(\int_{\Phi(Q,\cdot)} (e_i + \nabla w_{e_i}(y,\cdot))^T A_{per}\left(\Phi^{-1}(y,\cdot)\right) e_j \, dy\right), \quad (4.32)$$

where for any $p \in \mathbb{R}^d$, w_p is the corrector defined by (4.30).

We will not give the proof of Theorems 4.3 and 4.4, since it is contained in Blanc et al. (2006), and to some extent in Blanc et al. (2007b). Let us only point out that, as in the standard ergodic case, the central points are:

- the div-curl lemma;
- the convergence of rescaled functions, which here is closely related to the property $\varepsilon \Phi\left(\frac{x}{\varepsilon}\right) \longrightarrow \int_Q \mathbb{E}(\nabla \Phi) \cdot x$ in $L^\infty_{\text{loc}}(\mathbb{R}^d)$, almost surely. This in particular gives:

$$A_{per}\left(\Phi^{-1}\left(\frac{x}{\varepsilon}\right)\right) \overset{*}{\rightharpoonup} \mathbb{E}\left(\int_Q \det(\nabla \Phi(y,\cdot))dy\right)^{-1}$$
$$\times \mathbb{E}\left(\int_{\Phi(Q,\cdot)} A_{per}\left(\Phi^{-1}(x,\cdot)\right) dx\right).$$

Note that many extensions of the results of the present section are possible, with, for instance, A_{per} being stationary instead of periodic, or with a notion of continuous ergodicity. We refer to Blanc et al. (2007b) for the details.

Let us finally point out that in the particular case when $\Phi(x) = x + \eta\Psi(x)$ with η small, the above results may be expanded as powers of η: the first order is clearly the periodic case, and the second one may be computed in a much simpler way than (4.32). Indeed, its computation only involves solutions to *periodic* corrector problems, rather then *stationary* ones, which are much simpler to compute. All this is detailed in Blanc et al. (2007b).

4.5 Infinite sets of points

This section is devoted to some considerations on the definition of average energies for infinite sets of points. Subsection 4.5.1 deals with the deterministic case, which was treated in Blanc et al. (2003), while Subsection 4.5.2 deals with a stochastic case, as treated in Blanc et al. (2007a).

As above, let us first have a look at the periodic case, then at the stationary case: consider thus a set of N particles of positions $\{X_i\}_{1 \le i \le N}$ in $\mathbb{R}^d$ interacting by a two-body potential W, which is supposed to be smooth, radially symmetric, and decay fast enough at infinity. Then, the energy per particle is defined by:

$$\mathcal{E}(\{X_i\}_{1 \le i \le N}) = \frac{1}{2N} \sum_{1 \le i \ne j \le N} W(X_i - X_j). \tag{4.33}$$

Assume that the set $\{X_i\}_{1 \le i \le N}$ is a subset of some lattice, say, $\mathbb{Z}^d$. Then, if $\{X_i\}_{1 \le i \le N}$ converges to $\mathbb{Z}^d$ in a suitable sense (this is the case, for instance, when $\{X_i\}_{1 \le i \le N} = \mathbb{Z}^d \cap B_{R_N}$ with R_N going to infinity), a simple computation proves that the energy (4.33) has a limit as N goes to infinity, namely:

$$\lim_{N \to +\infty} \mathcal{E}(\{X_i\}_{1 \le i \le N}) = \frac{1}{2} \sum_{j \in \mathbb{Z}^d \setminus \{0\}} W(j). \tag{4.34}$$

Of course, this simple computation carries through to the case of a stationary sequence X_i. Indeed, using the same stochastic setting as in Subsection 4.3.3, if we assume that the sequence $\{X_i\}_{i \in \mathbb{Z}^d}$, now indexed by $\mathbb{Z}^d$ rather than $\mathbb{N}$, is stationary, i.e:

$$\forall i, j \in \mathbb{Z}^d, \quad X_{i+j}(\omega) = X_i(\tau_j \omega), \tag{4.35}$$

almost surely, than we have the following convergence:

$$\lim_{N \to +\infty} \left(\frac{1}{(2N+1)^d} \sum_{X_i \in (2N+1)Q} \sum_{X_j \in (2N+1)Q, j \ne i} W(X_i - X_j) \right)$$
$$= \mathbb{E} \left(\sum_{j \in \mathbb{Z}^d \setminus \{0\}} W(X_j - X_0) \right), \tag{4.36}$$

almost surely. The proof is rather simple: noticing that the inner sum in the right-hand side is almost equal to:

$$\sum_{i \in \mathbb{Z}^d \setminus \{0\}} W(X_i - X_j) := F_i(\omega),$$

one easily proves that F_i is stationary in the sense of (4.35). Hence, applying the ergodic theorem (Theorem 4.2), one finds (4.36).

As in Section 4.3, these two particular cases (periodic and stationary) will serve as a guideline in the following subsections.

4.5.1 The deterministic case

The question considered in Blanc et al. (2003) is the following: what are the most general (deterministic) hypotheses which allow us to carry out the same kind of computation as (4.34). The following answer was given:

Definition 4.6 *We shall say that a set of points $\{X_i\}_{i \in \mathbb{N}}$ is admissible if it satisfies the following:*

(H1) $\sup_{x \in \mathbb{R}^d} \#\{i \in \mathbb{N} \ / \ |x - X_i| < 1\} < +\infty;$

(H2) $\exists R > 0$ *such that* $\inf_{x \in \mathbb{R}^d} \#\{i \in \mathbb{N} \ / \ |x - X_i| < R\} > 0;$

(H3) for any $n \in \mathbb{N}$, the following limit exists:

$$\lim_{R \to \infty} \frac{1}{|B_R|} \sum_{X_{i_0} \in B_R} \cdots \sum_{X_{i_n} \in B_R} \delta_{(X_{i_0} - X_{i_1}, \ldots X_{i_0} - X_{i_n})}(h_1, \ldots, h_n)$$
$$= l^n(h_1, \ldots, h_n), \tag{4.37}$$

and is a nonnegative uniformly locally bounded measure.

We use here the convention that if $n = 0$, l^0 is the constant function equal to:

$$l^0 = \lim_{R \to \infty} \frac{1}{|B_R|} \#\{i \in \mathbb{N} \ / \ X_i \in B_R\}. \tag{4.38}$$

It is proved in Blanc et al. (2003) that if $\{X_i\}_{i \in \mathbb{N}}$ is an admissible set of points, then a formula similar to (4.34) holds:

$$\lim_{R \to +\infty} \left(\frac{1}{\#(B_R \cap \{X_i\}_{i \in \mathbb{N}})} \sum_{X_i \neq X_j \in B_R} W(X_i - X_j) \right) = \frac{1}{l^0} \int_{\mathbb{R}^3} W(h) l^1(dh). \tag{4.39}$$

Actually, for this special case of two-body interactions, one only needs hypothesis (H3) for $n = 0, 1$. Indeed, these are exactly the convergence one needs in the computation leading to (4.39). However, since Blanc et al. (2003) aimed at dealing with more general models, such as N-body energies, hypotheses (H1)–(H2)–(H3) were naturally derived: the first one corresponds to the fact that there is no cluster of particles, the second one that there is no big hole in the position configuration, and the third one should be seen as the existence of an n-body average correlation between the X_i. As pointed out in Blanc et al. (2003), these hypotheses are not logically linked.

Given Definition 4.6, one can define the corresponding functional spaces:

Definition 4.7 *Let $\{X_i\}_{i\in\mathbb{N}}$ be an admissible set, and denote by $\mathcal{A}(\{X_i\})$ the vector space generated by the functions of the form:*

$$f(x) = \sum_{i_1\in\mathbb{N}}\sum_{i_2\in\mathbb{N}}\cdots\sum_{i_n\in\mathbb{N}} \varphi(x - X_{i_1}, x - X_{i_2}, \ldots, x - X_{i_n}), \qquad (4.40)$$

with $\varphi \in C_c^\infty(\mathbb{R}^{3n})$. Then, for any $k \in \mathbb{N}$ and any $p \in [1,+\infty)$, we denote by $\mathcal{A}^{k,p}(\{X_i\})$, or simply $\mathcal{A}^{k,p}$ when there is no ambiguity, the closure of $\mathcal{A}(\{X_i\})$ for the norm $\|\cdot\|_{W^{k,p}_{\text{unif}}}$.

When $k = 0$, we use the notation $\mathcal{A}^p$ for $\mathcal{A}^{0,p}$. The closure of $\mathcal{A}$ for the norm $\|\cdot\|_{L^\infty(\mathbb{R}^3)}$ being a set of continuous functions, we will denote it by $\mathcal{A}_c$. We will call $\mathcal{A}^\infty$ the closure for the $L^\infty(\mathbb{R}^d)$ norm of the space of functions of the form (4.40), with $\varphi \in L^\infty(\mathbb{R}^d)$ having compact support.

Note that $\mathcal{A}^{k,p}$ is the closure for the $W^{k,p}_{\text{unif}}$ norm of the algebra generated by functions of the form:

$$f(x) = \sum_{i\in\mathbb{N}} \varphi(x - X_i), \quad \varphi \in \mathcal{D}(\mathbb{R}^d).$$

Moreover, in the particular case of a periodic lattice $\{X_i\}_{i\in\mathbb{N}}$, $\mathcal{A}^{k,p}(\{X_i\}_{i\in\mathbb{N}})$ is the algebra of periodic functions with the appropriate period and regularity.

A direct consequence of the above definitions is the following

Lemma 4.2 (Blanc et al. 2003) *Let $\{X_i\}_{i\in\mathbb{N}}$ be an admissible (in the sense of Definition 4.6) set of points. Then, for any $f \in \mathcal{A}^{k,p}$, the following limit exists:*

$$\langle f \rangle := \lim_{R\to\infty} \fint_{B_R} f.$$

In addition, in the special case of an f of the form (4.40), we have:

$$\langle f \rangle = \int_{\mathbb{R}^d}\int_{\mathbb{R}^{d(n-1)}} \varphi(x, x - h_1, \ldots, x - h_{n-1})dl^{n-1}(h_1, \ldots, h_{n-1})dx. \qquad (4.41)$$

4.5.2 Random deformation of periodic lattices

Being inspired by the stochastic setting introduced in Section 4.4, let us now consider a set of points $\{X_i\}_{i\in\mathbb{Z}^d}$, which is the deformation of a periodic lattice by a stationary diffeomorphism Φ. More precisely, we assume that Φ satisfies (4.27)–(4.28)–(4.29), and we define:

$$\forall i \in \mathbb{Z}^d, \quad X_i(\omega) = \Phi(i, \omega). \tag{4.42}$$

The relation between Definition 4.6 and the notion of stationary diffeomorphism is best illustrated by:

Proposition 4.1 (Blanc et al. 2003) *Let Φ be a stationary diffeomorphism, i.e a diffeomorphism satisfying (4.27)–(4.28)–(4.29). Let the set $\{X_i(\omega)\}_{i \in \mathbb{Z}^d}$ be defined by (4.42). Then, $\{X_i\}_{i \in \mathbb{Z}^d}$ satisfies (H1)–(H2)–(H3) of Definition 4.6, almost surely.*

This proposition allows us to assert that the average energy of this (stochastic) set of points may be defined in the same way as in (4.36). In addition, it is also possible to construct the algebras $\mathcal{A}^{k,p}$ defined in Subsection 4.5.1. This is done in:

Proposition 4.2 *Let Φ be a stationary diffeomorphism, i.e a diffeomorphism satisfying (4.27)–(4.28)–(4.29). Let the set $\{X_i(\omega)\}_{i \in \mathbb{Z}^d}$ be defined by (4.42). Define $\mathcal{A}$ as the vector space generated by the functions of the form:*

$$f(x, \omega) = \sum_{i_1 \in \mathbb{Z}^d} \sum_{i_2 \in \mathbb{Z}^d} \cdots \sum_{i_n \in \mathbb{Z}^d} \varphi(x - X_{i_1}(\omega), x - X_{i_2}(\omega), \ldots, x - X_{i_n}(\omega)), \tag{4.43}$$

with $\varphi \in C_c^\infty(\mathbb{R}^{3n})$. Denote by $\mathcal{A}^{k,p}$ the closure of $\mathcal{A}$ for the $L^1\left(\Omega, W_{\text{unif}}^{k,p}(\mathbb{R}^d)\right)$ norm. Then for any $f \in \mathcal{A}^{k,p}$, the following limit exists almost surely:

$$\langle f \rangle = \lim_{R \to \infty} \fint_{B_R} f.$$

In addition, $\langle f \rangle$ does not depend on ω, and:

$$f\left(\frac{x}{\varepsilon}, \omega\right) \xrightarrow[\varepsilon \to 0]{*} \langle f \rangle, \quad \text{almost surely}. \tag{4.44}$$

Let us point out that the hypotheses (H1)–(H2)–(H3) *do not* imply the convergence (4.44). This property is closely linked with stationarity, which is a form of translation invariance. Note indeed that in hypothesis (H3), the origin plays a special role, which prevents any form of translation invariance from being implied by this assumption.

Remark 4.3 *The proof of Proposition 4.2, is based on the following property: $\forall f \in \mathcal{A}^{k,p}$, there exists $g \in W_{\text{unif}}^{k,p}\left(\mathbb{R}^d, L^1(\Omega)\right)$ such that g is stationary, i.e:*

$$\forall j \in \mathbb{Z}^d, \quad \forall x \in \mathbb{R}^d, \quad g(x + j, \omega) = g(x, \tau_j \omega),$$

almost surely, and:

$$\forall x \in \mathbb{R}^d, \quad f(x, \omega) = g\left(\Phi^{-1}(x, \omega), \omega\right),$$

almost surely.

We are now in position to relate Subsection 4.5.2 and the homogenization setting discussed in Section 4.4. We recall Φ is a stationary diffeomorphism (i.e Φ satisfies (4.27)–(4.28)–(4.29)), and the set $\{X_i\}_{i\in\mathbb{Z}^d}$ is defined by (4.42), that is:

$$\forall i \in \mathbb{Z}^d, \quad X_i(\omega) = \Phi(i,\omega).$$

In addition, the algebras $\mathcal{A}^{k,p}$ are defined as in Proposition 4.2. Hence, if we consider a matrix $A \in \mathcal{A}^\infty(\{X_i\}_{i\in\mathbb{Z}^d})$, Remark 4.3 implies that there exists a stationary matrix B such that:

$$A(x,\omega) = B\left(\Phi^{-1}(x,\omega),\omega\right).$$

Consequently, Theorems 4.3 and 4.4 apply to the present case, giving:

Theorem 4.5 *Let $A \in \mathcal{A}^\infty(\{X_i\}_{i\in\mathbb{Z}^d})$ be a square matrix which satisfies (4.8)–(4.9). Then for any $p \in \mathbb{R}^d$, the system:*

$$\begin{cases} -\operatorname{div}\left[A(y,\omega)(p+\nabla w_p)\right] = 0, \\ w_p(y,\omega) = \tilde{w}_p\left(\Phi^{-1}(y,\omega),\omega\right), \quad \nabla \tilde{w}_p \text{ is stationary in the sense of (4.19)} \\ \mathbb{E}\left(\int_{\Phi(Q)} \nabla w_p(y,\cdot)dy\right) = 0, \end{cases} \quad (4.45)$$

has a solution in $\{w \in L^2_{\text{loc}}(\mathbb{R}^d, L^2(\Omega)), \nabla w \in L^2_{\text{unif}}(\mathbb{R}^d, L^2(\Omega))\}$. In addition, this solution is unique up to the addition of a (random) constant.

Theorem 4.6 *Let $\mathcal{D}$ be a bounded smooth open subset of $\mathbb{R}^d$, and let $f \in H^{-1}(\mathcal{D})$. Let A satisfy the hypotheses of Theorem 4.5. Then the solution $u_\varepsilon(x,\omega)$ of:*

$$\begin{cases} -\operatorname{div}\left(A\left(\frac{x}{\varepsilon},\omega\right)\nabla u\right) = f & \text{in } \mathcal{D}, \\ u = 0 & \text{on } \partial\mathcal{D} \end{cases} \quad (4.46)$$

satisfies the following properties:

(i) $u_\varepsilon(x,\omega)$ converges to some $u_0(x)$ strongly in $L^2(\mathcal{D})$ and weakly in $H^1(\mathcal{D})$, almost surely;

(ii) the function u_0 is a solution to the homogenized problem:

$$\begin{cases} -\operatorname{div}(A^*\nabla u) = f & \text{in } \mathcal{D}, \\ u = 0 & \text{on } \partial\mathcal{D}. \end{cases} \quad (4.47)$$

In (4.47), the homogenized matrix A^* is defined by:

$$A^*_{ij} = \det\left(\mathbb{E}\left(\int_Q \nabla\Phi(z,\cdot)dz\right)\right)^{-1}$$
$$\times \mathbb{E}\left(\int_{\Phi(Q,\cdot)} (e_i + \nabla w_{e_i}(y,\cdot))^T A(y,\cdot) e_j \, dy\right), \quad (4.48)$$

where for any $p \in \mathbb{R}^d$, w_p is the corrector defined by the system (4.45).

We end this section with two remarks:

Remark 4.4 *The above considerations on the random lattices may be generalized further. In particular, only the increments $X_i - X_j$ are present in the energy. Therefore it is possible to study the thermodynamic limit in the case of stationary increments. Furthermore, the energy should be a function of the (infinite) set of points as a whole. It is thus natural to study the thermodynamic limit for a form of stationarity involving only the set $\ell = \{X_i\}$, that is, $\ell(\tau_k \omega) = \ell(\omega) - k$. All the above considerations apply to this case* mutatis mutandis. *We refer to Blanc et al. (2007a,b) for the details.*

Remark 4.5 *It is also possible to define a deterministic setting for non periodic homogenization, in the spirit of the thermodynamic limit study of Subsection 4.5.1. However, one in general adds a form of translation invariance to the hypothesis (H3) in order to be able to do so. We refer to Nguetseng (1989, 2003, 2004a,b) for studies of this kind.*

References

Adams, R. A. (1975). *Sobolev Spaces*. Pure and Applied Mathematics, Vol. 65. Academic Press [A subsidiary of Harcourt Brace Jovanovich, Publishers], New York-London.

Allaire, G. (1992). Homogenization and two-scale convergence. *SIAM J. Math. Anal.* **23**(6), 1482–1518.

Babuška, I. (1976). Homogenization approach in engineering. In *Computing Methods in Applied Sciences and Engineering (Second Internat. Sympos., Versailles, 1975), Part 1*, pp. 137–53. Lecture Notes in Econom. and Math. Systems, Vol. 134. Springer: Berlin.

Bensoussan, A., Lions, J.-L. and Papanicolaou, G. (1978). *Asymptotic Analysis for Periodic Structures*, Volume 5 of *Studies in Mathematics and its Applications*. North-Holland Publishing Co: Amsterdam.

Blanc, X., Le Bris, C. and Lions, P.-L. (2003). A definition of the ground state energy for systems composed of infinitely many particles. *Comm. Partial Differential Equations* **28**(1-2), 439–75.

Blanc, X., Le Bris, C. and Lions, P.-L. (2006). Une variante de la théorie de l'homogénéisation stochastique des opérateurs elliptiques. *C. R. Math. Acad. Sci. Paris* **343**(11-12), 717–24.

Blanc, X., Le Bris, C. and Lions, P.-L. (2007a). Stochastic homogenization and random lattices. *J. Math. Pures Appl.* **88**(1), 34–63.

Blanc, X., Le Bris, C. and Lions P.-L. (2007b). The energy of some microscopic stochastic lattices. *Arch. Ration. Mech. Anal.* **184**(2), 303–39.

Brezis, H. (1983). *Analyse fonctionnelle*. Collection Mathématiques Appliquées pour la Maîtrise. [Collection of Applied Mathematics for the Master's Degree]. Paris: Masson. Théorie et applications. [Theory and applications].

Cioranescu, D., Damlamian, A. and Griso, G. (2002). Periodic unfolding and homogenization. *C. R. Math. Acad. Sci. Paris* **335**(1), 99–104.

Cioranescu, D. and Donato, P. (1999). *An Introduction to Homogenization*, Volume 17 of *Oxford Lecture Series in Mathematics and its Applications*. The Clarendon Press Oxford University Press: New York.

Gilbarg, D. and Trudinger, N. S. (2001). *Elliptic Partial Differential Equations of Second Order*. Classics in Mathematics. Springer-Verlag: Berlin. Reprint of the 1998 edition.

Jikov, V. V., Kozlov, S. M. and Oleĭnik, O. A. (1994). *Homogenization of Differential Operators and Integral Functionals*. Springer-Verlag: Berlin. Translated from the Russian by G. A. Yosifian [G. A. Iosif′ yan].

Kallenberg, O. (2002). *Foundations of modern probability*. 2nd ed. Probability and Its Applications. xvii, 638 p. Springer: New York, NY.

Krengel, U. (1985). *Ergodic Theorems*, Volume 6 of *de Gruyter Studies in Mathematics*. Walter de Gruyter & Co: Berlin. With a supplement by Antoine Brunel.

Lukkassen, D., Nguetseng, G. and Wall, P. (2002). Two-scale convergence. *Int. J. Pure Appl. Math.* **2**(1), 35–86.

Murat, F. (1978). Compacité par compensation. *Ann. Scuola Norm. Sup. Pisa Cl. Sci. (4)* **5**(3), 489–507.

Murat, F. and Tartar, L. (1997). H-convergence. In *Topics in the Mathematical Modelling of Composite Materials*, Volume 31 of *Progr. Nonlinear Differential Equations Appl.*, pp. 139–173. Birkhäuser Boston: Boston, MA.

Nguetseng, G. (1989). A general convergence result for a functional related to the theory of homogenization. *SIAM J. Math. Anal.* **20**(3), 608–23.

Nguetseng, G. (2003). Homogenization structures and applications. I. *Z. Anal. Anwendungen* **22**(1), 73–107.

Nguetseng, G. (2004a). Homogenization in perforated domains beyond the periodic setting. *J. Math. Anal. Appl.* **289**(2), 608–28.

Nguetseng, G. (2004b). Homogenization structures and applications. II. *Z. Anal. Anwendungen* **23**(3), 483–508.

Persson, L. E., Persson, L., Svanstedt, N. and Wyller, J. (1993). *The Homogenization Method*. Lund: Studentlitteratur. An introduction.

Shiryaev, A. N. (1996). *Probability* (Second ed.), Volume 95 of *Graduate Texts in Mathematics*. Springer-Verlag: New York. Translated from the first (1980) Russian edition by R. P. Boas.

Tartar, L. (1979). Compensated compactness and applications to partial differential equations. In *Nonlinear Analysis and Mechanics: Heriot-Watt*

Symposium, Vol. IV, Volume 39 of *Res. Notes in Math.*, pp. 136–212. Pitman: Boston, Mass.

Tartar, L. (1997). Estimations of homogenized coefficients. In *Topics in the Mathematical Modelling of Composite Materials*, Volume 31 of *Progr. Nonlinear Differential Equations Appl.*, pp. 9–20. Birkhäuser Boston: Boston, MA.

Visintin, A. (2007). Two-scale convergence of some integral functionals. *Calc. Var. Partial Differential Equations* **29**(2), 239–65.

5

VALIDITY AND NON-VALIDITY OF PROPAGATION OF CHAOS

Karsten Matthies and Florian Theil

Abstract

We develop a novel rigorous approach to analyse the validity of continuum approximations for deterministic interacting particle systems. Some of our ideas have been used earlier in the context of annihilating Brownian spheres (Sznitman 1991). We study the Boltzmann–Grad limit of ballistic annihilation, a topic which has has received considerable attention in the physics literature. Due to the deterministic nature of the evolution it is possible that correlations build up and the mean-field approximation by the Boltzmann equation breaks down. We find a sharp condition on the initial distribution which ensures the validity of the Boltzmann equation and demonstrate the failure of the mean-field theory if the condition is violated.

The derivation of the continuum models of mathematical physics from atomistic descriptions is a longstanding and fundamental problem. One of the most notorious challenges is the question whether the Hamiltonian nature of the fundamental laws of motion (quantum mechanics, Newtonian mechanics) is compatible with the fact that the second principle of thermodynamics postulates that macroscopic systems are irreversible. An illustration of this question is provided by deterministic hard ball dynamics with random initial states. For high particle densities and suitably scaled diameters it is expected that the time-evolution of the density is close to the solution of the Boltzmann equation:

$$\partial_t f + v \cdot \partial_u f = \int_{\mathbb{R}^d \times S^{d-1}} (f(u,\tilde{v})f(u,\tilde{v}') - f(u,v)f(u,v'))\left((v-v') \cdot \nu\right)_+ d\nu' \, d\nu,$$

(5.1)

where $\tilde{v}, \tilde{v}'$ are obtained from v, v' by exchanging the respective components of v and v' in direction ν, that is:

$$\tilde{v} = v + (v'-v) \cdot \nu \nu, \quad \tilde{v}' = v' + (v-v') \cdot \nu \nu,$$

and $f_t(u,v)$ is the density of presence at time t of particles at locations u with velocity v, see Spohn (1991).

An important concept which sheds some light on the connection between the Boltzmann equation and hard ball dynamics is the propagation of chaos. This means that the distribution $p_N(u_1, v_1 \ldots, u_N, v_N, t)$ of N particles will lose its product structure for nonzero time t. However, the marginal distribution of the first k particles should be very close to a product measure when the total number of particles N is large. A classical method to establish propagation of chaos is to express the evolution of k-particle marginals in terms of the $k+1$-particle marginals. This strategy is implemented in the BBGKY hierarchy. The weakness of this approach consists in the fact that establishing convergence of the resulting series is hard in many cases. O. Lanford succeeded in proving that in the case of hard ball dynamics the series that corresponds to the BBGKY hierarchy converges for small times to a solution of the Boltzmann equation (Lanford 1975). Unfortunately it cannot be shown that the time interval where the series is known to converge is larger than a small fraction of the mean free time, regardless of the initial data. This problem was partially overcome by Illner and Pulvirenti (1989) who managed to obtain a global result if the positions are in $\mathbb{R}^d$ and the initial density is sufficiently small. However, currently there is no result which covers the case where both data and time are large. It is arguable that the justification of the Boltzmann equation (5.1) as a scaling limit of deterministic evolution constitutes a part of Hilbert's sixth problem (Hilbert 2000).

In Lang and Nguyen (1980) the same strategy is applied to the simpler problem of coagulation. Here the spheres move along Brownian paths and two intact spheres annihilate each other if the distance between the centres drops below a. Although the series generated by the BBGKY hierarchy does not converge globally in time, Lang and Nguyen were able to give a rigorous justification of the corresponding Boltzmann equation by restarting the procedure at small positive time.

In this paper we consider kinetic annihilation, another simplification of hard ball dynamics which keeps two central features of the original evolution: the initial state is random, the evolution is deterministic. We assume that the initial phase space positions in the phase space $\mathbb{T}^d \times \mathbb{R}^d$ ($\mathbb{T}^d$ is the unit torus) form a Poisson point process with some intensity $\mu \in M(\mathbb{T}^d \times \mathbb{R}^d)$. As long as they are intact the centres of the spheres move along straight lines with constant velocity. When the centres of two spheres, which are still intact, come within distance a, then both spheres are destroyed.

We will consider the asymptotic behaviour of the system in the limit where the diameter a of the particles tends to 0 and the total intensity $n = \mu(\mathbb{T}^d \times \mathbb{R}^d)$ is linked to a by the Boltzmann–Grad relation:

$$na^{d-1} = 1. \tag{5.2}$$

The central question in this paper is whether for small values of a the many-body evolution can be described by the gainless Boltzmann equation:

$$\partial_t f + v \cdot \partial_u f = Q_-[f, f], \tag{5.3}$$

where $Q_-[f,g](v) = -\kappa_d f(v) \int_{\mathbb{R}^d} \mathrm{d}g(v') \, |v-v'|$ is the loss term of the hard-sphere collision kernel of the Boltzmann equation (5.1) and κ_d is the volume of the $d-1$

dimensional unit-ball. For the sake of simplicity we will restrict ourselves to the case where the initial density f_0 does not depend on u, in this case the transport term $v \cdot \partial_u f$ in equation (5.3) vanishes and $f_t(u,v) = f_t(v)$.

Solutions f of the Boltzmann equation (5.3) have two different but closely related interpretations. The macroscopic interpretation involves the empirical quantity $b(U,t) = \#(\omega(t) \cap U)$ where $U \subset \mathbb{T}^d \times \mathbb{R}^d$ is arbitrary (measurable) and $\omega(t) \subset \mathbb{T}^d \times \mathbb{R}^d$ is the realization of the phase space positions of the particles at time t. Equation (5.3) is valid in the macroscopic sense if $b(U,t)/n$ converges to $\int_U du\, df_t(v)$ in probability as a tends to 0.

A microscopic interpretation is based on the single-particle marginals of the N-particle distribution. The gainless homogeneous Boltzmann equation is valid in the microscopic sense if:

$$\lim_{a \to 0} \text{Prob}((u_1(t), v_1(t)) \in U \text{ and particle 1 is intact at time } t)$$
$$= \frac{1}{f_0(\mathbb{R}^d)} \int_U du\, df_t(v).$$

Since the distribution of the N particles is invariant under permutation it is irrelevant which particle index we use to define the microscopic validity.

It is well known that microscopic validity and a simple bound on correlations implies macroscopic validity. Ballistic annihilation has been studied extensively in the physics literature, see Elskens and Frisch (1985); Piasecki (1995); Droz et al. (1995, 2002); and Coppex et al. (2003). Kinetic annihilation dynamics can be used to model growth and coarsening of surfaces, see Krug and Spohn (1988).

This paper contains an outline of the essential steps of the first mathematical proof of the microscopic validity of the Boltzmann equation as a scaling limit of kinetic annihilation (Theorem 5.1). The main idea is to determine the probability distribution of objects that are adapted to the system under consideration. In this case we work with marked trees that record the history of potential collisions which a tagged particle experiences. The trees have the property that their expected size remains finite as a tends to 0. This idea has been used earlier in Sznitman (1991) in the context of Brownian spheres. Note however that our results differ significantly from those in Sznitman (1991).

First of all, we are working with a *deterministic* evolution. The Boltzmann equation emerges because of random initial conditions. We obtain a limiting measure in the *phase space* $\mathbb{T}^d \times \mathbb{R}^d$, not in *position* space $\mathbb{T}^d$. Since we consider initial distributions which are u-independent we end up with measures on the velocity space $\mathbb{R}^d$.

Secondly, for a large subset of trees we obtain a very simple, explicit representation formula for the distribution of the trees (5.38). Thanks to this formula we are able to establish explicit $o(1)$-bounds of the total variation difference between empirical distribution $\hat{P}$ and the limiting measure P as a tends to 0.

Thirdly, since the only source of stochasticity are the initial values, it is less obvious that the initial chaos is propagated to such a degree that the limiting evolution can be described by a simple mean-field theory which leads to the Boltzmann equation. We obtain novel necessary conditions on the absence of certain concentrations in the initial density (5.7) which are sufficient for the validity of the Boltzmann equation. A counter-example (Theorem 5.2) demonstrates that, if the concentrations are present in the initial density, the mean-field theory that underlies the Boltzmann equation is not consistent with the many body evolution. This shows on the one hand, that our condition is actually sharp, and on the other hand, that a previously published justification of the Boltzmann equation by Droz et al. (2002) requires the additional assumption that the initial velocity density is absolutely continuous with respect to the Lebesgue measure.

Since the Boltzmann equation is insensitive to these concentrations it is impossible to derive condition (5.7) from the mean-field theory itself.

Failure of the Boltzmann approximation of high-dimensional deterministic many-body systems has been previously observed in the case of ballistic annihilation for for $d = 1$, see Elskens and Frisch (1985) and for discrete velocity models of collisional dynamics, see Ushiyama (1988). In both cases the failure of the mean-field theory can be traced back to the the finiteness of the set of possible directions. Our analysis shows that the buildup of correlations is actually caused by concentrations in the initial distributions, not by the specifics of the evolution.

5.1 Validity

We consider N particles with initial values $(u_0(i), v_0(i)) \in \mathbb{T}^d \times \mathbb{R}^d$, $i = 1\ldots N$, which evolve by Newtonian dynamics:

$$u(i, t = 0) = u_0(i), \quad v(i, t = 0) = v_0(i),$$
$$\dot{u}(i,t) = v(i,t), \quad \dot{v}(i,t) = 0. \tag{5.4}$$

For each $t \in [0, \infty)$, $i \in \{1\ldots N\}$ with $z_i = (u_i, v_i)$, there exists a unique scattering state $\beta^{(a)}(i,t) \in \{0,1\}$. Here $\beta^{(a)}(i,t) = 1$ means that the ith particle has not collided with another particle up to time t. The scattering state satisfies the implicit relation:

$$\beta^{(a)}(i,t) = \begin{cases} 1 & \text{if } \text{dist}(z_i, z_{i'}, s) \geq a\beta^{(a)}(i', s) \text{ for all } s \in [0,t),\ i' \neq i, \\ 0 & \text{else} \end{cases} \tag{5.5}$$

with a modified distance function to ignore initial intersections:

$$\text{dist}((u,v),(u',v'),s) = |u - u' + s(v - v')| + a\chi_{[0,a]}(|u - u'|), \tag{5.6}$$

where $|.|$ is the metric on the torus $\mathbb{T}^d = \mathbb{R}^d/\mathbb{Z}^d$ and χ_A is the characteristic function of A.

Definition 5.1 (Tagged Poisson point processes) *Let Ω be a measure space. The random variable $z \in \bigcup_{N=0}^{\infty} \Omega^N$ is a realization of a Poisson point process (PPP) with density $\mu \in M_+(\Omega)$ if:*

$$\mathrm{Prob}(z \in \Omega^N) = e^{-\mu(\Omega)} \frac{\mu(\Omega)^N}{N!}, \quad \mathrm{law}(z_i) = \mu/\mu(\Omega),$$

and $z_1, \ldots, z_N$ are independent. Realizations of the tagged Poisson point process (tPPP) are obtained by adding an independent random variable z_0 with law $\mu/\mu(\Omega)$, i.e. for symmetric $A \subset \bigcup_{N=1}^{\infty} \Omega^N$ one obtains that:

$$\mathrm{Prob}_{\mathrm{tPPP}}((z_0, z) \in A) = \frac{1}{\mu(\Omega) e^{\mu(\Omega)}} \sum_{N=0}^{\infty} \frac{1}{N!} \int_{A \cap \Omega^{N+1}} \mathrm{d}\mu(z_0) \ldots \mathrm{d}\mu(z_N).$$

Note that the tagged PPP is a symmetric point process. The motivation for working with this process is that the realizations of the tagged PPP without the tagged particle form a PPP and we obtain a very simple explicit formula for the distribution of trees, see (5.38), hence the complexity of the proof can be reduced. On the other hand, it seems that the formulae for the joint distribution of two trees are much more complicated, therefore we will only make statements which concern the law of a single, tagged particle.

Theorem 5.1 *(Justification of the gainless homogeneous Boltzmann equation) Let $f_0 \in PM_+(\mathbb{R}^d), d \geq 2$ be a momentum density with finite second moment ($\int_{\mathbb{R}^d} \mathrm{d} f_0(v)(1 + |v|^2) < \infty$) which does not concentrate mass on lines:*

$$f_0(v + \mathbb{R} w) = 0 \quad \text{for all } v, w \in \mathbb{R}^d. \tag{5.7}$$

Assume that the intensity of the tagged PPP is $\mu = n(\mathbf{1}_{\mathbb{T}^d} \otimes f_0)$, with n given by (5.2), then:

$$\lim_{a \to 0} \mathrm{Prob}_{\mathrm{tPPP}} \left(z(0, t) \in A \text{ and } \beta^{(a)}(0, t) = 1 \right) = \int_A \mathrm{d}u \, \mathrm{d} f_t(v), \tag{5.8}$$

where $z(0, t), \beta(0, t)$ are position and status of the tagged particle at time t and $f : [0, \infty) \to M_+(\mathbb{R}^d)$ solves the gainless homogeneous Boltzmann equation:

$$\partial_t f = Q_-[f, f], \quad f(t = 0) = f_0. \tag{5.9}$$

The assumption that $\int_{\mathbb{R}^d} \mathrm{d} f_0(v) = 1$ is not necessary. We make it because it simplifies the notation in the proof.

Assumption (5.7) does not exclude the possibility that f_0 is concentrated on lower dimensional subsets, for example the uniform distribution on the sphere S^{d-1} is admissible, i.e., f_0 satisfies:

$$\int \varphi(v) \, \mathrm{d} f_0(v) := \frac{1}{\mathcal{H}^{d-1}(S^{d-1})} \int_{S^{d-1}} \varphi(v) \, \mathrm{d}\mathcal{H}^{d-1}(v), \tag{5.10}$$

for all test-functions $\varphi \in C_c(\mathbb{T}^d \times \mathbb{R}^d)$, where $\mathcal{H}^d$ is the d-dimensional Hausdorff-measure.

5.2 Effective descriptions

5.2.1 The hierarchy of evolutions

Instead of expanding f_t into a power-series in t and matching coefficients, in a first step, we replace the initial value problem (5.9) by an infinite system using general initial distribution without concentrations:

$$\dot{f}_k = Q_-[f_{k-1}, f_k], \quad f_{t=0,k} = f_0. \tag{5.11}$$

Since Q_- is quadratic, for fixed k the integro-differential equation (5.11) is in fact linear and nonautonomous. We can therefore work with the mathematically much more convenient mild formulation. The differential equation completely decouples in v and the equation for each v is a scalar linear nonautonomous ODE, which can be directly integrated to:

$$f_{t,k} = \exp(-\int_0^t L[f_{s,k-1}]\,\mathrm{d}s)f_0, \tag{5.12}$$

where $L[f](v) = \kappa_d \int \mathrm{d}f(v')\,|v-v'|$. We observe that $\mathrm{d}f_{t,k}(v)$ is absolutely continuous with respect to $\mathrm{d}f_0(v)$ due to the decoupling in v.

Lemma 5.1 Let $f_0 \in M_{(1+|v|)^2}$ then f_k converges in $C_\rho^0([0,\infty), M_{1+|v|})$ to f for some $\rho > 0$ and $f \in C^1([0,\infty), M_{1+|v|})$ is the unique solution of (5.9).

By $M_{1+|v|}$ and $M_{(1+|v|)^2}$ we mean the set of Radon measures on $\mathbb{R}^d$ with first and second moment, C_ρ denotes the continuous functions on $\mathbb{R}^+$ which grow not faster than $e^{\rho t}$. The proof of Lemma 5.1 together with a precise definition of the function spaces is standard.

Now we have to translate this idea into the context of deterministic manybody dynamics. To limit the complexity of the notation we will from now on assume that everything except the constants depends on a without displaying the dependency. For every realization of the N-body evolution the random variable $\beta(i,t) \in \{0,1\}$, which encodes the scattering state of particle $i \in \{1\ldots N\}$ at time $t \in [0,\infty)$ satisfies the implicit relation (5.5). The computation of the scattering state β can be simplified by introducing a hierarchy of artificial evolutions indexed by $k \in \mathbb{N}$. We assume that the initial values of the particles at all levels are identical. The particles at level $k=1$ are simply transported and do not interact with anything. The particles at level $k > 1$ interact only with the particles at level $k-1$, but not with each other. For each $k \in \mathbb{N}$ and $i \in \{1\ldots N\}$ the scattering state $\beta_k(i,t) \in \{0,1\}$ is defined in the following way:

$$\beta_k(i,t) = \begin{cases} 1 & \text{if } \mathrm{dist}(z_i, z_{i'}, s) \geq a\beta_{k-1}(i', s) \text{ for all } s \in [0,t), i' \neq i, \\ 0 & \text{else}, \end{cases} \tag{5.13}$$

$$\beta_1(i) \equiv 1, \tag{5.14}$$

with dist as in (5.6).

Remark 5.1 *While the determination of the collision-state $\beta(i,t)$ is a complicated problem, the state $\beta_k(i,t)$ emerges via a very simple calculation from $\beta_{k-1}(\cdot,t)$.*

Lemma 5.2 *For all realizations of the initial conditions $\omega \in \bigcup_{N=1}^{\infty}(\mathbb{T}^d \times \mathbb{R}^d)^N$ both $\beta_k(i,t)$ and $\beta(i,t)$ are well defined and:*

$$\lim_{k \to \infty} \beta_k(i,t) = \beta(i,t) \tag{5.15}$$

pointwise in i and uniformly in t.

5.2.2 The concept of marked trees

The translation of the N-body evolution into scattering states β is greatly facilitated by the concept of trees. In the collision tree with root (u,v) we will collect information of collisions and potential collisions up to time t for a particle with initial data u,v.

As an example assume that $N = 4$ and consider the scenario in Fig. 5.1 where the letters A, B, C, D are the labels of the four particles, the empty circles are the initial positions and the arrows are the initial velocities. Consequentially the arrow-tips indicate the positions of the particles at time $t=1$. To determine whether a certain particle has been scattered before time $t=1$ it suffices to analyse the associated collision tree which is constructed as follows: the particle of interest is the root with initial data (u,v). The particles which are potentially scattered by the root are added as nodes, i.e., a particle with initial data (u',v') is added, if $|u+sv-(u'+sv')| \leq a$ for some $s \in [0,t]$. This procedure is recursively

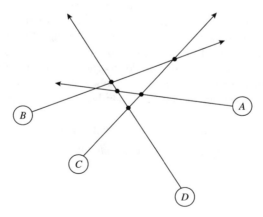

FIG. 5.1: Initial positions and velocities of four particles. The bullets indicate the positions where the particles are potentially scattered. The shown configuration is not very likely and consequentially the collision trees are quite complex.

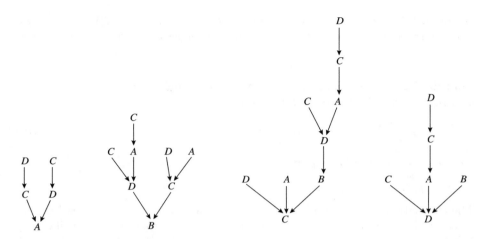

FIG. 5.2: Collision trees of the four particles with initial positions and collision structure given in Fig. 5.1. At time $t = 1$ particles C and D have been scattered, particles A and B have not. Note that the labels of the particles which generate the potential scattering events are only included in the picture in order to illustrate the translation of Fig. 5.1 into collision trees. The scattering state of the particle at the root is completely determined by the tree structure, the labels of the tree nodes are irrelevant. For example, the tree of particle B does not contain enough information to decide whether particle A is scattered.

applied to every node but we consider only potential scattering events which are upstream, i.e., before the event which is responsible for adding the leaf. The four collision trees associated to the scenario in Fig. 5.1 are shown in Fig. 5.2. The extraction of the collision trees amounts to a significant reduction of the complexity of the problem. In general, the number of potential scattering events (bullets) is proportional to N but thanks to the Boltzmann–Grad-scaling (5.2) the number of nodes in the individual trees is a Poissonian random number with an intensity which is asymptotically independent of N and grows exponentially with t, see Proposition 5.1.

We now convert the example into a general concept.

Definition 5.1 Let $\mathbb{N} = \{1, 2, \ldots\}$. The height of a node (or multi-index) $l \in \mathbb{N}^i$ is defined by $|l| := i$, the parent node of $l \in \mathbb{N}^i$ is $\bar{l} = (l_1, \ldots, l_{i-1})$. Let $\mathcal{F} = \bigcup_{i=1}^{\infty} \mathbb{N}^i$ be the set of multi-indices. We say that $m \subset \mathcal{F}$ is a tree with root ($m \in \mathcal{T}$), if:

1. $\#m < \infty$,
2. $m \cap \mathbb{N} = \{1\}$,
3. $\bar{l} \in m$ for all $l \in m \setminus \mathbb{N}$,
4. $l - 1 \in m$ for all $l \in m$ such that $l \neq (*, \ldots, *, 1)$,

where $l-1 = l-(0,\ldots,0,1)$. We say that a tree m has at most height k ($m \in \mathcal{T}_k$) if $m \cap \mathbb{N}^{k+1} = \emptyset$.

Let $Y = \{(u, v, s, \nu) \in \mathbb{T}^d \times \mathbb{R}^d \times [0, \infty) \times S^{d-1}\}$ be the space of initial values and collision parameters. The set of marked trees is given by:

$$\mathcal{MT} = \left\{ (m, \phi) \,\bigg|\, m \in \mathcal{T}, \phi: m \to Y \text{ with the property } s_l \in [s_{l-1}, s_{\bar{l}}] \right.$$
$$\left. \text{and } \nu_l = \tfrac{1}{a}(u_{\bar{l}} - u_l + s_l(v_{\bar{l}} - v_l)) \text{ for all } l \in m \setminus \mathbb{N} \right\},$$

where $s_{(*,\ldots*,0)} = 0$. For each skeleton $m \in \mathcal{T}$ we define the set:

$$\mathcal{E}(m) = \{(\tilde{m}, \phi) \in \mathcal{MT} \mid \tilde{m} = m\}, \tag{5.16}$$

which contains all trees with skeleton m. We stipulate a strict order of the set of nodes m:

$$l < l' \text{ if either } |l| < |l'| \text{ or } (|l| = |l'| \text{ and } \bar{l} < \bar{l}') \text{ or } (\bar{l} = \bar{l}' \text{ and } l_{|l|} < l'_{|l|}). \tag{5.17}$$

This order is induced by the link between the collision time and the indices $l \in m$. For example, $\{(1), (1,1), (1,2), (1,3), (1,1,1), (1,1,2)\} \in \mathcal{T}_3$, but $\{(1), (2,1)\}$ is not a tree skeleton. The assumption $s_l \in [s_{l-1}, s_{\bar{l}}]$ implies that for all nontrivial permutations $\pi \in S_{\#m} \setminus \text{Id}$ (S_n is the set of permutations of n symbols) and all trees $\Phi = (m, \phi) \in \mathcal{MT}$ the permuted tree $\Phi^\pi = (m, \phi^\pi)$ with $\phi_l^\pi = \phi_{\pi(l)}$ is not a tree in the sense of Definition 5.1.

The value ν_1 has no relevance. To circumvent this problem we fix a point $\nu^* \in (S^{d-1})$, define:

$$\mathcal{MT}^* = \{\Phi \in \mathcal{MT} \mid \nu_1 = \nu^*\}.$$

We will in future denote $\mathcal{MT}^*$ by $\mathcal{MT}$.

It is clear from the definition that for each tree $m \in \mathcal{T}$ there exists a function $r: m \to \mathbb{N} \cup \{0\}$ which counts the number of direct successors, i.e:

$$r_l = \#\{l' \in m \mid \bar{l}' = l\}.$$

Remark 5.2 *Graph theoretical description of collisions in a hard-sphere gas can lead to many different graphs, which are not necessarily trees. The advantage of our definition is that this graph will always be a tree. Particles might appear several times in a tree, as in Fig. 5.2. This will not destroy the tree structure, as these are due to different collision events. Multiple collisions, which are well-defined in our setting, can lead to identical branches within the tree, but the definition $\mathcal{T}$ will discriminate between these and the graph of collisions is still a tree.*

The scattering state $\beta: m \to \{0, 1\}$ is determined uniquely by the skeleton, i.e., the labels of the particles are immaterial, but the actual computation is not completely trivial. The most important aspect of the computation of β is that the scattering information flows from the leaves to the root, i.e., the scattering state of a node is completely determined by the state of the nodes above, the nodes below are irrelevant.

We will construct now two families of probability measures $P_{t,k}, \hat{P}_{t,k} \in PM(\mathcal{MT})$. The empirical distribution $\hat{P}_{t,k}$ is induced by the many-body dynamics and will be constructed recursively in Section 5.2.4. The mean-field distribution $P_{t,k}$ is given by an explicit formula (5.18). The link between $P_{t,k}$ and $\hat{P}_{t,k}$ is provided by the set of good trees $\mathcal{G}(a) \subset \mathcal{MT}$ (Definition 5.3) which has the properties that restriction of $\hat{P}_{t,k}$ on $\mathcal{G}(a) \cap \mathcal{MT}$ converges to $P_{t,k}$ and $P_{t,k}(\mathcal{G}(a))$ goes to 1 as a tends to 0 (Proposition 5.2).

This is the crucial step which eventually yields the justification of the mean-field theory. In other words, the main task consists in analysing the mean-field measure $P_{t,k}$, the empirical distribution $\hat{P}_{t,k}$ enters only when we prove that $P_{t,k}$ is consistent with $\hat{P}_{t,k}$.

5.2.3 The mean-field distribution $P_{t,k}$

We construct now the mean-field distribution of trees $P_{t,k} \in PM(\mathcal{MT})$. Let $\Omega \subset \mathcal{MT}$ and $t \in [0, \infty)$. The mean-field probability that the observed tree is in Ω is given by:

$$P_{t,k}(\Omega) = \sum_{m \in \mathcal{T}_k} \int_{\Omega \cap \mathcal{E}(m)} e^{-\sum_{j<k} \Gamma_j(\Phi)} \, d\lambda^m(\phi) \tag{5.18}$$

where $\mathcal{E}(m)$ was defined in (5.16) and

$$\Gamma_j(\Phi) = \sum_{l \in m, |l|=j} \gamma_l(\Phi),$$

$\gamma_l(\Phi) = \int_0^{s_l} L[f_0](v_l) \, ds' = s_l L[f_0](v_l) \geq 0$ is the collision rate of particle l,

$$\lambda^m(\phi) = \mu(z_1) \otimes \delta(s_1 - t) \otimes \prod_{l \in m \setminus \mathbb{N}} \left[((v_l - v_{\bar{l}}) \cdot \nu_l)_+ \chi_{[s_{l-1}, s_{\bar{l}}]}(s_l) \, df_0(v_l) \, d\nu_l \, ds_l \right],$$

(5.19)

$\mu(u, v) = \mathbf{1}_{\mathbb{T}^d}(u) \otimes f_0(v).$

Remark 5.3

1. Note that the positions u_l are completely determined by (u_1, v_1) and $(v_l, s_l, \nu_l)_{l \in m \setminus \{1\}}$. Since we have assumed that ν_1 is fixed, the value of $P_{t,k}(\Omega)$ is well-defined.
2. It is noteworthy that the measures $P_{t,k}$ depend on time only via the parameter t. In other words, time plays the role of a parameter which propagates through the tree and qualifies the local branching structure.
3. For some event $\Omega \subset \mathcal{MT}_k$ the probability $P_{t,k'}(\Omega)$ is independent of k' if $k' > k$. Equivalently, $P_{t,k_1}(\Omega \cap \mathcal{E}(m)) = P_{t,k_2}(\Omega \cap \mathcal{E}(m))$, if the height of m is strictly smaller than $\min\{k_1, k_2\}$.

We can simplify the measure $P_{t,k}$ by integrating over the collision parameters $\nu_l \in S^{d-1}$, $l \in m$. Let $\hat{Y} = \mathbb{R}^d \times [0, \infty)$ be the reduced set of collision data. For every $\Omega \subset \mathcal{T}(\hat{Y})$ we find that when still denoting the collision data as ϕ:

$$\bar{P}_{t,k}(\Omega) = \sum_{m \in \mathcal{T}_k} \int_{\Omega \cap \mathcal{E}(m)} \mathrm{d}\bar{\lambda}^m(\phi)\, e^{-\sum_{j<k} \Gamma_j(\Phi)} \qquad (5.20)$$

with

$$\bar{\lambda}^m(\phi) = f_0(v_1) \otimes \delta(s_1 - t) \otimes \prod_{l \in m \setminus \mathbb{N}} \left[\kappa_d\, |v_l - v_{\bar{l}}|\, \chi_{[s_{l-1}, s_{\bar{l}}]}(s_l) \,\mathrm{d} f_0(v_l)\, \mathrm{d} s_l \right].$$

The measures $P_{t,k}$ have the remarkable property that the expectation of certain random variables can be computed exactly or estimated accurately.

Proposition 5.1 *For a tree $m \in \mathcal{T}$ the number of non-root nodes is given by $R(m) = \sum_{r \in m} r_l = \#m - 1$. The expected value of R satisfies the estimate uniformly in k:*

$$\mathbb{E}(R) \leq K_{\mathrm{ini}} \exp(\kappa_d K_{\mathrm{ini}} t), \qquad (5.21)$$

with $K_{\mathrm{ini}} = \int_{\mathbb{R}^d} \mathrm{d} f_0(v)\, (1 + |v|)^2$ and:

$$P_{t,k+1}(v_1 \in \Omega \text{ and } \beta(1,t) = 1) = \mathrm{d} f_{t,k}(v) \qquad (5.22)$$

holds, where $f_{t,k}$ is the solution of system (5.12).

The proof relies on the recursive structure of the definition of the measure $P_{t,k}$ and can be found in Matthies and Theil (2008).

5.2.4 The empirical distribution $\hat{P}_{t,k}$

We return now to the hierarchy of many body evolutions described in Section 5.2.1. The initial values of the particles form a random set $\omega \subset \mathbb{T}^d \times \mathbb{R}^d$ and it is assumed that the law of ω is the Poisson point process with density μ, where $\mu = n \mathbf{1}_{\mathbb{T}^d} \otimes f_0$. Hence, the size $n = \#\omega$ is Poissonian random variable with intensity n. As explained in Section 5.2.2, the family of probability measures $\hat{P}_{t,k} \in PM(\mathcal{MT})$ is the empirical distribution of the tree Φ which is generated by the many-body evolution and has a randomly chosen (tagged) particle as its root. This tree is only well defined if $N > 0$, i.e. ω is nonempty. For this reason we define $\hat{P}_{t,k}(\Omega)$ as the conditional probability that the tree is contained in the set Ω, given that $N = \#\omega > 0$.

A particularly simple method of sampling from this conditional distribution consists in drawing a realization of ω according to the unconditioned Poisson point process, and an independent random variable $z \in \mathbb{T}^d \times \mathbb{R}^d$ with law $\mu(z) = \mathbf{1}_{\mathbb{T}^d}(u) \otimes f_0(v)$ which is the initial value of the tagged particle. It can be checked without difficulty that the joint distribution of ω and z is the previously defined conditional distribution.

The trees generated by this procedure are denoted by $\Phi(t,k) = (m(t,k), \phi) \in \mathcal{MT}_k$, where $m(t,k) \in \mathcal{T}_k$ is the skeleton and $\phi : m(t,k) \to Y$ specifies the initial values, the collision times and the impact parameters. The measures $\hat{P}_{t,k}$ are the image measure of $\mathrm{Prob}_{\mathrm{tPPP}}$ induced by the many-particle flows so that for each $\Omega \subset \mathcal{MT}$ we obtain:

$$\hat{P}_{t,k}(\Omega) := \mathrm{Prob}_{\mathrm{tPPP}}((m(t,k), \phi) \in \Omega). \tag{5.23}$$

The tree measures $\hat{P}_{t,k}$ are derived from $\mathrm{Prob}_{\mathrm{tPPP}}$, but $\mathrm{Prob}_{\mathrm{tPPP}}$ cannot be derived from $\hat{P}_{t,k}$.

By construction, for fixed ω the skeleton m is monotonously increasing in t and k, and for fixed $l \in m$ the data ϕ_l does not depend on t or k. This is equivalent to saying that the j-marginal of $\hat{P}_{t,k}$ (trees of height $j \leq k$) is given by $\hat{P}_{t,j}$, i.e.:

$$\hat{P}_{t,k}\left(\left(m(t,k) \cap (\bigcup_{i=1}^{j} \mathbb{N}^i), (\phi_l)_{|l| \leq j}\right) \in \Omega\right) = \hat{P}_{t,j}((m(t,j), (\phi_l)_{|l| \leq j}) \in \Omega) \tag{5.24}$$

for all $\Omega \subset \mathcal{MT}_j$, $k \geq j$.

We will use formula (5.24) to construct an alternative characterization of $\hat{P}_{t,k}$ which reflects the iterative process that underlies the definition of $m(t,k)$. Using this alternative characterization one can easily establish total-variation bounds for $P_{t,k} - \hat{P}_{t,k}$. Since the time t is arbitrary but fixed we will often write $\hat{P}_k$ instead of $\hat{P}_{t,k}$.

Let $(m', \phi') \in \mathcal{MT}_{k-1}$ and let $\hat{P}_k(\,\cdot\,|(m', \phi')) \in PM(\mathcal{MT}_k)$ be the conditional distribution of $\hat{P}_k$ in the sense that:

$$\hat{P}_k(\Omega \,|\, (m', \phi')) := \hat{P}_k\Big((m(k), \phi) \in \Omega \,|\, m \cap \mathbb{N}^j = m' \cap \mathbb{N}^j \text{ for all } j \in \{1 \ldots k-1\}$$

$$\text{and } \phi_l = \phi'_l \text{ for all } l \in m \text{ such that } |l| < k\Big).$$

Formula (5.24), which characterizes the j-marginals of $\hat{P}_{t,k}$, yields the following recurrence relation for $\hat{P}_k$:

$$\hat{P}_k(\Omega) = \int_{\mathcal{MT}_{k-1}} \mathrm{d}\hat{P}_{k-1}(\Phi') \, \hat{P}_k(\Omega \,|\, \Phi'). \tag{5.25}$$

Repeating this step $k-1$ times we obtain the following iterative representation of $\hat{P}_k$:

$$\hat{P}_k(\Omega) = \int_{\mathcal{MT}_1} \mathrm{d}P_1(\Phi_1) \int_{\mathcal{MT}_2} \mathrm{d}\hat{P}_2(\Phi_2 \,|\, \Phi_1)$$
$$\ldots \int_{\mathcal{MT}_{k-1}} \mathrm{d}\hat{P}_{k-1}(\Phi_{k-1} \,|\, \Phi_{k-2}) \, \hat{P}_k(\Omega \,|\, \Phi_{k-1}), \tag{5.26}$$

where:

$$P_1(z_1) = \frac{1}{n}\mu(z_1) \in PM\left(\mathbb{T}^d \times \mathbb{R}^d\right) \tag{5.27}$$

is the distribution of initial value for the root particle.

5.2.5 Convergence of $\hat{P}_k$ to P_k

Having constructed an iterative characterization of $\hat{P}_k$ we will now show that it is very similar to the mean field measure P_k in a precise way. The key is to identify the mechanisms by which the two probability distributions fail to be equal. In this part of the paper we will work with the phase-space representation of the trees: $z_l = (u_l, v_l) \in \mathbb{T}^d \times \mathbb{R}^d$.

Remark 5.4 *There are only two reasons why $\hat{P}_k$ may fail to coincide with P_k in the limit $a \to 0$:*

1. *The cylinders which are covered by the paths of the particles might contain self-intersections due to the periodic boundary conditions: $v - v' \in R(t, a)$ with:*

$$R(t, a) = \left\{ v \in \mathbb{R}^d \; \middle| \; \min\left\{ |s\, v - \xi| \; \middle| \; s \in [0, t], \xi \in \mathbb{Z}^d \setminus \{0\} \right\} \leq a \right\}. \quad (5.28)$$

2. *Particles might appear at different positions within a tree, i.e., the map $z : m \to \mathbb{T}^d \times \mathbb{R}^d$ might be not injective.*

The set $R(t, a)$, which can easily seen to be nonempty, is relevant due to periodic boundary conditions, which will lead to self-intersections of the cylinders. This happens, if $v - v_j$ is sufficiently close to a velocity v^*, where the components of $v_1^*, \ldots, v_d^*$ are rationally dependent, i.e. $\eta \cdot v^* \in \mathbb{Z}$ with $\eta \in \mathbb{Z}^d$, but only if $|\eta| \leq t$. The effect is not present in a setting where $(u, v) \in \mathbb{R}^d \times \mathbb{R}^d$.

The second effect is caused by the notorious recollisions. These dependencies disappear as the diameter a tends to zero.

Motivated by Remark 5.4 we define the set of 'good' trees.

Definition 5.3 *For each $a_0 > 0$ the set of 'good' trees $\mathcal{G}(a_0) \subset \mathcal{MT}$ consists of those trees $(m, \phi) \in \mathcal{MT}$ with the property that for all $0 < a \leq a_0$ and all $l \in m$:*

$$v_{\bar{l}} - v_l \in \mathbb{R}^d \setminus R(t, a) \quad \text{(all parent-child-pairs are non-resonant),} \quad (5.29)$$

$$z_l \notin \bigcup_{\substack{l' \leq l \\ l' \neq l}} C_{l'} \quad \text{(no particle appears more than once in the tree),}$$

$$(5.30)$$

where we associate to each node $l \in m$ the set of colliding initial values:

$$C_l = \left\{ z' \in \mathbb{T}^d \times \mathbb{R}^d \; \middle| \; \min_{s' \in [0, s_l]} |\text{dist}(z_l, z', s')| \leq a \right\},$$

and dist *as in (5.6) ignores overlap in the initial data.*

Note that $\mathcal{G}(a_0) \subset \mathcal{MT}$ is a family of sets which increases monotonically with decreasing a_0. An elementary calculation yields that for all $v' \in \mathbb{R}^d \setminus (v_l + R(t, a))$

$$n \, \mathcal{H}^d \left(C_l \cap (\mathbb{T}^d \times \{v'\}) \right) = \kappa_d |v_l - v'| s_l. \quad (5.31)$$

The significance of $\mathcal{G}(a_0)$ is given by the following results:

$$\liminf_{a_0 \to 0} \inf_k P_k(\mathcal{G}(a_0)) = 1, \qquad (5.32)$$

$$\limsup_{a \to 0} \left\{ \left| \hat{P}_k(\Omega) - P_k(\Omega) \right| \; \Big| \; k \in \mathbb{N}, \; \Omega \subset \mathcal{G}(a_0) \right\} = 0 \text{ for fixed } a_0, \qquad (5.33)$$

which are given in Proposition 5.2. For this we need a more explicit characterization of the distributions $\hat{P}_k(\cdot \mid \Phi_{k-1})$ and $\hat{P}_k(\cdot)$.

We recall the following fundamental independence-principle of Poisson-point processes which allows us to compute certain conditional probabilities explicitly.

Lemma 5.3 *Let the random set $\omega \subset \mathbb{T}^d \times \mathbb{R}^d$ be distributed according to a Poisson point process with density μ, $\bar{\mathcal{C}}, \mathcal{C} \subset \mathbb{T}^d \times \mathbb{R}^d$ and $A \subset \bigcup_{r=0}^{\infty} (\mathcal{C} \setminus \bar{\mathcal{C}})^r$ be symmetric. Then we obtain the following formula for the conditional probability of the event A:*

$$\text{Prob}_{\text{tPPP}} \left(\text{tup}(\omega \cap \mathcal{C}) \in A \mid \omega \cap \bar{\mathcal{C}} = \emptyset \right) = \exp\left(-\mu(\mathcal{C} \setminus \bar{\mathcal{C}})\right) \sum_{r=0}^{\infty} \frac{1}{r!} \int_{A \cap \mathcal{C}^r} d\mu^r(z), \qquad (5.34)$$

where $\mu^r = \underbrace{\mu \otimes \ldots \otimes \mu}_{r \text{ terms}}$ and $\text{tup}(\{z_1, \ldots, z_r\}) = (z_1, \ldots, z_r)$.

To apply Lemma 5.3 we have to work with the phase-space representation of trees. Owing to the decomposition $\Omega = \dot{\bigcup}_{m \in \mathcal{T}} \mathcal{E}(m) \cap \Omega$ we can assume that $\Omega \subset \mathcal{E}(m)$ for some $m \in \mathcal{T}$.

Note that for a general tree $\Phi = (m, \phi) \in \mathcal{MT}$ the number of nodes $\#m$ can be bigger than the number of particles involved in the collisions, i.e. it is possible that the map $z : m \to \mathbb{T}^d \times \mathbb{R}^d$ is not injective and $z_l = z_{l'}$ for some pair $l, l' \in m$, $l \neq l'$. This scenario corresponds to a bad tree where the same particle appears twice in the tree. For this reason we restrict our attention to sets Ω which are subsets of $\mathcal{G}(a)$. The excluded set has nonzero probability, however we will show that the probability of $\mathcal{MT} \setminus \mathcal{G}(a)$ tends with a to 0. By construction for all trees in Ω the map $l \mapsto z_l$ is injective.

The order defined by (5.17) induces a representation of the events $\Omega \subset \mathcal{MT}$ in phase-space coordinates:

$$A(\Omega) \subset (\mathbb{T}^d \times \mathbb{R}^d)^{\#m}.$$

In the same spirit one obtains a one-to-one correspondence between the initial values of particles associated with the tree-nodes at height k and subsets of $(\mathbb{T}^d \times \mathbb{R}^d)^{\#m \cap \mathbb{N}^k}$:

$$Z_k = (z_l)_{|l|=k} \in (\mathbb{T}^d \times \mathbb{R}^d)^{\#(m \cap \mathbb{N}^k)}.$$

We will also need the conditional events:

$$A_k(\Omega, \Phi) = \left\{ Z_k \in (\mathbb{T}^d \times \mathbb{R}^d)^{\#(m \cap \mathbb{N}^k)} \mid (Z_k, \Phi) \in \Omega \right\},$$

where $\Phi \in \mathcal{MT}_{k-1}$ and $(Z_k, \Phi) \in \mathcal{MT}_k$ is the tree obtained by attaching the leaves Z_k to the topmost nodes of Φ.

Recall that the density of the Poisson point process which generates the initial positions of the particles is given by $N\mu$ where:

$$\int d\mu(z)\, \varphi(z) = \int_{\mathbb{R}^d} df_0(v) \int_{\mathbb{T}^d} du\, \varphi(u, v)$$

for every test-function $\varphi \in C_c(\mathbb{T}^d \times \mathbb{R}^d)$.

Before applying Lemma 5.3 we have to specify the sets $\mathcal{C}$ and $\bar{\mathcal{C}}$. Fix $a_0 > 0$ and let $\Phi \in \mathcal{MT} \cap \mathcal{G}(a_0)$. We are interested in the distribution of those trees which coincide with Φ up to level k. Clearly, the initial positions of the particles at height $k+1$ are contained in the set:

$$\mathcal{C}_k(\Phi) := \bigcup_{l \in m \cap \mathbb{N}^k} C_l \subset \mathbb{T}^d \times \mathbb{R}^d,$$

with $\Phi = (m, \phi)$. In order to apply formula (5.34) we have to identify the conditioning of the distribution $\omega \cap \mathcal{C}_k(\Phi)$. Define the collection of cylinders:

$$\bar{\mathcal{C}}_k(\Phi) := \bigcup_{|l| < k} C_l \subset \mathbb{T}^d \times \mathbb{R}^d$$

which contains those initial values that would affect the lower nodes. By construction the information on the point process ω that we have accumulated so far is given by $\omega \cap \bar{\mathcal{C}}_k(\Phi) = \{z_l \mid |l| \leq k\}$. Furthermore, since $\Phi \in \mathcal{G}(a_0)$ we have that $\omega \cap \mathcal{C}_k(\Phi) \cap \bar{\mathcal{C}}_k(\Phi) = \emptyset$. This implies that for each $\Omega \subset \mathcal{MT} \cap \mathcal{G}(a_0)$ and $\Phi \in \mathcal{MT}_k \cap \mathcal{G}(a_0)$ that

$$\hat{P}_{k+1}(\Omega \mid \Phi) = \text{Prob}_{\text{tPPP}}(\mathcal{C}_k(\Phi) \cap \omega \in \text{sym}(A_k(\Omega, \Phi)) \mid \mathcal{C}_k(\Phi) \cap \bar{\mathcal{C}}_k(\Phi) \cap \omega = \emptyset).$$

where $\text{sym}(A)$ is the symmetrization of the set A, i.e. $(z_1, \ldots, z_n) \in \text{sym}(A)$ if there exists a permutation $\pi \in S_n$ such that $(z_{\pi(1)}, \ldots, z_{\pi(n)}) \in A$; in particular $A \subset \text{sym}(A)$. This is the crucial step where the complicated dependency on the past of the many-body evolution is reduced to a simple conditional expectation of the Poisson point process. Since:

$$A(\Omega, \Phi) \cap \underbrace{\bar{\mathcal{C}}_k(\Phi) \times \ldots \times \bar{\mathcal{C}}_k(\Phi)}_{r \text{ terms}} = \emptyset$$

for each r we can use formula (5.34) and deduce that:

$$\hat{P}_{k+1}(\Omega \mid \Phi) = e^{-\hat{\Gamma}_k(\Phi)} \frac{1}{r!} \int_{\text{sym}(A_{k+1}(\Omega, \Phi))} d\mu^r(Z_{k+1})$$

where
$$\hat{\Gamma}_k(\Phi) = \mu(\hat{\mathcal{C}}_k(\Phi)) \tag{5.35}$$
and $\hat{\mathcal{C}}(k) = \mathcal{C}_k(\Phi) \setminus \bar{\mathcal{C}}_k(\Phi)$. Recall the convention that the value of the integral over $(\mathbb{T}^d \times \mathbb{R}^d)^0$ is 1.

Since each nontrivial permutation of the labels $l \in m$ destroys the tree structure we obtain that if $z_\pi \in A$ and $z \in A$, then necessarily π is the identity transformation, i.e., $z_\pi = z$. This implies that if we replace in the above formula sym(A) by the nonsymmetric set A we have to drop the term $\frac{1}{r!}$.

$$\hat{P}_{k+1}(\Omega \mid \Phi) = e^{-\hat{\Gamma}_k(\Phi)} \int_{A_{k+1}(\Omega,\Phi)} d\mu^r(Z_{k+1}). \tag{5.36}$$

Plugging the expression (5.36) for the conditional expectation $\hat{P}_{k+1}(\cdot \mid \Phi)$ into equation (5.26) yields that:

$$\hat{P}_k(\Omega) = \int_{\mathbb{T}^d \times \mathbb{R}^d} dP_1(\phi_1(Z_1)) \, e^{-\hat{\Gamma}_1(\Phi_1(Z_1))} \int_{(\mathbb{T}^d \times \mathbb{R}^d)^{r_2}} \mu^{r_2}(\Phi_2(Z_2))$$
$$\ldots e^{-\hat{\Gamma}_{k-1}(\Phi_{k-1}(Z_1 \ldots Z_{k-1}))} \int_{A_k(\Omega,\Phi_{k-1}(Z_1 \ldots Z_{k-1}))} d\mu^{r_k}(Z_k)$$
$$= \sum_{m \in \mathcal{T}_k} \int_{A(\Omega)} d\mu^{\#m}(z) \, e^{-\sum_{j<k} \hat{\Gamma}_j(\Phi(z))}. \tag{5.37}$$

The intermediate step in the computation above relies on the additional assumption that $m \in \mathcal{T}_k \setminus \mathcal{T}_{k-1}$. In general we have to be more careful concerning the domains of integration, but the the final formula is unaffected.

We return now to the collision representation of the trees. This means that the variables $(z_l)_{l \in m}$ are replaced by $(u_1, v_1) \times (s_l, \nu_l, v_l)_{l \in m \setminus \{1\}}$. The determinant of the derivative of this transformation is given by:

$$\det D_\Phi z(\Phi) = \prod_{l \in m \setminus \{1\}} \left(a^{d-1}[\nu_l \cdot (v_l - v_{\bar{l}})]_+\right).$$

Thus changing coordinates in the integrals we obtain that for each $m \in \mathcal{T}$:

$$\int_{A(\Omega)} e^{-\sum_{j<k} \hat{\Gamma}_j(\Phi(z))} d\mu^{\#m}(z)$$
$$= \int_\Omega dP_1(z_1) \, e^{-\sum_{j<k} \hat{\Gamma}_j(\Phi)}$$
$$\prod_{l \in m \setminus \{1\}} \left(N \, df_0(v_l) \, d\nu_l \, ds_l \, \chi_{[s_{l-1}, s_{\bar{l}}]}(s_l) \, a^{d-1} \left[(v_l - v_{\bar{l}}) \cdot \nu_l\right]_+\right)$$
$$\stackrel{(5.2)}{=} \int_\Omega dP_1(z_1) \, e^{-\sum_{j<k} \hat{\Gamma}_j(\Phi)} \prod_{l \in m \setminus \{1\}} \left(df_0(v_l) \, d\nu_l \, ds_l \, \chi_{[s_{l-1}, s_{\bar{l}}]}(s_l) \left[(v_l - v_{\bar{l}}) \cdot \nu_l\right]_+\right)$$
$$= \int_\Omega d\lambda^m(\phi) \, e^{-\sum_{j<k} \hat{\Gamma}_j(\Phi)}.$$

Hence we have shown that for all $\Omega \subset \mathcal{G}(a)$:

$$\hat{P}_k(\Omega) = \sum_{m \in \mathcal{T}_k} \int_{\Omega \cap \mathcal{E}(m)} e^{-\sum_{j<k} \hat{\Gamma}_j(\Phi)} \, d\lambda^m(\phi) \tag{5.38}$$

and:

$$P_k(\Omega) = \hat{P}_k(\Omega) + e_k(\Omega), \tag{5.39}$$

where the error has the form:

$$e_k(\Omega) = \sum_{m \in \mathcal{T}_k} \int_{\Omega \cap \mathcal{E}(m)} d\lambda^m(\phi) \left(e^{-\sum_{j<k} \Gamma_j(\Phi)} - e^{-\sum_{j<k} \hat{\Gamma}_j(\Phi)} \right). \tag{5.40}$$

Since $\hat{\Gamma}_j(\Phi) \leq \Gamma_j(\Phi)$ the difference $e_k(\cdot)$ is a nonnegative measure.

Now we are in a good position to prove that equations (5.32) and (5.33) hold.

Proposition 5.2 (Similarity of $\hat{P}_k$ and P_k) *Let $\mathcal{G}(a)$ the set of good trees from Definition 5.3, and $\Omega \subset \mathcal{G}(a_0)$. Then equations (5.32) and (5.33) hold.*

The proof requires elementary but tedious estimates of sets $C_l \cap C_{l'}$ and can be found in Matthies and Theil (2008).

Proof of Theorem 5.1

We first demonstrate that the distribution of a single tagged particle satisfies the Boltzmann equation. Let $A \subset \mathbb{T}^d \times \mathbb{R}^d$ and define $\Omega(A) \subset \mathcal{MT}$ by:

$$\Omega(A) = \{\Phi \in \mathcal{MT} \mid \beta_1(m) = 1 \text{ and } z_1 \in A\}.$$

With this notation we obtain that for every $a_0 > 0$:

$$\left| \lim_{a \to 0} \lim_{k \to \infty} \hat{P}_{t,k}(\Omega) - \int_A du \, df_t(v) \right|$$

$$\stackrel{\text{Lem. 5.1}}{=} \lim_{a \to 0} \lim_{k \to \infty} \left| \hat{P}_{t,k}(\Omega) - \int_A du \, df_{t,k-1}(v) \right|$$

$$\stackrel{\text{Prop. 5.1}}{=} \lim_{a \to 0} \lim_{k \to \infty} \left| \hat{P}_{t,k}(\Omega) - P_{t,k}(\Omega) \right|$$

$$= \lim_{a \to 0} \lim_{k \to \infty} \Big| \hat{P}_{t,k}(\Omega \cap \mathcal{G}(a_0)) - P_{t,k}(\Omega \cap \mathcal{G}(a_0))$$

$$- P_{t,k}(\Omega \setminus \mathcal{G}(a_0)) + \hat{P}_{t,k}(\Omega \setminus \mathcal{G}(a_0)) \Big|$$

$$\stackrel{(5.33)}{\leq} \lim_{a \to 0} \lim_{k \to \infty} P_{t,k}(\mathcal{MT} \setminus \mathcal{G}(a_0)) + \lim_{a \to 0} \lim_{k \to \infty} \hat{P}_{t,k}(\mathcal{MT} \setminus \mathcal{G}(a_0))$$

Now using equation (5.33) again for $\tilde{\Omega} := \mathcal{MT} \cap \mathcal{G}(a_0)$ and that $\hat{P}_{t,k}$ and $P_{t,k}$ are probability measures, we also obtain that $\lim_{a \to 0} \hat{P}_{t,k}(\mathcal{MT} \setminus \mathcal{G}(a_0)) = P_{t,k}(\mathcal{MT} \setminus$

$\mathcal{G}(a_0)$). Now proceeding:

$$\leq 2 \lim_{k \to \infty} P_{t,k}(\mathcal{MT} \setminus \mathcal{G}(a_0)),$$

we send now a_0 to 0, apply (5.32) and obtain that $\lim_{a_0 \to 0} \lim_{k \to \infty} P_{t,k}(\mathcal{MT} \setminus \mathcal{G}(a_0)) = 0$, hence $\lim_{a \to 0} \lim_{k \to \infty} \hat{P}_{t,k}(\Omega) = \int_A \mathrm{d}u \, \mathrm{d}f_t(v)$.

The proof of Theorem 5.1 is complete. □

5.3 Concentrations and non-validity

We illustrate now that the mean-field theory does not capture the many-particle dynamics if the initial distribution f_0 exhibits strong concentrations. To simplify the long calculations in the proof we assume that $d = 2$, but similar results are expected to hold in the case $d = 3$.

Theorem 5.2 *Let $v \in \mathbb{R}^2$ be nonresonant ($\alpha \cdot v \notin \mathbb{Z}$ for all $\alpha \in \mathbb{Z}^d$) such that $|v| = 1$ and set $f_0 = \frac{1}{2}(\delta(\cdot - v) + \delta(\cdot + v))$. If $\hat{Q}(t) = \lim_{a \to 0} \lim_{k \to \infty} \hat{P}_{t,k}(\beta_1 = 1)$ denotes the empirical probability that a tagged particle does not collide, then:*

$$\lim_{t \to 0} \frac{1}{t^3} \left(\hat{Q}(t) - \int_{\mathbb{R}^2} \mathrm{d}f_t(v) \right) = \frac{1}{9}, \tag{5.41}$$

where $f_t = \frac{1}{1+t} f_0$ is the unique solution of the Boltzmann equation (5.9) which satisfies the initial condition $f_{t=0} = f_0$.

A numerical simulation for $N = 100000$ particles (Fig. 5.3) illustrates the prediction (5.41). The proof is based on a simple but lengthy calculation and can be found in Matthies and Theil (2008).

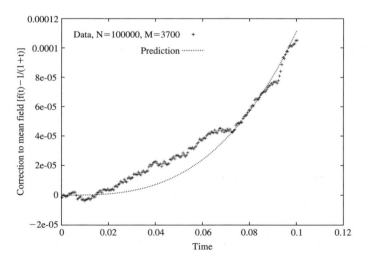

FIG. 5.3: Comparison between the empirical probability of colliding and the mean-field prediction. The dashed line is the cubic parabola $t \mapsto \frac{1}{9}t^3$, the signs '+' mark the difference between the number of non-collided particles at time t divided by N and the mean-field prediction $\frac{1}{1+t}$.

References

Coppex, F., Droz, M., Piasecki, J., Trizac, E. and Wittwer P. (2003). Some exact results for Boltzmann's annihilation dynamics. *Phys. Rev. E* **79**, 21103.

Droz, M., Frachebourg, L., Piasecki, J. and Rey, P.-A. (1995). Ballistic annihilation kinetics for a multivelocity one-dimensional ideal gas. *Phys. Rev. E* **51**, 5541–48.

Droz, M., Piasecki, J. and Trizak, E. (2002). Dynamics of ballistic annihilation. *Phys. Rev. E* **65**, 66111.

Elskens, Y. and Frisch H. (1985). Annihilation kinetics in the one-dimensional ideal gas. *Phys. Rev. A* **31**, 3812–16.

Hilbert, D. (2000). Mathematical problems (reprinted from *Bull. Amer. Math. Soc.* **8** (1902), pp 437–79). *Bull. Amer. Math. Soc. (N.S.)* **37**, 407–36.

Illner, R. and Pulvirenti, M. (1989). Global validity of the Boltzmann equation for two- and three-dimensional gas in vacuum. Erratum and improved result. *Comm. Math. Phys* **121**(1), 143–46.

Krug, J. and Spohn, H. (1988). Universality classes for deterministic surface growth. *Phys. Rev. A* **38**(8), 4271–83.

Lanford, O. (1975). Time evolution of large classical systems. In J. Moser (Ed.), *Dynamical Systems, Theory and Applications*, Volume 38 of *Lecture notes in Physics*, pp. 1–113. Springer Verlag: Berlin.

Lang, R. and Nguyen, X. (1980). Smoluchowski's theory of coagulation holds rigorously in the Boltzmann-Grad limit. *Z. Wahrs. Verw. Geb.* **54**, 227–80.

Matthies, K. and Theil, F. (2008).Validity and failure of the Boltzmann approximation of kinetic annihilation. submitted to *J. Nonlin. Sci.*.

Piasecki, J. (1995). Ballistic annihilation in a one-dimensional fluid. *Phys. Rev. E* **51**, 5535–40.

Spohn, H. (1991). *Large Scale Dynamics of Interacting Particles*. Texts and Monographs in Physics. Springer Verlag: Berlin.

Sznitman, A. (1991). Topics in the propagation of chaos. In P. Hennequin (Ed.), *Ecole d'Eté de Probabilités de Saint-Flour 1989*, Lecture Notes in Mathematics, pp. 165–251, Springer: Berlin.

Ushiyama, K. (1988). On the Boltzmann–Grad limit for the Broadwell model of the Boltzmann equation. *J. Stat. Phys.* **52**, 331–55.

C

APPLICATIONS IN PHYSICS

6

APPLICATIONS OF THE LACE EXPANSION TO STATISTICAL-MECHANICAL MODELS

Akira Sakai

6.1 Introduction

Synergetics is a common feature in interesting statistical-mechanical problems. One of the most important examples of synergetics is the emergence of a second-order phase transition and critical behaviour. It is rich and still far from fully understood. The reason why it is so difficult is due to the increase to infinity of the number of strongly correlated variables in the vicinity of the critical point. For example, the Ising model, which is a model for magnets, exhibits critical behaviour as the temperature T comes closer to its critical value T_c; the closer the temperature T is to T_c, the more spin variables cooperate with each other to attain the global magnetization. In this regime, neither standard probability theory for independent random variables nor naive perturbation techniques work. The lace expansion, which is the topic of this article, is currently one of the few approaches to rigorous investigation of critical behaviour for various statistical-mechanical models. We summarize here some of the most intriguing lace-expansion results for self-avoiding walk (SAW), percolation, and the Ising model. We also briefly explain the proof based on the latest version of bootstrapping argument (Borgs et al. 2005; Heydenneich et al. 2008; Slade 2006) and the first few stages of the derivation of the lace expansion for those three models.

6.1.1 *Models*

Let D be either the step distribution of the simple symmetric random walk (i.e., $D(x) = \frac{1}{2d}\mathbf{1}_{\{|x|=1\}}$, where $\mathbf{1}_{\{\,\dots\,\}}$ is the indicator function) or the following spread-out probability distribution on $\mathbb{Z}^d$. Given $L \in [1,\infty)$, we let h_L be a piecewise-continuous bounded probability distribution on $\mathbb{R}^d$ that respects the $\mathbb{Z}^d$-symmetry and satisfies $h_L(Lu) > 0$ for any unit vector $u \in \mathbb{Z}^d$ and $h_L(o) = 0$ at the origin $o = (0, \dots, 0)$. Then, we define:

$$D(x) = \frac{h_L(x/L)}{\sum_{y \in \mathbb{Z}^d} h_L(y/L)}, \qquad (6.1)$$

where $x/L = (x_1/L, \dots, x_d/L) \in \mathbb{Z}^d/L$. By the assumed bound on h_L, we have $D(o) = 0$ and $D(u) > 0$ for any unit vector $u \in \mathbb{Z}^d$. Moreover, by the Riemann-sum approximation, the denominator in (6.1) is $O(L^d)$, which plays a significant role in the analysis of the lace expansion.

We will use D as a microscopic coupling of the models on $\mathbb{Z}^d$. We refer to a model defined by $D(x) = \frac{1}{2d}\mathbf{1}_{\{|x|=1\}}$ as a nearest-neighbour model and to a model defined by (6.1) as a spread-out model. Furthermore, we refer to a model defined by h_L with bounded domain as a finite-range model, to a model with $\sum_x |x|^{2+\epsilon} h_L(x) < \infty$ for some $\epsilon > 0$ as a finite-variance model, and to a model with $h_L(x) \asymp |x|^{-d-\alpha}$ (i.e., $|x|^{d+\alpha} h_L(x)$ is bounded away from zero and infinity) for large $|x|$ as a long-range model with index $\alpha > 0$. Obviously, finite-range models and long-range models with index $\alpha > 2$ are finite-variance models.

6.1.1.1 Self-avoiding walk (SAW)
We denote by $\vec{w}_n$ an ordered sequence $(w_0, w_1, \ldots, w_n)$ of sites in $\mathbb{Z}^d$, and say that $\vec{w}_n$ is an SAW path if $w_i \neq w_j$ for $i \neq j$. We define the SAW two-point function by:

$$G_p^{\text{SAW}}(x,y) = \sum_{n=0}^{\infty} p^n \sum_{\substack{\vec{w}_n = (x,\ldots,y) \\ \text{SAW}}} \prod_{j=1}^{n} D(w_j - w_{j-1}), \qquad (6.2)$$

where, and in the rest of this article, we interpret the $n=0$ term as $\delta_{x,y}$. Since this is invariant under translation, we may suppress the notation to $G_p^{\text{SAW}}(y-x) := G_p^{\text{SAW}}(o, y-x)$. If the self-avoiding constraint is absent from (6.2), $G_p^{\text{SAW}}(x)$ is simply reduced to the random-walk Green's function:

$$S_p(x) = \sum_{n=0}^{\infty} p^n \sum_{\vec{w}_n = (o,\ldots,x)} \prod_{j=1}^{n} D(w_j - w_{j-1}) \equiv \sum_{n=0}^{\infty} p^n D^{*n}(x), \qquad (6.3)$$

where D^{*n} is the n-fold convolution of D, and $p^0 D^{*0}(x) = \delta_{x,o}$ by convention. If S_p is summable, then its Fourier transform can be solved as:

$$\hat{S}_p(k) := \sum_{x \in \mathbb{Z}^d} e^{ik \cdot x} S_p(x) = \frac{1}{1 - p\hat{D}(k)} = \frac{1}{1 - p + p(1 - \hat{D}(k))}. \qquad (6.4)$$

6.1.1.2 Percolation
A pair $\{x,y\}$ of sites in $\mathbb{Z}^d$ is called a bond, which is either occupied with probability $pD(y-x) \in [0,1]$ or vacant with probability $1 - pD(y-x)$, independently of the other bonds. Since D is a probability distribution, the percolation parameter p is the average number of occupied bonds per site. We define $\{x \longleftrightarrow y\}$ to be the event that x is connected to y, i.e., if either $x=y$ or there is an SAW path of occupied bonds from x to y. We define the percolation two-point function by:

$$G_p^{\text{perc}}(x,y) = \mathbb{P}_p(x \longleftrightarrow y), \qquad (6.5)$$

where $\mathbb{P}_p$ is the probability distribution of the bond variables. We will denote the expectation against $\mathbb{P}_p$ by $\mathbb{E}_p$. Similarly to $G_p^{\text{SAW}}(y-x)$, we may simply denote $G_p^{\text{perc}}(x,y)$ by $G_p^{\text{perc}}(y-x)$.

6.1.1.3 *Ising model* Let Λ be a finite subset of $\mathbb{Z}^d$ containing the origin o, e.g., a d-dimensional hypercube centred at o. At each site $x \in \Lambda$, there is a spin variable φ_x that takes values either $+1$ or -1. The energy of the system is defined by:

$$H_\Lambda(\varphi) = -\sum_{\{x,y\}\subset\Lambda} J_{x,y}\varphi_x\varphi_y, \qquad (6.6)$$

where $\varphi := \{\varphi_x\}_{x\in\Lambda}$ is a spin configuration and $\{J_{x,y}\}_{x,y\in\mathbb{Z}^d}$ is a collection of microscopic spin-spin couplings. We assume that $J_{x,y} = J_{o,y-x} \geq 0$ (i.e., translation-invariant and ferromagnetic) and that $\sum_x J_{o,x} < \infty$. Given the inverse temperature $\beta \geq 0$, we let:

$$p = \sum_{x\in\mathbb{Z}^d} \tanh(\beta J_{o,x}), \qquad D(x) = \frac{1}{p}\tanh(\beta J_{o,x}). \qquad (6.7)$$

We use p as the parameter of the model, since it is increasing in $\beta < \infty$ given $\{J_{o,x}\}_{x\in\mathbb{Z}^d}$. We define the Ising two-point function $G_p^{\text{Ising}}(x,y)$ as the unique infinite-volume limit of the thermal average $\langle\varphi_x\varphi_y\rangle_{p;\Lambda}$:

$$\langle\varphi_x\varphi_y\rangle_{p;\Lambda} = \frac{\sum_{\varphi\in\{\pm 1\}^\Lambda} \varphi_x\varphi_y\, e^{-\beta H_\Lambda(\varphi)}}{\sum_{\varphi\in\{\pm 1\}^\Lambda} e^{-\beta H_\Lambda(\varphi)}}, \qquad G_p^{\text{Ising}}(x,y) = \lim_{\Lambda\uparrow\mathbb{Z}^d} \langle\varphi_x\varphi_y\rangle_{p;\Lambda}, \qquad (6.8)$$

where the uniqueness is assured by the second Griffiths inequality (e.g., Fernández et al. 1992). We may denote $G_p^{\text{Ising}}(x,y)$ by $G_p^{\text{Ising}}(y-x)$, due to its translation-invariance (e.g., Bodineau 2006).

6.1.2 Phase transition

Let $d \geq 2$. It is known that there is a model-dependent critical point $p_c \in (0,\infty)$ such that the susceptibility:

$$\chi_p := \|G_p\|_1 \equiv \sum_{x\in\mathbb{Z}^d} G_p(x) \qquad (6.9)$$

is finite if $p < p_c$, and infinite if $p > p_c$. For percolation, in particular, χ_p is the expected number of sites connected from the origin o: $\chi_p = \mathbb{E}_p[\sum_x \mathbb{1}_{\{o\leftrightarrow x\}}]$. We denote p_c for each model by p_c^{SAW}, p_c^{perc} or p_c^{Ising} if necessary. By (6.2)–(6.3) and (6.5), $1 \leq p_c^{\text{SAW}} \leq p_c^{\text{perc}}$; by (6.70) below, $p_c^{\text{Ising}} \geq 1$. For percolation and the Ising model, p_c can also be characterized by the positivity of the percolation probability (i.e., the probability of existence of an infinite cluster of occupied bonds at the origin) (Aizenman and Barsky 1987) or the spontaneous magnetization (i.e., the infinite-volume limit of $\langle \varphi_o \rangle^+_{p;\Lambda}$ under the plus-boundary condition) (Aizenman et al. 1987).

For the finite-range models, in particular, the finiteness of χ_p is equivalent to exponential decay of correlation: for every $p < p_c$, there are $C_p < \infty$ (which is 1 for percolation and the Ising model) and $m_p > 0$ such that:

$$G_p(x) \leq C_p\, e^{-m_p \|x\|_\infty} \qquad (x \in \mathbb{Z}^d). \tag{6.10}$$

This is a result of partial removal of the self-avoiding constraint (Madras and Slade 1993) or of the Simon–Lieb inequality (Lieb 1980; Simon 1980).

6.1.3 Critical exponents and their universality

The susceptibility χ_p is infinite not only for $p > p_c$, but also in the limit $p \uparrow p_c$ (e.g., Aizenman 1982; Aizenman and Newman 1984; Madras and Slade 1993). In addition, m_p in (6.10) tends to zero as $p \uparrow p_c$. It is expected that there are model-dependent critical exponents γ and η such that:

$$\chi_p \underset{p \uparrow p_c}{\approx} (p_c - p)^{-\gamma}, \qquad \hat{G}_{p_c}(k) := \lim_{p \uparrow p_c} \hat{G}_p(k) \underset{|k| \downarrow 0}{\approx} \left(1 - \hat{D}(k)\right)^{-1+\eta}, \tag{6.11}$$

in some appropriate sense of limits (e.g., convergence of the log-ratio of the left-hand side to the right-hand side). Other observables, such as the percolation probability and the spontaneous magnetization, are also expected to exhibit power-law behaviour near $p = p_c$ characterized by their own critical exponents. To prove existence of these critical exponents and identify their values is one of the most important problems in statistical physics. So far, there is no general proof of existence. The values of the critical exponents for two-dimensional models can be identified assuming that scaling limits of interfaces and critical cluster boundaries of the models are represented by the Schramm–Loewner Evolution (e.g., Werner 2004). The assumption has been proved affirmative for some special cases, such as nearest-neighbour site percolation on the triangular lattice (e.g., Camia and Newman 2007).

The most important prediction about the critical exponents is their universality. The critical exponents are believed to depend only on d (and possibly on α for the long-range models with index $\alpha \leq 2$) and to be insensitive to the microscopic details, such as the value of the spread-out parameter L. For example, the value of γ^{Ising} for the nearest-neighbour Ising model is believed to be equal to that for the finite-variance Ising model for every $d \geq 2$, independently of $L < \infty$. It is natural to believe so, because the critical exponents characterize macroscopic physics, which should not be very much affected by the difference in the microscopic coupling D of the finite-variance models (the long-range models with index $\alpha \leq 2$ may not be in the same universality class). Therefore, the critical exponents are considered to possess the intrinsic nature of the models.

As described below, the lace-expansion approach provides some evidence to believe in universality, in high dimensions.

6.1.4 Mean-field behaviour above the upper-critical dimension

It is the interaction among constituents (such as spins) of each model that makes investigation of critical behaviour difficult. For SAW, for example, if the effect of the self-avoiding constraint were negligible, then we could approximate $G_p^{\mathrm{SAW}}(X)$ by the random-walk Green's function $S_p(x)$ and expect that $p_c^{\mathrm{SAW}} \simeq 1$, $\gamma^{\mathrm{SAW}} \simeq 1$ and $\eta^{\mathrm{SAW}} \simeq 0$, due to (6.4). Here, we call the critical behaviour described by the underlying random walk as the mean-field behaviour.

However, the mean-field approximation is problematic, because of the following two reasons. (i) By ignoring the interaction among constituents, p_c decreases towards the mean-field value 1, so we cannot really observe the behaviour around the true p_c. (ii) In lower dimensions, the critical exponents are expected to take on non-mean-field values; e.g., $(\gamma^{\mathrm{SAW}}, \eta^{\mathrm{SAW}}) = (\frac{43}{32}, \frac{5}{24})$, $(\gamma^{\mathrm{perc}}, \eta^{\mathrm{perc}}) = (\frac{43}{18}, \frac{5}{24})$ and $(\gamma^{\mathrm{Ising}}, \eta^{\mathrm{Ising}}) = (\frac{7}{4}, \frac{1}{4})$ in two-dimensions (see Fernández et al. 1992; Madras and Slade 1993; and Smirnov and Werner 2001 and references therein). Let d_c denote the model-dependent threshold dimension above which the critical exponents take on their mean-field values. If $d_c = \infty$, it is hopeless to use the mean-field approximation to investigate the critical behaviour.

There is a sufficient condition for the mean-field behaviour. Let:

$$B_p = (G_p * G_p)(o) \equiv \sum_{x \in \mathbb{Z}^d} G_p(x)^2, \tag{6.12}$$

$$T_p = (G_p * G_p * G_p)(o) \equiv \sum_{x,y \in \mathbb{Z}^d} G_p(x)\, G_p(x,y)\, G_p(y), \tag{6.13}$$

and say that the bubble (resp., triangle) condition holds if $B_{p_c} < \infty$ (resp., $T_{p_c} < \infty$). It is known that γ and various other critical exponents take on their mean-field values if the bubble condition holds for SAW and the Ising model and if the triangle condition holds for percolation (see Fernández et al. 1992; Gimmett 1999; Madras and Slade 1993 and references therein). This is because, for example (see Madras and Slade 1993):

$$\frac{(\chi_p^{\mathrm{SAW}})^2}{B_p^{\mathrm{SAW}}} \leq \frac{\mathrm{d}(p\chi_p^{\mathrm{SAW}})}{\mathrm{d}p} \leq (\chi_p^{\mathrm{SAW}})^2 \qquad (0 < p < p_c^{\mathrm{SAW}}). \tag{6.14}$$

We obtain $\gamma^{\mathrm{SAW}} = 1$ by integrating this differential inequality under the bubble condition (and using the monotonicity of B_p^{SAW} in p).

On the other hand, if there is a parameter change $\mu_p \in [0,1]$ with $\mu_{p_c} = 1$ such that $G_p \asymp S_{\mu_p}$ uniformly in $p < p_c$, then $B_{p_c} < \infty$ if $d > 4$ and $T_{p_c} < \infty$ if $d > 6$ for the finite-variance models, and $B_{p_c} < \infty$ if $d > 2\alpha$ and $T_{p_c} < \infty$ if $d > 3\alpha$ for the long-range models with index $\alpha \leq 2$. Therefore, we are led to expect that the mean-field prediction is correct in high dimensions.

As explained in the next section, the lace expansion proves this consistency for the nearest-neighbour models in sufficiently high dimensions and for the spread-out models with $L \gg 1$ above the model-dependent upper-critical dimension.

Before closing this section, we stress that the exponential decay (6.10) does not prove the bubble/triangle condition. For example, since $m_p \to 0$ as $p \uparrow p_c$ as mentioned earlier:

$$B_{p_c} = \lim_{p \uparrow p_c} \sum_{x \in \mathbb{Z}^d} G_p(x)^2 \leq 1 + \lim_{p \uparrow p_c} \sum_{r=1}^{\infty} O(r^{d-1}) e^{-2m_p r} = 1 + \lim_{p \uparrow p_c} O(m_p^{-d}) = \infty.$$
(6.15)

6.2 Results of the lace expansion

In this section, we summarize some of the most intriguing results of the lace expansion.

The lace-expansion approach was initiated by Brydges and Spencer (1985) for investigation of nearest-neighbour weakly SAW for $d > 4$. Since then, it has been applied successfully to various finite-variance statistical-mechanical models, such as strictly SAW for $d > 4$, lattice trees and lattice animals for $d > 8$, percolation for $d > 6$, oriented percolation and the contact process for $d > 4$. See Slade (2006) for the references up to the year 2005. It has been growing to cover other models, such as finite-range self-interacting random-walk models (van der Hofstad and Holmes 2006; Holmes and Sakai 2007) and the finite-range Ising model for $d > 4$ (Sakai 2007). Analysis of the lace expansion has also been improved to deal with the long-range models with index $\alpha > 0$ (Chen and Sakai forthcoming; Heydenreich et al. forthcoming).

In general, the lace expansion for $G_p(x)$ is the following recursion equation (assuming its convergence):

$$G_p(x) = \Delta_p(x) + (pD * G_p)(x) + (\Pi_p * G_p)(x),$$
(6.16)

where $\{\Delta_p(x)\}_{x \in \mathbb{Z}^d}$ and $\{\Pi_p(x)\}_{x \in \mathbb{Z}^d}$ are the model-dependent expansion coefficients. If $\Delta_p(x) = \delta_{x,o}$ and $\Pi_p(x) = 0$, then (6.16) is reduced to the recursion equation for the random-walk Green's function $S_p(x)$:

$$S_p(x) = \delta_{o,x} + (pD * S_p)(x).$$
(6.17)

It is therefore natural to expect that, assuming some nice properties of Δ_p and Π_p, we may find a suitable parameter change $\mu_p \in [0,1]$ such that $G_p \asymp S_{\mu_p}$ for $p \leq p_c$.

6.2.1 Main results

Before showing the results below, we first introduce some notation. For the finite-variance models, we let $a_d = \frac{d}{2}\pi^{-d/2}\Gamma(\frac{d}{2}-1)$ and $\sigma^2 = \sum_x |x|^2 D(x)$, hence $S_1(x) \sim \frac{a_d}{\sigma^2}|x|^{2-d}$ (see, e.g., Hara 2008; Hara et al. 2003), and let:

$$d_c = \begin{cases} 4 & \text{(SAW and the Ising model)}, \\ 6 & \text{(percolation)}. \end{cases} \quad (6.18)$$

For the long-range models with index $\alpha \leq 2$, we let:

$$d_c = \begin{cases} 2\alpha & \text{(SAW and the Ising model)}, \\ 3\alpha & \text{(percolation)}. \end{cases} \quad (6.19)$$

Let:

$$\lambda = \begin{cases} d^{-1} & \text{(nearest-neighbour models)}, \\ L^{-d} & \text{(spread-out models)}. \end{cases} \quad (6.20)$$

Theorem 6.1 (Hara and Slade 1990, 1992; Heydenreich et al. 2008)
Let $d > d_c$. For the nearest-neighbour models with $d \gg 1$ and the spread-out models with $L \gg 1$

$$\hat{G}_p(k) = \frac{1+O(\lambda)}{\chi_p^{-1} + p(1-\hat{D}(k))} \quad (6.21)$$

holds uniformly in $p < p_c$ and $k \in [-\pi, \pi]^d$. Consequently, $\gamma = 1$ and $\eta = 0$, and various other critical exponents take on their respective mean-field values.

Theorem 6.2 (Hara 2008; Hara et al. 2003; Sakai 2007) *Let $d > d_c$. For the nearest-neighbour models with $d \gg 1$ and the finite-range spread-out models with $L \gg 1$*

$$G_{p_c}(x) \sim \frac{K a_d}{\sigma^2}|x|^{2-d} \quad \text{as } |x| \to \infty, \quad (6.22)$$

where $K \in (0, \infty)$ is a model-dependent constant.

Theorem 6.3 (Van der Hofstad and Sakai 2005) *Let $d > d_c$. For the finite-range models, as $L \to \infty$:*

$$p_c = 1 + \begin{cases} \sum_{n=2}^{\infty} D^{*n}(o) + O(\lambda^2) & (SAW), \\ D^{*2}(o) + \sum_{n=3}^{\infty} \frac{n+1}{2} D^{*n}(o) + O(\lambda^2) & (percolation), \end{cases} \quad (6.23)$$

where $\sum_{n=2}^{\infty} D^{*n}(o)$ and $D^{*2}(o) + \sum_{n=3}^{\infty} \frac{n+1}{2} D^{*n}(o)$ are both $O(\lambda)$.

Remark 6.1

1. *Theorems 6.1–6.2 hold for nearest-neighbour SAW as soon as $d \geq 5$ (Hara 2008; Hara and Slade 1992).*

2. Theorem 6.2 may be extended to the finite-variance models, following the argument in Hara (2008).
3. Similar asymptotic results to (6.23) have been obtained for spread-out oriented percolation, where $p_c - 1 \sim \frac{1}{2}\sum_{n=2}^{\infty} D^{*2n}(o)$ (Chen and Sakai forthcoming; van der Hofstad and Sakai 2005), and for the finite-range contact process, where $p_c - 1 \sim \sum_{n=2}^{\infty} D^{*n}(o)$ (van der Hofstad and Sakai 2005). It is expected that p_c^{Ising} obeys the same asymptotics as p_c^{SAW} in (6.23), with a different error term.

6.2.2 Idea of the proof

Since the lace expansion (6.16) is a convolution equation, a natural approach to investigate it is to take its Fourier transform and solve the resulting equation for $\hat{G}_p(k)$. For SAW, in particular, $\hat{G}_p^{\text{SAW}}(k)$ is considered as the Fourier–Laplace transform of the n-step SAW two-point function $Z_n^{\text{SAW}}(x)$ (= the sum in (6.2) over SAWs $\vec{w}_n = (o, \ldots, x)$), where p is the fugacity. Therefore, one way to investigate $\hat{Z}_n(k)$ for fixed n is to use the Tauberian theorem to find the coefficient of p^n in $\hat{G}_p^{\text{SAW}}(k)$. Another way is to solve a recursion equation for $\hat{Z}_n^{\text{SAW}}(k)$, which is a version of (6.16) for fixed n, by induction on n. To investigate asymptotic behaviour of $G_p(x)$ as $|x| \to \infty$, we may reorganize (6.16) to approximate $G_p(x)$ by $\kappa S_\mu(x)$ for some $\kappa \in (0, \infty)$ and $\mu \in [0, 1]$. See Slade (2006) and references therein for more details of those approaches. Here, we focus on the proof of Theorem 6.1.

First, we heuristically explain the required properties of Δ_p and Π_p. Suppose that Δ_p and Π_p are absolutely summable, hence $\hat{\Delta}_p(k)$ and $\hat{\Pi}_p(k)$ exist (we assume more below). Then, by solving the Fourier transform of (6.16) for $\hat{G}_p(k)$:

$$\hat{G}_p(k) = \hat{\Delta}_p(k) + p\hat{D}(k)\hat{G}_p(k) + \hat{\Pi}_p(k)\hat{G}_p(k)$$
$$= \frac{\hat{\Delta}_p(k)}{1 - p\hat{D}(k) - \hat{\Pi}_p(k)} = \frac{\hat{\Delta}_p(k)}{\hat{\Delta}_p(o)\chi_p^{-1} + p(1 - \hat{D}(k)) + \hat{\Pi}_p(o) - \hat{\Pi}_p(k)},$$
(6.24)

where we have used $\hat{\Delta}_p(o)\chi_p^{-1} \equiv \hat{\Delta}_p(o)\hat{G}_p(o)^{-1} = 1 - p - \hat{\Pi}_p(o)$. We obtain (6.21) if there are p-independent constants $C, C' < \infty$ such that:

$$\left.\begin{array}{l}|\hat{\Delta}_p(k) - 1|\\|\hat{\Pi}_p(k)|\end{array}\right\} \leq C\lambda, \qquad |\hat{\Pi}_p(o) - \hat{\Pi}_p(k)| \leq C'\lambda p\underbrace{\big(1 - \mu_p \hat{D}(k)\big)}_{\hat{S}_{\mu_p}(k)^{-1}}, \qquad (6.25)$$

where:

$$\mu_p = 1 - \chi_p^{-1} \in [0, 1] \qquad (6.26)$$

is a parameter change to compare $\hat{G}_p(k)$ with $\hat{S}_{\mu_p}(k)$ (cf., $f_p^{(1)}$ in (6.28) below); due to this definition, $\hat{G}_p(o) = \hat{S}_{\mu_p}(o)$ for any $p < p_c$. The first inequality in (6.25) means, in a weak sense, that $\Delta_p(x) \simeq \delta_{x,o}$ and $\Pi_p(x) \simeq 0$ (cf., the discussion around (6.16)–(6.17)).

The above heuristic argument shows that (6.25) implies (6.21). To complete the proof of Theorem 6.1, it suffices to prove the converse statement: (6.21) implies (6.25). In fact, the actual proof is based on bootstrapping argument. To explain its latest version (Borgs et al. 2005; Heydenreich et al. forthcoming; Slade 2006), we define:

$$f_p = p \vee f_p^{(1)} \vee f_p^{(2)} \geq 0, \tag{6.27}$$

where:

$$f_p^{(1)} = \sup_{k \in [-\pi,\pi]^d} \frac{\hat{G}_p(k)}{\hat{S}_{\mu_p}(k)}, \tag{6.28}$$

$$f_p^{(2)} = \sup_{k,l \in [-\pi,\pi]^d} \frac{\hat{S}_{\mu_p}(k) \left|\hat{G}_p(l) - \frac{1}{2}(\hat{G}_p(l+k) + \hat{G}_p(l-k))\right|}{100 \sum_{(j,j')=(0,\pm 1),(1,-1)} \hat{S}_{\mu_p}(l+jk) \hat{S}_{\mu_p}(l+j'k)}. \tag{6.29}$$

The function $f_p^{(2)}$ is to compare the Fourier transform of $(1 - \cos(k \cdot x))G_p(x)$, which is:

$$\sum_{x \in \mathbb{Z}^d} e^{il \cdot x}(1 - \cos(k \cdot x))G_p(x) = \hat{G}_p(l) - \frac{1}{2}(\hat{G}_p(l+k) + \hat{G}_p(l-k)), \tag{6.30}$$

with a constant $(= 100)$ multiple of:

$$\hat{S}_{\mu_p}(k)^{-1} \sum_{(j,j')=(0,\pm 1),(1,-1)} \hat{S}_{\mu_p}(l+jk) \hat{S}_{\mu_p}(l+j'k). \tag{6.31}$$

It has been proved (Borgs et al. 2005; Heydenreich et al. forthcoming; Slade 2006) (and we will demonstrate a part of (i) in Section 6.3) that:

(i) $f_p \leq 3$ implies (6.25) for every $p < p_c$ if $d > d_c$ and $\lambda \ll 1$;
(ii) $f_p \leq 3$ and (6.25) imply the stronger bound $f_p \leq 2$ for every $p < p_c$ if $d > d_c$ and $\lambda \ll 1$;
(iii) f_p is continuous in $p < p_c$, with $f_0 = 1$.

Because of (i)–(ii), $f_p \notin (2, 3]$ for every $p < p_c$. With the help of (iii), $f_p \leq 2$ for all $p < p_c$, hence (6.25) indeed holds uniformly in $p < p_c$. This proves Theorem 6.1.

6.3 Lace expansion and bounding diagrams

In this section, we explain (the first few stages of) the derivation of the lace expansion (6.16) for each model. We note here that there are many identities of the form (6.16) depending on Δ_p and Π_p. Among those identities, the lace expansion is the one with optimal Δ_p and Π_p, which means that Δ_p and Π_p are absolutely summable for $p \leq p_c$ (cf., the mean-field approximation, which is valid only for $p < 1$) and $d > d_c$. We briefly explain that Δ_p and Π_p are bounded by geometrical series of nonzero bubbles for SAW and the Ising model and of nonzero triangles for percolation, where the nonzero bubble/triangle is $O(\lambda)$ if $d > d_c$.

6.3.1 Expansion for self-avoiding walk

In this section, we use inclusion-exclusion to derive the expansion (6.16) for $G_p^{\text{SAW}}(x)$, with $\Delta_p^{\text{SAW}}(x) = \delta_{x,o}$ and:

$$\Pi_p^{\text{SAW}}(x) = \sum_{N=1}^{\infty} (-1)^N \Pi_{p;N}^{\text{SAW}}(x) \equiv -\bigcirc_{x=o} + \bigoplus_{o}^{x} - \triangle_{ox} + \cdots, \tag{6.32}$$

where each $\Pi_{p;N}^{\text{SAW}}(x) \geq 0$ consists of N vertices and $2N - 1$ lines; unlabelled vertices in $\Pi_{p;N}^{\text{SAW}}(x)$ for $N \geq 3$ are summed over $\mathbb{Z}^d$; unslashed lines correspond to nonzero SAW paths (or loop when $N = 1$), while slashed lines in $\Pi_{p;N}^{\text{SAW}}(x)$ for $N \geq 3$ correspond to SAW paths which may have zero length. Some lines in each term are mutually avoiding, others are not. For example, all the lines in $\Pi_{p;2}^{\text{SAW}}(x)$ are mutually avoiding, which is not the case for $\Pi_{p;3}^{\text{SAW}}(x)$. Refer to, e.g., Madras and Slade (1993) for the subtlety of this mutual avoidance.

Before showing the derivation of the lace expansion, we demonstrate how $f_p \leq 3$ implies the inequalities in (6.25) for Π_p^{SAW} if $d > d_c^{\text{SAW}}$ and $\lambda \ll 1$. First, by ignoring the mutual avoidance among consisting SAW paths, $|\hat{\Pi}_p^{\text{SAW}}(k)|$ is bounded as:

$$|\hat{\Pi}_p^{\text{SAW}}(k)| \leq \sum_{x \in \mathbb{Z}^d} \left(\bigcirc_{x=o} + \tilde{G}_p^{\text{SAW}}(x)^3 + \tilde{G}_p^{\text{SAW}}(x) \left((\tilde{G}_p^{\text{SAW}})^2 * (\tilde{G}_p^{\text{SAW}})^2 \right)(x) + \cdots \right), \tag{6.33}$$

where $\tilde{G}_p^{\text{SAW}}(x) = G_p^{\text{SAW}}(x) - \delta_{x,o}$, which obeys the trivial bound:

$$\tilde{G}_p^{\text{SAW}}(x) \leq (pD * G_p^{\text{SAW}})(x) \leq pD(x) + p^2(D^{*2} * G_p^{\text{SAW}})(x). \tag{6.34}$$

If $p \vee f_p^{(1)} \leq 3$ and $d > \frac{1}{2}d_c^{\text{SAW}}$, then the sum of the first term in (6.33) is bounded as:

$$\sum_{x \in \mathbb{Z}^d} \bigcirc_{x=o} \leq p^2(D^{*2} * G_p^{\text{SAW}})(o) = p^2 \int_{[-\pi,\pi]^d} \frac{d^d k}{(2\pi)^d} \hat{D}(k)^2 \hat{G}_p^{\text{SAW}}(k)$$

$$\leq 3^3 \int_{[-\pi,\pi]^d} \frac{d^d k}{(2\pi)^d} \hat{D}(k)^2 \hat{S}_{\mu_p}(k) \leq 3^3 \int_{[-\pi,\pi]^d} \frac{d^d k}{(2\pi)^d} \frac{\hat{D}(k)^2}{1 - \hat{D}(k)} = O(\lambda), \tag{6.35}$$

where we have used the random-walk estimates (see, e.g., Slade (2006) for the nearest-neighbour model, and Chen and Sakai (2007), van der Hofstad and Sakai (2007), and van der Hofstad and Slade (2002) for the spread-out model). For the sum of the second term in (6.33), we use:

$$\sum_{x \in \mathbb{Z}^d} \tilde{G}_p^{\text{SAW}}(x)^3 \leq \|\tilde{G}_p^{\text{SAW}}\|_\infty \, (\tilde{G}_p^{\text{SAW}})^{*2}(o), \tag{6.36}$$

where $\|\tilde{G}_p^{\text{SAW}}\|_\infty = O(\lambda)$ if $d > \frac{1}{2}d_c^{\text{SAW}}$, due to (6.34), $\|D\|_\infty = O(\lambda)$ and a similar analysis to (6.35). Moreover, if $d > d_c^{\text{SAW}}$:

$$(\tilde{G}_p^{\text{SAW}})^{*2}(o) \leq p^2(D * G_p^{\text{SAW}})^{*2}(o) = p^2 \int_{[-\pi,\pi]^d} \frac{d^d k}{(2\pi)^d} \hat{D}(k)^2 \hat{G}_p^{\text{SAW}}(k)^2$$

$$\leq 3^4 \int_{[-\pi,\pi]^d} \frac{d^d k}{(2\pi)^d} \hat{D}(k)^2 \hat{S}_{\mu_p}(k)^2 \leq 3^4 \int_{[-\pi,\pi]^d} \frac{d^d k}{(2\pi)^d} \frac{\hat{D}(k)^2}{(1 - \hat{D}(k))^2} = O(\lambda). \tag{6.37}$$

Therefore, $\|\Pi_{p;2}^{\text{SAW}}\|_1 \leq O(\lambda)^2$. By similar manipulation of supremums and sums, we can show $\|\Pi_{p;N}^{\text{SAW}}\|_1 \leq O(\lambda)^N$ if $d > d_c^{\text{SAW}}$ (see, e.g., Madras and Slade 1993), hence the series (6.33) converges and is $O(\lambda)$ if $\lambda \ll 1$. This proves the first inequality in (6.25).

For the second inequality in (6.25), we use $f_p^{(2)}$ as well. For example:

$$\hat{\Pi}_{p;2}^{\text{SAW}}(o) - \hat{\Pi}_{p;2}^{\text{SAW}}(k) = \sum_{x \in \mathbb{Z}^d} (1 - \cos(k \cdot x)) \Pi_{p;2}^{\text{SAW}}(x)$$

$$\leq \sum_{x \in \mathbb{Z}^d} (1 - \cos(k \cdot x)) \tilde{G}_p^{\text{SAW}}(x)^3$$

$$\leq \left(\sup_{x \in \mathbb{Z}^d} (1 - \cos(k \cdot x)) G_p^{\text{SAW}}(x) \right) \underbrace{(\tilde{G}_p^{\text{SAW}})^{*2}(o)}_{pO(\lambda)}, \tag{6.38}$$

where, by (6.30) and $f_p^{(2)} \leq 3$:

$$\begin{aligned}
&\left(1 - \cos(k \cdot x)\right) G_p^{\mathrm{SAW}}(x) \\
&\leq \int_{[-\pi,\pi]^d} \frac{\mathrm{d}^d l}{(2\pi)^d} \left| \hat{G}_p^{\mathrm{SAW}}(l) - \tfrac{1}{2}\left(\hat{G}_p^{\mathrm{SAW}}(l+k) + \hat{G}_p^{\mathrm{SAW}}(l-k)\right) \right| \\
&\leq \sum_{(j,j')=(0,\pm 1),(1,-1)} \int_{[-\pi,\pi]^d} \frac{\mathrm{d}^d l}{(2\pi)^d} \frac{300\, \hat{S}_{\mu_p}(k)^{-1}}{(1 - \hat{D}(l+jk))(1 - \hat{D}(l+j'k))},
\end{aligned} \quad (6.39)$$

which is $O(1)\hat{S}_{\mu_p}(k)^{-1}$ if $d > d_c^{\mathrm{SAW}}$, hence $\hat{\Pi}_{p;2}^{\mathrm{SAW}}(o) - \hat{\Pi}_{p;2}^{\mathrm{SAW}}(k) \leq O(\lambda)p\hat{S}_{\mu_p}(k)^{-1}$. Similarly, we can show $\hat{\Pi}_{p;N}^{\mathrm{SAW}}(o) - \hat{\Pi}_{p;N}^{\mathrm{SAW}}(k) \leq O(\lambda)^N p\hat{S}_{\mu_p}(k)^{-1}$ if $d > d_c^{\mathrm{SAW}}$ (Heydenreich et al. forthcoming; Slade 2006), which yields the second inequality in (6.25) if $\lambda \ll 1$.

Derivation of the lace expansion for $G_p^{\mathrm{SAW}}(x)$. We only explain the first two stages of the derivation of the expansion (6.16). To complete the expansion, we refer to, e.g., Madras and Slade (1993).

The first stage of the expansion is to isolate the zero-step walk and identify the position of w_1 for the nonzero SAW paths $\vec{w}_n = (o, w_1, \ldots, x)$. Denoting these nonzero SAW paths by $\vec{w}_n = (o, \vec{w}'_{n-1})$, we obtain:

$$\begin{aligned}
G_p^{\mathrm{SAW}}(x) &= \delta_{x,o} + \sum_{n=1}^{\infty} p^n \sum_{y \in \mathbb{Z}^d} \sum_{\substack{\vec{w}_n = (o,y,\ldots,x) \\ \mathrm{SAW}}} \prod_{j=1}^{n} D(w_j - w_{j-1}) \\
&= \delta_{x,o} + \sum_{y \in \mathbb{Z}^d} pD(y) \sum_{n=1}^{\infty} p^{n-1} \sum_{\substack{\vec{w}'_{n-1} = (y,\ldots,x) \\ \mathrm{SAW}}} \prod_{j=1}^{n-1} D(w'_j - w'_{j-1}) \mathbf{1}_{\{\vec{w}'_{n-1} \not\ni o\}}.
\end{aligned} \quad (6.40)$$

Since $\vec{w}'_{n-1}$ is an SAW path, the last indicator is 1 if and only if $\vec{w}_n = (o, \vec{w}'_{n-1})$ is an SAW path. If this indicator is absent, then the sum over n is simply equal to $G_p^{\mathrm{SAW}}(y,x)$; the correction is the contribution from $\mathbf{1}_{\{\vec{w}'_{n-1} \ni o\}} \equiv 1 - \mathbf{1}_{\{\vec{w}'_{n-1} \not\ni o\}}$, due to inclusion-exclusion. Therefore:

$$G_p^{\mathrm{SAW}}(x) = \delta_{x,o} + (pD * G_p^{\mathrm{SAW}})(x) - R_{p;1}^{\mathrm{SAW}}(x), \quad (6.41)$$

with:

$$R_{p;1}^{\mathrm{SAW}}(x) = \sum_{y \in \mathbb{Z}^d} pD(y) \sum_{n=0}^{\infty} p^n \sum_{\substack{\vec{w}_n = (y,\ldots,o,\ldots,x) \\ \mathrm{SAW}}} \prod_{j=1}^{n} D(w_j - w_{j-1}), \quad (6.42)$$

Applications of the lace expansion to statistical-mechanical models   135

where we have replaced $n-1$ in (6.40) by $n' \in \mathbb{Z}_+ := \{0,1,\ldots\}$ and then removed all the primes from the expression. The derivation of the full expansion is completed if we can show $R^{\text{SAW}}_{p;1}(x) = -(\Pi^{\text{SAW}}_p * G^{\text{SAW}}_p)(x)$.

The second stage of the expansion is to split an SAW path $\vec{w}_n = (y,\ldots,o,\ldots,x)$ in (6.42) into two subwalks according to the 'time' t at which $\vec{w}_n$ is at o. Let $\vec{w}_n = (\vec{w}_t, \vec{w}'_{n-t})$, where $\vec{w}_t = (y,\ldots,o)$ and $\vec{w}'_{n-t} = (o,\ldots,x)$. Then, we obtain:

$$R^{\text{SAW}}_{p;1}(x) = \sum_{y \in \mathbb{Z}^d} pD(y) \sum_{t=0}^{\infty} p^t \sum_{\substack{\vec{w}_t=(y,\ldots,o) \\ \text{SAW}}} \prod_{i=1}^{t} D(w_i - w_{i-1})$$

$$\times \sum_{n=t}^{\infty} p^{n-t} \sum_{\substack{\vec{w}'_{n-t}=(o,\ldots,x) \\ \text{SAW}}} \prod_{j=1}^{n-t} D(w'_j - w'_{j-1}) 1_{\{\vec{w}_t \cap \vec{w}'_{n-t} = \{o\}\}}. \quad (6.43)$$

Since $\vec{w}_t$ and $\vec{w}'_{n-t}$ are SAW paths, the last indicator is 1 if and only if $\vec{w}_n = (\vec{w}_t, \vec{w}'_{n-t})$ is an SAW path. If we ignore this indicator, the first and second lines become independent, yielding $(\Pi^{\text{SAW}}_{p;1} * G^{\text{SAW}}_p)(x)$, where $\Pi^{\text{SAW}}_{p;1}(x) \equiv \bigcirc_{x=o}$ is the self-avoiding loop at the origin. The correction is the contribution from $1_{\{\vec{w}_t \cap \vec{w}'_{n-t} \supsetneq \{o\}\}} \equiv 1 - 1_{\{\vec{w}_t \cap \vec{w}'_{n-t} = \{o\}\}}$. Therefore:

$$R^{\text{SAW}}_{p;1}(x) = (\Pi^{\text{SAW}}_{p;1} * G^{\text{SAW}}_p)(x) - R^{\text{SAW}}_{p;2}(x), \quad (6.44)$$

where, by replacing $n-t$ in (6.43) by $s \in \mathbb{Z}_+$:

$$R^{\text{SAW}}_{p;2}(x) = \sum_{y \in \mathbb{Z}^d} pD(y) \sum_{t=0}^{\infty} p^t \sum_{\substack{\vec{w}_t=(y,\ldots,o) \\ \text{SAW}}} \prod_{i=1}^{t} D(w_i - w_{i-1})$$

$$\times \sum_{s=0}^{\infty} p^s \sum_{\substack{\vec{w}'_s=(o,\ldots,x) \\ \text{SAW}}} \prod_{j=1}^{s} D(w'_j - w'_{j-1}) 1_{\{\vec{w}_t \cap \vec{w}'_s \supsetneq \{o\}\}}. \quad (6.45)$$

This completes the second stage of the expansion.

To obtain the higher-order expansion coefficient $\Pi^{\text{SAW}}_{p;N}(x)$, we look at the last SAW path in the higher-order remainder $R^{\text{SAW}}_{p;N}(x)$, e.g., $\vec{w}'_s$ in $R^{\text{SAW}}_{p;2}(x)$, and see when for the first time the condition in the indicator, e.g., $\vec{w}_t \cap \vec{w}'_s \supsetneq \{o\}$ in (6.45), is satisfied. Let $u \in \{0, 1\ldots, s\}$ be that 'first time' and split the path $\vec{w}'_s$ into $\vec{w}'_s = (\vec{w}'_u, \vec{w}''_{s-u})$. If we forget that $\vec{w}'_u$ and $\vec{w}''_{s-u}$ are mutually avoiding, we obtain $(\Pi_{p;N}^{\text{SAW}} * G_p^{\text{SAW}})(x)$; the correction becomes $-R_{p;N+1}^{\text{SAW}}(x)$. See, e.g., Madras and Slade (1993) for more details of this inclusion-exclusion argument.

6.3.2 Expansion for percolation

Here and in Section 6.3.3, we explain the derivation of the following expansion for percolation and the Ising model[1]:

$$G_p(x) = \Delta_p(x) + (\Delta_p * pD * G_p)(x) \equiv \Delta_p(x) + (pD * G_p)(x) \\ + \big(\underbrace{(\Delta_p - \delta) * pD}_{\Pi_p} * G_p\big)(x), \qquad (6.46)$$

where $\Delta_p(x)$ is the alternating series of the model-dependent N^{th}-order expansion coefficients $\Delta_{p;N}(x) \geq 0$:

$$\Delta_p(x) = \sum_{N=0}^{\infty} (-1)^N \Delta_{p;N}(x). \qquad (6.47)$$

For percolation (Hara and Slade 1990), $\Delta_p^{\text{perc}}(x)$ may be depicted as:

$$\Delta_p^{\text{perc}}(x) = o \bigcirc x \;-\; o \Diamond x \;+\cdots, \qquad (6.48)$$

where each line corresponds to an occupied path that may have zero length (the horizontal bold line in the second term has an occupied bond at the left end, hence it is nonzero). The first term represents the probability that o is doubly-connected to x, i.e., either $x = o$ or there are at least two bond-disjoint occupied paths from o to $x \neq o$. The higher-order terms may be interpreted in a similar way. For example, in the second term, the thinner lines are mutually bond-disjoint occupied paths, and so are the bold lines. The difference in thickness is to represent nested expectations; the bold lines are on a probability space that is different from the one the thinner lines are defined on, and these lines are coupled in order to satisfy certain geometrical conditions. The higher-order terms are also defined by using nested expectations. See, e.g., Hara and Slade (1990) for more details.

Before going into the derivation of the expansion (6.46), we demonstrate how to use $f_p \leq 3$ to obtain (6.25) if $d > d_c^{\text{perc}}$ and $\lambda \ll 1$. The key is the BK inequality (van der Berg and Kesten 1985), by which we can prove that the probability of occurrence of bond-disjoint connections is bounded from above by the product

---

[1] To keep this article as simple and intuitive as possible, we pretend here and in Section 6.3.3 that the expansion (6.46) is complete without a remainder, with the series representation (6.47) for $\Delta_p(x)$. See Hara and Slade (1990) and Sakai (2007) for the lace expansion up to any ℓ^{th} order, with $\Delta_p(x) = \sum_{N=0}^{\ell} (-1)^N \Delta_{p;N}(x)$ and a remainder.

of the probability of occurrence of each connection. For example, by the BK inequality:

$$\delta_{x,o} \leq \; o\!\!\bigcirc\!\!\!\bigcirc\!\!x \;\leq G_p^{\text{perc}}(x)^2 = \delta_{x,o} + \tilde{G}_p^{\text{perc}}(x)^2, \tag{6.49}$$

where each $\tilde{G}_p^{\text{perc}}(x) := G_p^{\text{perc}}(x) - \delta_{x,o}$ is bounded, again by the BK inequality, as:

$$\tilde{G}_p^{\text{perc}}(x) \leq \sum_{v \in \mathbb{Z}^d} \mathbb{P}_p\Big(\{o,v\} \text{ is occupied, } v \longleftrightarrow x \text{ without using } \{o,v\}\Big)$$
$$\leq (pD * G_p^{\text{perc}})(x) \leq pD(x) + p^2(D^{*2} * G_p^{\text{perc}})(x). \tag{6.50}$$

Therefore, by the same computation as in (6.37), we obtain $|\hat{\Delta}_{p;0}^{\text{perc}}(k) - 1| \leq O(\lambda)$ if $d > \frac{2}{3} d_{\text{c}}^{\text{perc}}$. Similarly, by the BK inequality and taking supremums and sums:

$$\|\Delta_{p;1}^{\text{perc}}\|_1 \leq \sum_{u,v \in \mathbb{Z}^d} G_p^{\text{perc}}(u)\, G_p^{\text{perc}}(u,v)\, G_p^{\text{perc}}(v)$$
$$\times \sum_{y,z \in \mathbb{Z}^d} (pD * G_p^{\text{perc}})(v,y)\, G_p^{\text{perc}}(y,z)\, G_p^{\text{perc}}(z,u) \sum_{x \in \mathbb{Z}^d} G_p^{\text{perc}}(y,x)\, G_p^{\text{perc}}(x,z)$$
$$\leq (G_p^{\text{perc}})^{*3}(o)\, \|pD * (G_p^{\text{perc}})^{*3}\|_\infty\, \|(G_p^{\text{perc}})^{*2}\|_\infty. \tag{6.51}$$

Since $f_p^{(1)} \leq 3$ and $d > d_{\text{c}}^{\text{perc}}$, both $(G_p^{\text{perc}})^{*3}(o)$ and $\|(G_p^{\text{perc}})^{*2}\|_\infty$ are $O(1)$. Moreover, by (6.50) and $G_p^{\text{perc}}(x) \geq \delta_{x,o}$ and using $p \vee f_p^{(1)} \leq 3$ and the random-walk estimates mentioned below (6.35):

$$\|pD * (G_p^{\text{perc}})^{*3}\|_\infty \leq p\|D\|_\infty + 3p^2 \|D^{*2} * (G_p^{\text{perc}})^{*3}\|_\infty$$
$$\leq 3\|D\|_\infty + 3^6 \int_{[-\pi,\pi]^d} \frac{d^d k}{(2\pi)^d} \frac{\hat{D}(k)^2}{(1 - \hat{D}(k))^3} = O(\lambda), \tag{6.52}$$

if $d > d_{\text{c}}^{\text{perc}}$, hence $\|\Delta_{p;1}^{\text{perc}}\|_1 \leq O(\lambda)$. By similar manipulations of supremums and sums, we can show $\|\Delta_{p;N}^{\text{perc}}\|_1 \leq O(\lambda)^N$ for $N \geq 2$ if $d > d_{\text{c}}^{\text{perc}}$ (see, e.g., Heydenreich et al. forthcoming). As a result, if $\lambda \ll 1$:

$$|\hat{\Delta}_p^{\text{perc}}(k) - 1| \leq |\hat{\Delta}_{p;0}^{\text{perc}}(k) - 1| + \sum_{N=1}^\infty \|\Delta_{p;N}^{\text{perc}}\|_1 = O(\lambda), \tag{6.53}$$

which also implies $|\hat{\Pi}_p^{\text{perc}}(k)| \leq O(\lambda)$. This proves the first inequality in (6.25).

For the second inequality in (6.25), we first note that:

$$\hat{\Pi}_p^{\text{perc}}(o) - \hat{\Pi}_p^{\text{perc}}(k) = p\Big(\big(\hat{\Delta}_p^{\text{perc}}(o) - 1\big) - \big(\hat{\Delta}_p^{\text{perc}}(k) - 1\big)\hat{D}(k)\Big)$$
$$= p\Big(\big(\hat{\Delta}_p^{\text{perc}}(o) - \hat{\Delta}_p^{\text{perc}}(k)\big) + \big(\hat{\Delta}_p^{\text{perc}}(k) - 1\big)\big(1 - \hat{D}(k)\big)\Big). \tag{6.54}$$

Since $1 - \hat{D}(k) \leq 2\hat{S}_{\mu_p}(k)^{-1}$, it thus suffices to show $|\hat{\Delta}_p^{\text{perc}}(o) - \hat{\Delta}_p^{\text{perc}}(k)| \leq O(\lambda)\hat{S}_{\mu_p}(k)^{-1}$. However, we only show here that $\hat{\Delta}_{p;0}^{\text{perc}}(o) - \hat{\Delta}_{p;0}^{\text{perc}}(k) \leq O(1)\hat{S}_{\mu_p}(k)^{-1}$ (the wanted $O(\lambda)$ term can be extracted using (6.50); Heydenreich et al. forthcoming). By (6.49) and (6.30) and using $f_p^{(1)} \vee f_p^{(2)} \leq 3$, we obtain:

$$\hat{\Delta}_{p;0}^{\text{perc}}(o) - \hat{\Delta}_{p;0}^{\text{perc}}(k) = \sum_{x \in \mathbb{Z}^d} \big(1 - \cos(k \cdot x)\big)\Delta_{p;0}^{\text{perc}}(x) \leq \sum_{x \in \mathbb{Z}^d} \big(1 - \cos(k \cdot x)\big)G_p^{\text{perc}}(x)^2$$
$$= \int_{[-\pi,\pi]^d} \frac{\mathrm{d}^d l}{(2\pi)^d} \hat{G}_p^{\text{perc}}(l)\Big(\hat{G}_p^{\text{perc}}(l) - \tfrac{1}{2}\big(\hat{G}_p^{\text{perc}}(l+k) + \hat{G}_p^{\text{perc}}(l-k)\big)\Big)$$
$$\leq \sum_{(j,j')=(0,\pm 1),(1,-1)} \int_{[-\pi,\pi]^d} \frac{\mathrm{d}^d l}{(2\pi)^d} \frac{900\,\hat{S}_{\mu_p}(k)^{-1}}{(1 - \hat{D}(l))(1 - \hat{D}(l+jk))(1 - \hat{D}(l+j'k))}, \tag{6.55}$$

which is $O(1)\hat{S}_{\mu_p}(k)^{-1}$ if $d > d_c^{\text{perc}}$, as required. The higher-order term $\hat{\Delta}_{p;N}^{\text{perc}}(o) - \hat{\Delta}_{p;N}^{\text{perc}}(k)$ obeys a diagrammatic bound that contains N nonzero triangles, which is bounded by $O(\lambda)^N \hat{S}_{\mu_p}(k)^{-1}$ for $N \geq 1$ if $d > d_c^{\text{perc}}$ (Heydenreich et al. forthcoming), yielding the second inequality in (6.25) if $\lambda \ll 1$.

Derivation of the lace expansion for $G_p^{\text{perc}}(x)$. Here, we only explain the first stage of the derivation of the expansion (6.46). See Hara and Slade (1990) for the completion of the expansion.

First, we introduce the following notions and notation (which will also be used for the Ising model):

Definition 6.1

1. For $V \subset \mathbb{Z}^d$, we denote by $\mathbb{B}_V$ the set of bonds whose vertices are both in V. Given a bond configuration $\boldsymbol{n} \equiv \{n_b\}$ (where $n_b = 1_{\{b \text{ is occupied}\}}$), we say that x is $\boldsymbol{n}$-connected to y in V, denoted $x \xleftrightarrow{\boldsymbol{n}} y$ in V, if either $x = y \in V$ or there is a path from x to y consisting of bonds $b \in \mathbb{B}_V$ with $n_b > 0$.

2. For an event E (= a set of bond configurations) and a bond b, the event $\{E$ without using $b\}$ is the set of bond configurations $\bm{n} \in E$ such that the new configuration obtained by changing n_b is also in E. Given a bond configuration $\bm{n}$, we let:

$$\mathcal{C}_{\bm{n}}^b(x) = \{y : x \underset{\bm{n}}{\longleftrightarrow} y \text{ without using } b\}. \tag{6.56}$$

3. We denote the head and tail of a directed bond b by $\bar{b}$ and $\underline{b}$, respectively: $b = (\underline{b}, \bar{b})$. Given a bond configuration $\bm{n}$, we say that a directed bond b is pivotal for $x \underset{\bm{n}}{\longleftrightarrow} y$ from x, if $x \underset{\bm{n}}{\longleftrightarrow} \underline{b}$ without using b and $\bar{b} \underset{\bm{n}}{\longleftrightarrow} y$ in $\mathbb{Z}^d \setminus \mathcal{C}_{\bm{n}}^b(x)$. Let:

$$\texttt{piv}_{\bm{n}}(x, y) = \{b : b \text{ is pivotal for } x \underset{\bm{n}}{\longleftrightarrow} y \text{ from } x\}. \tag{6.57}$$

If $x \underset{\bm{n}}{\longleftrightarrow} y$ and $\texttt{piv}_{\bm{n}}(x, y) = \varnothing$, we say that x is $\bm{n}$-doubly connected to y, denoted $x \underset{\bm{n}}{\Longleftrightarrow} y$.

4. As usual, we abbreviate $\{\bm{n} : \mathcal{C}_{\bm{n}}^b(x) = V\}$ to $\{\mathcal{C}^b(x) = V\}$, and $\{\bm{n} : b \in \texttt{piv}_{\bm{n}}(x, y)\}$ to $\{b \in \texttt{piv}(x, y)\}$, etc.

We now begin with the derivation of the lace expansion for $G_p^{\text{perc}}(x) \equiv \mathbb{P}_p(o \longleftrightarrow x)$. First we note that, for each bond configuration $\bm{n}$, there are two possibilities for $o \underset{\bm{n}}{\longleftrightarrow} x$: either $\texttt{piv}_{\bm{n}}(o, x) = \varnothing$ or $\texttt{piv}_{\bm{n}}(o, x) \neq \varnothing$. If $\texttt{piv}_{\bm{n}}(o, x) \neq \varnothing$, we take its first element b (so that $o \underset{\bm{n}}{\Longleftrightarrow} \underline{b}$). Then, we have:

$$\{o \longleftrightarrow x\} = \underbrace{\{o \longleftrightarrow x, \texttt{piv}(o, x) = \varnothing\}}_{\{o \Longleftrightarrow x\}} \dot{\cup} \bigcup_b \{o \longleftrightarrow x, b \in \texttt{piv}(o, x), o \Longleftrightarrow \underline{b}\}, \tag{6.58}$$

where, by definition, the event subject to the big union over b can be written as:

$$\{b \text{ is occupied}\} \cap \{o \Longleftrightarrow \underline{b} \text{ without using } b\} \cap \{\bar{b} \longleftrightarrow x \text{ in } \mathbb{Z}^d \setminus \mathcal{C}^b(o)\}. \tag{6.59}$$

Let $p_b = pD(\bar{b} - \underline{b}) \equiv \mathbb{P}_p(b \text{ is occupied})$. Since $\{b \text{ is occupied}\}$ is independent of the other two events in (6.59), we obtain:

$$G_p^{\text{perc}}(x) = \Delta_{p;0}^{\text{perc}}(x) + \sum_b p_b \mathbb{E}_p \left[\mathbb{1}_{\{o \Longleftrightarrow \underline{b}\}} \mathbb{1}_{\{\bar{b} \longleftrightarrow x \text{ in } \mathbb{Z}^d \setminus \mathcal{C}^b(o)\}} \right], \tag{6.60}$$

where the 'without using b' condition has been ignored, as $\mathbb{1}_{\{\bar{b} \longleftrightarrow x \text{ in } \mathbb{Z}^d \setminus \mathcal{C}^b(o)\}}$ is always zero on the event $\{\bar{b} \in \mathcal{C}^b(o)\} \supset \{o \Longleftrightarrow x\} \setminus \{o \Longleftrightarrow x \text{ without using } b\}$.

Next we investigate the expectation in (6.60). For notational convenience, we will drop the subscript p from $\mathbb{P}_p$ and $\mathbb{E}_p$. First, by conditioning on the cluster $\mathcal{C}^b(o)$, we can formally write:

$$\mathbb{E}\Big[1_{\{o\Longleftrightarrow \underline{b}\}}1_{\{\overline{b}\longleftrightarrow x \text{ in } \mathbb{Z}^d\setminus \mathcal{C}^b(o)\}}\Big] = \sum_{V\subset \mathbb{Z}^d}\mathbb{E}\Big[1_{\{\mathcal{C}^b(o)=V, o\Longleftrightarrow \underline{b}\}}1_{\{\overline{b}\longleftrightarrow x \text{ in } \mathbb{Z}^d\setminus V\}}\Big]$$

$$= \sum_{V\subset \mathbb{Z}^d}\sum_{(l,m)}\mathbb{P}(l,m)1_{\{\mathcal{C}_l^b(o)=V,\,o\underset{l}{\Longleftrightarrow}\underline{b}\}}1_{\{\overline{b}\underset{m}{\longleftrightarrow}x\}},$$

(6.61)

where (l,m) is a bond configuration on $\mathbb{B}_{\mathbb{Z}^d}$, with $m \in \{0,1\}^{\mathbb{B}_{\mathbb{Z}^d\setminus V}}$ and $l \in \{0,1\}^{\mathbb{B}_{\mathbb{Z}^d}\setminus \mathbb{B}_{\mathbb{Z}^d\setminus V}}$, and we have ignored the unnecessary 'in $\mathbb{Z}^d\setminus V$' condition for such an m. Let:

$$\mathbb{P}_{\mathbb{Z}^d\setminus V}(m) = \prod_{b\in\mathbb{B}_{\mathbb{Z}^d\setminus V}}\big(p_b\delta_{m_b,1} + (1-p_b)\delta_{m_b,0}\big)$$

(6.62)

and let $\tilde{\mathbb{P}}_V$ be such that $\mathbb{P}(l,m) = \tilde{\mathbb{P}}_V(l)\,\mathbb{P}_{\mathbb{Z}^d\setminus V}(m)$. Then, by multiplying and dividing (6.61) by $\sum_{m'}\mathbb{P}_{\mathbb{Z}^d\setminus V}(m')$ (which is always 1 for percolation), we obtain

$$(6.61) = \sum_{V\subset\mathbb{Z}^d}\sum_{(l,m')}\underbrace{\tilde{\mathbb{P}}_V(l)\,\mathbb{P}_{\mathbb{Z}^d\setminus V}(m')}_{\mathbb{P}(l,m')}1_{\{\mathcal{C}_l^b(o)=V,\,o\underset{l}{\Longleftrightarrow}\underline{b}\}}\underbrace{\frac{\sum_m \mathbb{P}_{\mathbb{Z}^d\setminus V}(m)1_{\{\overline{b}\underset{m}{\longleftrightarrow}x\}}}{\sum_{m'}\mathbb{P}_{\mathbb{Z}^d\setminus V}(m')}}_{\mathbb{P}_{\mathbb{Z}^d\setminus V}(\overline{b}\longleftrightarrow x)}$$

$$= \sum_{V\subset\mathbb{Z}^d}\mathbb{E}\Big[1_{\{\mathcal{C}^b(o)=V,\,o\Longleftrightarrow\underline{b}\}}\mathbb{P}_{\mathbb{Z}^d\setminus V}(\overline{b}\longleftrightarrow x)\Big]$$

$$= \mathbb{E}\Big[1_{\{o\Longleftrightarrow\underline{b}\}}\mathbb{P}_{\mathbb{Z}^d\setminus \mathcal{C}^b(o)}(\overline{b}\longleftrightarrow x)\Big].$$

(6.63)

Of course, the above derivation using $\mathbb{P}(l)$ for $l \in \{0,1\}^{\mathbb{B}_{\mathbb{Z}^d}}$ is formal and rather blunt. However, in Section 6.3.3, we will find similarity in the derivation of an Ising version of this identity.

Finally, by substituting (6.63) back to (6.60), we arrive at:

$$G_p^{\text{perc}}(x) = \Delta_{p;0}^{\text{perc}}(x) + \sum_b \Delta_{p;0}^{\text{perc}}(\underline{b})\,p_b\,G_p^{\text{perc}}(\overline{b},x) - R_{p;1}^{\text{perc}}(x),$$

(6.64)

where:

$$R_{p;1}^{\text{perc}}(x) = \sum_b \mathbb{E}\Big[1_{\{o\Longleftrightarrow\underline{b}\}}\,p_b\big(\mathbb{P}(\overline{b}\longleftrightarrow x) - \mathbb{P}_{\mathbb{Z}^d\setminus \mathcal{C}^b(o)}(\overline{b}\longleftrightarrow x)\big)\Big].$$

(6.65)

This completes the first stage of the expansion.

Applications of the lace expansion to statistical-mechanical models 141

At every stage of the further expansion, we have to investigate the difference between the full two-point function $G_p^{\text{perc}}(y,x) \equiv \mathbb{P}(y \longleftrightarrow x)$ and its restricted version $\mathbb{P}_{\mathbb{Z}^d \setminus V}(y \longleftrightarrow x)$ for given $y \in \mathbb{Z}^d$ and $V \subset \mathbb{Z}^d$. However, since $\mathbb{P}_{\mathbb{Z}^d \setminus V}(y \longleftrightarrow x) = \mathbb{P}(y \longleftrightarrow x \text{ in } \mathbb{Z}^d \setminus V)$, the difference is actually equal to the probability of the event that y is connected to x in $\mathbb{Z}^d$, but not in $\mathbb{Z}^d \setminus V$, i.e., every connection from y to x has to go through a vertex in V. The key idea to derive the higher-order expansion coefficients is to see when for the first time the connection goes though V. If there are no pivotal bonds for that connection after going through V, this will be the contribution to $\Delta_{p;N}^{\text{perc}}(x)$ at the Nth stage of the expansion. On the other hand, if there are pivotal bonds for that connection after going through V, then we cut the structure at the first bond among those pivotal bonds, yielding $(\Delta_{p;N}^{\text{perc}} * pD * G_p^{\text{perc}})(x) - R_{p;N+1}^{\text{perc}}(x)$, where the correction term $R_{p;N+1}^{\text{perc}}(x)$ again contains the difference between a full two-point function and its restricted version. Therefore, we can repeat the same procedure and continue expanding indefinitely. See Hara and Slade (1990) for more details.

6.3.3 Expansion for the Ising model

In this section, we explain the derivation of the expansion (6.46) for the finite-volume Ising two-point function $\langle \varphi_o \varphi_x \rangle_\Lambda := \langle \varphi_o \varphi_x \rangle_{p;\Lambda}$ for $x, o \in \Lambda \subset \mathbb{Z}^d$:

$$\langle \varphi_o \varphi_x \rangle_\Lambda = \Delta_{p,\Lambda}^{\text{Ising}}(x) + \sum_{b \in \mathbb{B}_\Lambda} \Delta_{p,\Lambda}^{\text{Ising}}(\underline{b})\, p_b\, \langle \varphi_{\overline{b}} \varphi_x \rangle_\Lambda. \tag{6.66}$$

To obtain this identity, we use the so-called random-current representation, initiated by Griffiths et al. (1970). As explained below, it allows us to represent $\langle \varphi_o \varphi_x \rangle_\Lambda$ as a sort of percolation two-point function. Therefore, we can apply the inclusion-exclusion argument using pivotal bonds, as explained in the previous section. Moreover, the event defining the Nth-order expansion coefficient $\Delta_{p,\Lambda;N}^{\text{Ising}}(x)$ is identical to that for $\Delta_{p;N}^{\text{perc}}(x)$, for every $N \geq 0$. Then, why is d_c^{Ising} equal to d_c^{SAW}, not equal to d_c^{perc}, as in (6.18)–(6.19)? We will get back to this issue at the end of this section.

Now, we explain the random-current representation of $\langle \varphi_x \varphi_y \rangle_\Lambda$. We call $\boldsymbol{n} \equiv \{n_b\} \in \mathbb{Z}_+^{\mathbb{B}_\Lambda}$ a current configuration on $\mathbb{B}_\Lambda$ (cf., a bond configuration $\boldsymbol{n} \in \{0,1\}^{\mathbb{B}_\Lambda}$ for percolation). A vertex v is said to be a source of a current configuration $\boldsymbol{n}$ if $\sum_{b \ni v} n_b$ is an odd number, and we denote by $\partial \boldsymbol{n}$ the set of all sources of $\boldsymbol{n}$. Let:

$$W_\Lambda(\boldsymbol{n}) = \prod_{b \in \mathbb{B}_\Lambda} \frac{(\beta J_b)^{n_b}}{n_b!}. \tag{6.67}$$

Then, $\langle \varphi_x \varphi_y \rangle_\Lambda$ can be represented as (e.g., Griffiths et al. 1970):

$$\langle \varphi_x \varphi_y \rangle_\Lambda = \frac{\sum_{\partial \boldsymbol{n}=x\triangle y} W_\Lambda(\boldsymbol{n}) 1_{\{x \xleftrightarrow[\boldsymbol{n}]{} y\}}}{\sum_{\partial \boldsymbol{n}=\varnothing} W_\Lambda(\boldsymbol{n})} = \text{_____},\quad (6.68)$$

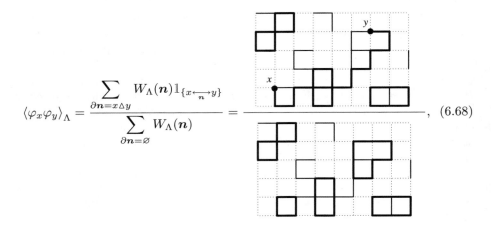

where $x \triangle y$ is an abbreviation for the symmetric difference $\{x\} \triangle \{y\}$: $x \triangle y = \{x, y\}$ if $x \neq y$, and $x \triangle y = \varnothing$ if $x = y$. In the above figures, each bold line-segment represents a bond with an odd current, while a thinner line-segment represents a bond with a positive-even current. Although the indicator function in the numerator is redundant because of the source constraint, we keep it here to emphasize that x is $\boldsymbol{n}$-connected to y. In fact, x is connected to y with bonds having odd currents. Using this fact, we have, for example:

$$\sum_{\partial \boldsymbol{n}=x\triangle y} W_\Lambda(\boldsymbol{n}) 1_{\{x \xleftrightarrow[\boldsymbol{n}]{} y\}} \leq \sum_{v \in \Lambda} \frac{\sum_{n_{x,v}:\text{odd}} (\beta J_{x,v})^{n_{x,v}}/n_{x,v}!}{\sum_{m_{x,v}:\text{even}} (\beta J_{x,v})^{m_{x,v}}/m_{x,v}!}$$

$$\times \sum_{\substack{\partial \boldsymbol{m}=v\triangle y \\ (m_{x,v}:\text{even})}} W_\Lambda(\boldsymbol{m}) 1_{\{v \xleftrightarrow[\boldsymbol{m}]{} y\}}$$

$$\leq \sum_{v \in \Lambda} \underbrace{\tanh(\beta J_{x,v})}_{p_{x,v} \equiv pD(v-x)} \langle \varphi_v \varphi_y \rangle_\Lambda \sum_{\partial \boldsymbol{n}=\varnothing} W_\Lambda(\boldsymbol{n}), \quad (6.69)$$

which implies in the infinite-volume limit (cf., (6.34) and (6.50))

$$\tilde{G}_p^{\text{Ising}}(x,y) := G_p^{\text{Ising}}(x,y) - \delta_{x,y} \leq (pD * G_p^{\text{Ising}})(x,y). \quad (6.70)$$

Repeated use of this inequality yields $G_p^{\text{Ising}}(x) \leq S_p(x)$ for any $p < 1$.

Derivation of the lace expansion for $\langle \varphi_o \varphi_x \rangle_\Lambda$. We only explain here the first stage of the derivation of the expansion (6.66), and refer to Sakai (2007) for the full expansion.

First, we define:

$$\Delta_{p,\Lambda;0}^{\text{Ising}}(x) = \frac{\sum_{\partial \boldsymbol{n}=o\Delta x} W_\Lambda(\boldsymbol{n}) \mathbf{1}_{\{o \underset{\boldsymbol{n}}{\Longleftrightarrow} x\}}}{\sum_{\partial \boldsymbol{n}=\varnothing} W_\Lambda(\boldsymbol{n})}. \tag{6.71}$$

Then, by (6.58) and (6.68), we obtain:

$$\langle \varphi_o \varphi_x \rangle_\Lambda = \Delta_{p,\Lambda;0}^{\text{Ising}}(x)$$
$$+ \sum_{b \in \mathbb{B}_\Lambda} \frac{\sum_{\partial \boldsymbol{n}=o\Delta x} W_\Lambda(\boldsymbol{n}) \mathbf{1}_{\{o \underset{\boldsymbol{n}}{\Longleftrightarrow} \underline{b} \text{ without using } b\}} \mathbf{1}_{\{n_b:\text{odd}\}} \mathbf{1}_{\{\bar{b} \underset{\boldsymbol{n}}{\longleftrightarrow} x \text{ in } \Lambda \setminus \mathcal{C}_{\boldsymbol{n}}^b(o)\}}}{\sum_{\partial \boldsymbol{n}=\varnothing} W_\Lambda(\boldsymbol{n})},$$
$$\tag{6.72}$$

where we have used the fact that $n_b > 0$ is an odd number if $o \underset{\boldsymbol{n}}{\longleftrightarrow} x$ and $b \in \text{piv}_{\boldsymbol{n}}(o, x)$.

Next, we investigate the numerator of the second term. First, by changing the parity of n_b as in (6.69), we obtain:

$$\sum_{\partial \boldsymbol{n}=o\Delta x} W_\Lambda(\boldsymbol{n}) \mathbf{1}_{\{o \underset{\boldsymbol{n}}{\Longleftrightarrow} \underline{b} \text{ without using } b\}} \mathbf{1}_{\{n_b:\text{odd}\}} \mathbf{1}_{\{\bar{b} \underset{\boldsymbol{n}}{\longleftrightarrow} x \text{ in } \Lambda \setminus \mathcal{C}_{\boldsymbol{n}}^b(o)\}}$$
$$= \frac{\sum_{n_b:\text{odd}} (\beta J_b)^{n_b}/n_b!}{\sum_{n_b:\text{even}} (\beta J_b)^{n_b}/n_b!} \sum_{\substack{\partial \boldsymbol{n}=o\Delta b \Delta x \\ (n_b:\text{even})}} W_\Lambda(\boldsymbol{n}) \mathbf{1}_{\{o \underset{\boldsymbol{n}}{\Longleftrightarrow} \underline{b} \text{ without using } b\}} \mathbf{1}_{\{\bar{b} \underset{\boldsymbol{n}}{\longleftrightarrow} x \text{ in } \Lambda \setminus \mathcal{C}_{\boldsymbol{n}}^b(o)\}}$$
$$= p_b \sum_{\partial \boldsymbol{n}=o\Delta b \Delta x} W_\Lambda(\boldsymbol{n}) \mathbf{1}_{\{o \underset{\boldsymbol{n}}{\Longleftrightarrow} \underline{b}\}} \mathbf{1}_{\{\bar{b} \underset{\boldsymbol{n}}{\longleftrightarrow} x \text{ in } \Lambda \setminus \mathcal{C}_{\boldsymbol{n}}^b(o)\}}, \tag{6.73}$$

where we have ignored the 'without using b' condition, due to the same reason as explained below (6.60); we have also dropped the 'n_b:even' condition, because it is impossible to satisfy the source constraint and the conditions in the indicators at the same time if n_b is an odd number. By conditioning on the cluster $\mathcal{C}^b(o)$, we have (cf., (6.61)):

$$(6.73) = p_b \sum_{V \subset \Lambda} \sum_{\partial \boldsymbol{n}=o\Delta b \Delta x} W_\Lambda(\boldsymbol{n}) \mathbf{1}_{\{\mathcal{C}_{\boldsymbol{n}}^b(o)=V,\, o \underset{\boldsymbol{n}}{\Longleftrightarrow} \underline{b}\}} \mathbf{1}_{\{\bar{b} \underset{\boldsymbol{n}}{\longleftrightarrow} x \text{ in } \Lambda \setminus V\}}$$
$$= p_b \sum_{V \subset \Lambda} \sum_{\substack{\partial \boldsymbol{l}=o\Delta \underline{b} \\ \partial \boldsymbol{m}=\bar{b}\Delta x}} \tilde{W}_V(\boldsymbol{l}) \, W_{\Lambda \setminus V}(\boldsymbol{m}) \mathbf{1}_{\{\mathcal{C}_{\boldsymbol{l}}^b(o)=V,\, o \underset{\boldsymbol{l}}{\Longleftrightarrow} \underline{b}\}} \mathbf{1}_{\{\bar{b} \underset{\boldsymbol{m}}{\longleftrightarrow} x\}}, \tag{6.74}$$

where $\boldsymbol{n} \equiv (\boldsymbol{l}, \boldsymbol{m})$ is a current configuration on $\mathbb{B}_\Lambda$, with $\boldsymbol{m} \in \mathbb{Z}_+^{\mathbb{B}_{\Lambda \setminus V}}$ and $\boldsymbol{l} \in \mathbb{Z}_+^{\mathbb{B}_\Lambda \setminus \mathbb{B}_{\Lambda \setminus V}}$, and $\tilde{W}_V$ is such that $W_\Lambda(\boldsymbol{l}, \boldsymbol{m}) = \tilde{W}_V(\boldsymbol{l}) \, W_{\Lambda \setminus V}(\boldsymbol{m})$. Here, in order for a joint current configuration (l, m) to satisfy all the conditions in the first line, l and m have to satisfy the required source constraints in the second line. Inspired by the computation in (6.63), we multiply and divide (6.74) by $\sum_{\partial m' = \varnothing} W_{\Lambda \setminus V}(m')$ (which is not 1 in general for the Ising model) and obtain:

$$(6.74) = p_b \sum_{V \subset \Lambda} \sum_{\partial(l,m') = o \triangle \underline{b}} \underbrace{\frac{\tilde{W}_V(l)\, W_{\Lambda \setminus V}(m')}{W_\Lambda(l,m')}}_{} 1_{\{C_l^b(o) = V,\, o \underset{l}{\Longleftrightarrow} \underline{b}\}}$$

$$\times \underbrace{\frac{\sum_{\partial m = \overline{b} \triangle x} W_{\Lambda \setminus V}(m) 1_{\{\overline{b} \underset{m}{\longleftrightarrow} x\}}}{\sum_{\partial m' = \varnothing} W_{\Lambda \setminus V}(m')}}_{\langle \varphi_{\overline{b}} \varphi_x \rangle_{\Lambda \setminus V}}$$

$$= \sum_{\partial n = o \triangle \underline{b}} W_\Lambda(n) 1_{\{o \underset{n}{\Longleftrightarrow} \underline{b}\}}\, p_b\, \langle \varphi_{\overline{b}} \varphi_x \rangle_{\Lambda \setminus C_n^b(o)}. \tag{6.75}$$

Finally, by substituting (6.75) back to (6.72), we arrive at:

$$\langle \varphi_o \varphi_x \rangle_\Lambda = \Delta^{\text{Ising}}_{p,\Lambda;0}(x) + \sum_{b \in \mathbb{B}_\Lambda} \Delta^{\text{Ising}}_{p,\Lambda;0}(\underline{b})\, p_b\, \langle \varphi_{\overline{b}} \varphi_x \rangle_\Lambda - R^{\text{Ising}}_{p,\Lambda;1}(x), \tag{6.76}$$

where:

$$R^{\text{Ising}}_{p,\Lambda;1}(x) = \sum_{b \in \mathbb{B}_\Lambda} \frac{\sum_{\partial n = o \triangle \underline{b}} W_\Lambda(n) 1_{\{o \underset{n}{\Longleftrightarrow} \underline{b}\}}\, p_b \left(\langle \varphi_{\overline{b}} \varphi_x \rangle_\Lambda - \langle \varphi_{\overline{b}} \varphi_x \rangle_{\Lambda \setminus C_n^b(o)} \right)}{\sum_{\partial n = \varnothing} W_\Lambda(n)}. \tag{6.77}$$

This completes the first stage of the expansion.

Similarly to percolation, at every stage of the further expansion, we have to investigate $\langle \varphi_y \varphi_x \rangle_\Lambda - \langle \varphi_y \varphi_x \rangle_{\Lambda \setminus V}$ for given $y \in \Lambda$ and $V \subset \Lambda$. However, since $\sum_{\partial n = \varnothing} W_A(n) \neq 1$ for any nontrivial $A \subset \mathbb{Z}^d$, we have:

$$\langle \varphi_y \varphi_x \rangle_{\Lambda \setminus V} \neq \frac{\sum_{\partial n = y \triangle x} W_\Lambda(n) 1_{\{y \underset{n}{\longleftrightarrow} x \text{ in } \Lambda \setminus V\}}}{\sum_{\partial n = \varnothing} W_\Lambda(n)}, \tag{6.78}$$

in contrast with the equality $\mathbb{P}_{\Lambda \setminus V}(y \longleftrightarrow x) = \mathbb{P}(y \longleftrightarrow x \text{ in } \Lambda \setminus V)$ for percolation. This makes it difficult to compare two-point functions on different sets of sites.

In Sakai (2007), we overcame this difficulty using the so-called source-switching lemma (Griffiths et al. 1970) and proved that:

$$\langle \varphi_y \varphi_x \rangle_\Lambda - \langle \varphi_y \varphi_x \rangle_{\Lambda \setminus V} = \frac{\displaystyle\sum_{\substack{\partial \boldsymbol{n} = y \Delta x \\ \partial \boldsymbol{m} = \varnothing}} W_\Lambda(\boldsymbol{n}) \, W_{\Lambda \setminus V}(\boldsymbol{m}) \mathbb{1}_{\{y \xleftrightarrow{m+n} x \text{ through } V\}}}{\displaystyle\sum_{\partial \boldsymbol{n} = \partial \boldsymbol{m} = \varnothing} W_\Lambda(\boldsymbol{n}) \, W_{\Lambda \setminus V}(\boldsymbol{m})}, \qquad (6.79)$$

where $\mathbb{1}_{\{y \xleftrightarrow{m+n} x \text{ through } V\}} = \mathbb{1}_{\{y \xleftrightarrow{m+n} x\}} - \mathbb{1}_{\{y \xleftrightarrow{m+n} x \text{ in } \Lambda \setminus V\}}$ for $\boldsymbol{m} + \boldsymbol{n} = \{m_b + n_b\} \in \mathbb{Z}_+^{\mathbb{B}_\Lambda}$. By this percolation representation, we can follow the same strategy as for percolation: find the first bond $b \in \texttt{piv}_{m+n}(y,x)$ such that $x \xleftrightarrow{m+n} \underline{b}$ though V. At the Nth stage of the expansion, the contribution from the case in which there are no such pivotal bonds is $\Delta_{p,\Lambda;N}^{\text{Ising}}(x)$, otherwise we have $\sum_b \Delta_{p,\Lambda;N}^{\text{Ising}}(\underline{b}) p_b \langle \varphi_{\overline{b}} \varphi_x \rangle_\Lambda - R_{p,\Lambda;N+1}^{\text{Ising}}(x)$, where the correction term $R_{p,\Lambda;N+1}^{\text{Ising}}(x)$ again contains the difference of the form (6.79). Then, we can repeat the same argument to continue the expansion indefinitely. See Sakai (2007) for more details.

Remark 6.2: *Since we have exploited the percolation structure of the random-current representation, the set of 'bond' configurations defining $\Delta_{p,\Lambda;N}^{\text{Ising}}(x)$ is identical to that for $\Delta_{p;N}^{\text{perc}}(x)$, for every $N \geq 0$. As explained in Section 6.3.2, in order to bound $\Delta_{p;N}^{\text{perc}}(x)$ for percolation, it suffices to know the geometrical structure of the relevant bond configurations. Then, the BK inequality does the remaining job: e.g., $\Delta_{p;0}^{\text{perc}}(x) \leq G_p^{\text{perc}}(x)^2$. However, if this was also the case for the Ising model, i.e., if there was an Ising version of the BK inequality such that $\Delta_{p,\Lambda;0}^{\text{Ising}}(x) \leq G_p^{\text{Ising}}(x)^2$, then $d_{\text{c}}^{\text{Ising}}$ would be equal to $d_{\text{c}}^{\text{perc}}$ (above which the triangle condition holds) rather than $d_{\text{c}}^{\text{SAW}}$ (above which the bubble condition holds, as required for the mean-field behaviour for the Ising model).*

In fact, instead of using visible clusters of positive-current bonds, we should use a graph $\mathbb{G}_\Lambda(\boldsymbol{n}) \equiv (\Lambda, E_\Lambda(\boldsymbol{n}))$ that is constructed from each current configuration $\boldsymbol{n} \in \mathbb{Z}_+^{\mathbb{B}_\Lambda}$ by joining $u, v \in \Lambda$ with $n_{u,v}$ labelled edges. Then, for each $\boldsymbol{n} \in \mathbb{Z}_+^{\mathbb{B}_\Lambda}$ defining $\Delta_{p,\Lambda;0}^{\text{Ising}}(x)$, there are at least three edge-disjoint paths from o to x in the graph $\mathbb{G}_\Lambda(\boldsymbol{n})$, as shown in the following example:

Example 6.1

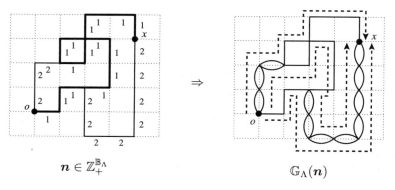

$\boldsymbol{n} \in \mathbb{Z}_+^{\mathbb{B}_\Lambda}$ $\qquad\qquad\qquad\qquad$ $\mathbb{G}_\Lambda(\boldsymbol{n})$

This observation leads us to predict that $\Delta^{\text{Ising}}_{p,\Lambda;0}(x)$ is bounded similarly to $\Delta^{\text{SAW}}_{p;1}(x)$. The prediction was proved affirmative in Sakai (2007): $\Delta^{\text{Ising}}_{p,\Lambda;0}(x) - \delta_{x,o} \leq \tilde{G}^{\text{Ising}}_p(x)^3$. It was also proved that the higher-order expansion coefficients for the Ising model obey similar diagrammatic bounds for the SAW expansion coefficients. This is why $d_c^{\text{Ising}} = d_c^{\text{SAW}}$, not $d_c^{\text{Ising}} = d_c^{\text{perc}}$.

Acknowledgements

The analysis of the lace expansion explained in this article is taken out of the recent joint work with Lung-Chi Chen, Markus Heydenreich and Remco van der Hofstad. I am grateful to them for the fruitful and enjoyable collaboration. I would also like to express my thanks to the Institute of Mathematics at Academia Sinica, the London Mathematical Society, and the Netherlands Organisation for Scientific Research (NWO) for supporting the travel expenses.

References

Aizenman, M. (1982). Geometric analysis of ϕ^4 fields and Ising models. *Comm. Math. Phys.*, **86**:1–48.

Aizenman, M. and Barsky, D. J. (1987). Sharpness of the phase transition in percolation models. *Comm. Math. Phys.*, **108**:489–526.

Aizenman, M., Barsky, D. J. and Fernández, R. (1987). The phase transition in a general class of Ising-type models is sharp. *J. Stat. Phys.*, **47**:343–74.

Aizenman, M. and Newman, C. M. (1984). Tree graph inequalities and critical behaviour in percolation models. *J. Stat. Phys.*, **36**:107–43.

van den Berg, J. and Kesten, H. (1985). Inequalities with applications to percolation and reliability. *J. Appl. Probab.*, **22**:556–69.

Bodineau, T. (2006). Translation invariant Gibbs states for the Ising model. *Probab. Theory Related Fields*, **135**:153–68.

Borgs, C., Chayes, J. T., van der Hofstad, R., Slade, G. and Spencer, J. (2005). Random subgraphs of finite graphs. II. The lace expansion and the triangle condition. *Ann. Probab.*, **33**:1886–944.

Brydges, D. and Spencer, T. (1985). Self-avoiding walk in 5 or more dimensions. *Comm. Math. Phys.*, **97**:125–48.

Camia, F. and Newman, C. M. (2007). Critical percolation exploration path and SLE_6: a proof of convergence. *Probab. Theory Related Fields*, **139**:473–519.

Chen, L.-C. and Sakai, A. (2008). Critical behaviour for long-range oriented percolation. I. *Probab. Theory Related Fields*, To appear.

Fernández, R., Fröhlich, J. and Sokal, A. D. (1992). *Random Walks, Critical Phenomena, and Triviality in Quantum Field Theory*. Springer, Berlin.

Griffiths, R. B., Hurst, C. A. and Sherman, S. (1970). Concavity of magnetization of an Ising ferromagnet in a positive external field. *J. Math. Phys.*, **11**:790–5.

Grimmett, G. (1999). *Percolation (2nd edition)*. Springer, Berlin.

Hara, T. (2008). Decay of correlations in nearest-neighbour self-avoiding walk, percolation, lattice trees and animals. *Ann. Probab.*, **36**:530–93.

Hara, T., van der Hofstad, R. and Slade, G. (2003). Critical two-point functions and the lace expansion for spread-out high-dimensional percolation and related models. *Ann. Probab.*, **31**:349–408.

Hara, T. and Slade, G. (1990). Mean-field critical behaviour for percolation in high dimensions. *Comm. Math. Phys.*, **128**:333–91.

Hara, T. and Slade, G. (1992). Self-avoiding walk in five or more dimensions. I. The critical behaviour. *Comm. Math. Phys.*, **147**:101–36.

Heydenreich, M., van der Hofstad, R. and Sakai, A. Mean-field behaviour for long- and finite-range Ising model, percolation and self-avoiding walk. In preparation.

van der Hofstad, R. and Holmes, M. (2006). An expansion for self-interacting random walks. Preprint.

van der Hofstad, R. and Sakai, A. (2005). Critical points for spread-out self-avoiding walk, percolation and the contact process above the upper critical dimensions. *Probab. Theory Related Fields*, **132**:438–70.

van der Hofstad, R. and Slade, G. (2002). A generalised inductive approach to the lace expansion. *Probab. Theory Related Fields*, **122**:389–430.

Holmes, M. and Sakai, A. (2007). Senile reinforced random walks. *Stochastic Process. Appl.*, **117**:1519–39.

Lieb, E. H. (1980). A refinement of Simon's correlation inequality. *Comm. Math. Phys.*, **77**:127–35.

Madras, N. and Slade, G. (1993). *The Self-Avoiding Walk*. Birkhäuser, Boston.

Sakai, A. (2007). Lace expansion for the Ising model. *Comm. Math. Phys.*, **272**: 283–344.

Simon, B. (1980). Correlation inequalities and the decay of correlations in ferromagnets. *Comm. Math. Phys.*, **77**:111–26.

Slade, G. (2006). The lace expansion and its applications. *Lecture Notes in Math.*, 1879.

Smirnov, S. and Werner, W. (2001). Critical exponents for two-dimensional percolation. *Math. Res. Lett.*, **8**:729–44.

Werner, W. (2004). Random planer curves and Schramm–Loewner evolutions. *Lecture Notes in Math.*, **1840**:107–95.

7

LARGE DEVIATIONS FOR EMPIRICAL CYCLE COUNTS OF INTEGER PARTITIONS AND THEIR RELATION TO SYSTEMS OF BOSONS

Stefan Adams

Abstract

Motivated by the Bose gas, we introduce certain combinatorial structures. We analyse the asymptotic behaviour of empirical shape measures and of empirical path measures of N Brownian motions with large deviations techniques. The rate functions are given as variational problems which we analyse. A symmetrized system of Brownian motions, that is, for any i, the terminal location of the i-th motion is affixed to the initial point of the $\sigma(i)$-th motion, where σ is a uniformly distributed random permutation of $1, \ldots, N$, is highly correlated and has to be formulated such that standard techniques can be applied. We review a novel spatial and a novel cycle structure approach for the symmetrized distributions of the empirical path measures. The cycle structure leads to a proof of a phase transition in the mean path measure.

7.1 Introduction

We study different aspects of combinatorial asymptotic large-N behaviour of distributions on the group $\mathfrak{S}_N$ of permutations of N elements and their cycles structures distributed on the set $\mathcal{P}_N$ of integer partitions of N. We combine this analysis with large deviations principles for certain empirical path measures of Brownian motions. We review two different approaches to analyse the large-N asymptotic of the mean path measure under symmetrized distributions. One is spatial structure of the symmetrization and the other one is the cycle structure for concatenations of Brownian bridges to Brownian bridges whose time horizons equal the cycle lengths.

The main focus is to derive variational problems whose analyses will provide deeper insight into the probabilistic asymptotic behaviour for large systems of Brownian motions. This combination of combinatorial studies, large deviations techniques, and variational analysis is novel and has its roots in the mathematical analysis of large systems of Bosons, and it is hence related to and carries forward the article Adams and König (2008a) in these proceedings. In Section 7.1.1 we review main features of studying systems of Bosons and in Section 7.1.2 we introduce probabilistic models and outline how these raise interesting combinatorial structures for permutations and integer partitions.

7.1.1 Motivation

The state of a large system of N identical quantum particles (subsystems) is described by the many-body wave function. Two many-body wave functions which result from each other by a permutation of the indices distinguishing the particles must describe the same state. Such a permutation can change the state vector (wave function) only by a numerical factor, and these factors must give a 1–dimensional representation of the permutation group $\mathfrak{S}_N$ of N elements. Hence, there are only two possible choices, -1 and $+1$. That is, wave functions are antisymmetric or symmetric under permutations. Due to Pauli's exclusion principle, systems of Fermions are described by antisymmetric wave functions. If the wave functions are symmetric, i.e., they are elements in the image of the projection $P_N^+\colon L^2(\mathbb{R}^{dN}) \to L_+^2(\mathbb{R}^{dN})$ of the N-particle Hilbert space $L^2(\mathbb{R}^{dN})$:

$$P_N^+(\Psi)(x) = \frac{1}{N!} \sum_{\sigma \in \mathfrak{S}_N} \Psi(x_{\sigma(1)}, \ldots, x_{\sigma(N)}),$$

one calls the quantum particles Bosons. The Bosons are well-known because they show a phenomenon known as Bose–Einstein condensation. It was predicted by Einstein (1925) on the basis of ideas of the Indian physicist Bose (1924) concerning the statistical description of the quanta of light: in a system of particles described by symmetric many-body wave functions and whose total number is conserved, there should be a temperature below which a finite fraction of all the particles 'condense' into the same one-particle state. Einstein's original prediction was for a noninteracting gas of particles. The predicted phase transition is associated with the condensation of atoms in the state of lowest energy and is the consequence of quantum statistical effects.

For a long time these predictions were considered as a curiosity of noninteracting gases and its statistics, called Bose statistics, and had no practical impact. But the ideal gas systems show that the above symmetrization generates correlations among the noninteracting particles. We review the mathematics concerned with this symmetrization and its relation to combinatorial studies of integer partitions and corresponding limit theorems. Our main objective is to derive variational formulae via large deviations principles for symmetrized systems of Brownian motions. We briefly motivate this ansatz in the following. N quantum particles are described by the N-particle Hamilton operator:

$$H_N = \sum_{i=1}^N \bigl(-\Delta_i + W(x_i) \bigr) + \sum_{1 \le i < j \le N} v(|x_i - x_j|), \qquad x_1, \ldots, x_N \in \mathbb{R}^d,$$

where the i-th Laplace operator, Δ_i, represents the kinetic energy of the i-th particle, and $W\colon \mathbb{R}^d \to [0,\infty]$ is the trap potential, and where the pair potential $v\colon \mathbb{R}_+ \to \mathbb{R}$ expresses the potential energy of two interacting particles. We do not specify here the assumptions on the trap potential W and on the pair potential v but refer to standard choices in Ruelle (1969) and Lieb et al. (2005). The ground state at zero temperature is the minimizer of the energy and is, due to the symmetry properties of the Hamilton operator, a symmetric N-particle wave function. The study of systems of Bosons at zero temperature is reviewed in the article Adams and König (2008a), more can be found in Lieb et al. (2005). Recent good references including experimental aspects are Pitaevskii and Stringari (2003) and Griffin et al. (1995).

To describe systems of Bosons at thermodynamic equilibrium with inverse temperature $\beta > 0$ one has to analyse the traces of the Boltzmann factor $\mathrm{e}^{-\beta H_N}$, like the free energy, or the pressure, where the trace is restricted to the subspace $L_+(\mathbb{R}^{dN})$ of symmetric N-particle wave functions. The trace class operator $\mathrm{e}^{-\beta H_N}$ is called the canonical ensemble for which the number of particles and the inverse temperature is fixed, see Khinchin (1960) and Thirring (1980) as standard references. The so-called quantum canonical partition function $Z_N^{(\mathrm{sym})}(\beta)$ is the trace of this operator, i.e.:

$$Z_N^{(\mathrm{sym})}(\beta) = \mathrm{Tr}_{L_+^2(\mathbb{R}^{dN})}\left(\mathrm{e}^{-\beta H_N}\right) = \mathrm{Tr}_{L^2(\mathbb{R}^{dN})}\left(P_N^+ \mathrm{e}^{-\beta H_N}\right).$$

However, these traces are very difficult to calculate because the spectral analysis of the Hamilton operator H_N with interaction is not known. We will discuss therefore only noninteracting Bosons in this article.

The genuine task of quantum statistical mechanics is to prove and analyse the thermodynamic limit $-\lim_{\Lambda \uparrow \mathbb{R}^d, N \to \infty} 1/\beta|\Lambda| \log Z_N^{(\mathrm{sym})}(\beta)$, which is the free energy, such that $N/|\Lambda| \to \rho \in (0, \infty)$, see Ruelle (1969) for a general introduction to the concept of thermodynamic limit. The quantum statistical mechanics of this ideal Bose gas is well understood (see for example Huang 1987). One can calculate the specific free energy in the thermodynamic limit as a function of the inverse temperature and the density. The Bose–Einstein condensation transition can be identified here as a singularity in the specific free energy for certain parameter values. The pressure of the ideal Bose gas at finite temperature can be calculated in the so-called grandcanonical ensemble, where the particle number is a Poissonian random variable. However, since Einstein's work in 1925 there has been no rigorous mathematical proof for interacting Bosons in the thermodynamic limit for finite density and positive temperature. The only exception is the proof of Bose–Einstein condensation on a lattice with hard-core exclusion and half filling, see Lieb et al. (2005). The main difficulties are the role of the symmetrization, the role of the interaction, and an appropriate definition/criterium of what Bose–Einstein is precisely. There have been three lines of rigorous mathematical attacks. One started with Landau (1941) and its description of superfluidity, which is considered as a Bose–Einstein condensation since London (1938), in terms of the spectrum of elementary excitations of the fluid. In 1947 Bogoliubov developed the first microscopic theory of interacting Bose gases, based on approximations of the Hamilton operator and the concept of Bose–Einstein condensation. This initiated several theoretical studies; a recent account on the state of the art can be found in Adams and Bru (2004a,b) and on its contribution to superfluidity theory in Adams and Bru (2004c).

A second line is devoted to the study of dilute interacting systems, compare the article Adams and König (2008a) for an overview on this. For these systems the first experimental realization of Bose–Einstein condensation was derived in 1995. This was followed by rigorous studies in a series of papers by Lieb et al., see Lieb et al. (2005) for more information.

The third line of attack focuses on probabilistic representations of traces and interacting Brownian motions. This started with Adams et al. (2006a) and Adams et al. (2006b) for dilute systems and in Adams and König (2007), Adams and Dorlas (2007), and Adams (2007) for symmetrized systems. In Section 7.1.2 below we outline this approach. This approach has two challenging task one is to deal with the interaction of the Brownian motions and the other one is to resolve the correlations due to the symmetrization. The symmetrization correlations are the main subject of this article, and we will outline several aspects of combinatorial and stochastic analysis related with these.

In order to understand Bose–Einstein condensation as a quantum phase transition one needs to study correlation functions. In quantum statistical mechanics correlations can be expressed as reduced traces of the Boltzmann factor (see Thirring (1980) or Bratteli and Robinson (1997)). The reduced one-particle density matrix defines an integral kernel for the corresponding operator. Penrose (1951) and Onsager and Penrose (1956) introduced the concept of the nondiagonal long-range order of the one-particle reduced density matrix (the integral kernel) and defined this as a criterion for Bose–Einstein condensation.

Let us make some remarks on related literature. Scaling limits for shape measures of integer partitions in $\mathcal{P}_N$ under uniform distribution are obtained in Vershik (1996). Large deviations from this limit behaviour are in Dembo et al. (2000), where large deviations principles for scaled shape measures for partitions as well as for strict partitions under uniform distributions are derived. Motivated by the statistics of combinatorial partitions, illustrated by Vershik in Vershik (1996), Benfatto et al. (2005) derived limit theorems for statistics of combinatorial partitions for the case of a mean field Bose gas in the grandcanonical ensemble. Here, in contrast to the canonical ensemble, only the mean of the particle number is fixed. Benfatto et al. (2005) are using Fourier analysis of the corresponding traces to derive a complete description of the statistics of short and long cycles. For a perturbed mean-field model the density of long cycles for a perturbed mean-field model is analysed in Dorlas et al. (2005).

7.1.2 *Systems of Bosons and Probabilistic models*

Feynman 1953 introduced the functional integration methods for traces, see de Witt and Storaeds (1970) and Bratteli and Robinson (1997) for details. Since the 1960s, interacting Brownian motions are generally used for probabilistic representations for these traces. The parameter β, which is interpreted as the inverse temperature of the system, is then the length of the time interval of the Brownian motions. Difficulties arise for systems of Bosons due to the symmetrization (see Bratteli and Robinson 1997) and the interaction.

Let $\Omega := \{\omega\colon [0,\infty) \to \mathbb{R}^d\colon \omega \text{ continuous}\}$ be the set of continuous functions $[0,\infty) \to \mathbb{R}^d$. The elements in Ω are called trajectories or paths and we denote by $\Omega_k = \{\omega\colon [0,k\beta] \to \mathbb{R}^d\colon \omega \text{ continuous}\}$, $k \in \mathbb{N}$, the set of paths for time horizon $[0,k\beta]$. We write Ω_β for Ω_1. We equip Ω (respectively Ω_k) with the topology of uniform convergence and with the corresponding Borel σ-field $\mathcal{B}$ (respectively $\mathcal{B}_k$). We consider N Brownian motions, $B^{(1)}, \ldots, B^{(N)}$, with time horizon $[0,\beta]$ as N random variables taking values in Ω_β. For the reader's convenience, we repeat the definition of a Brownian bridge measure; see the Appendix in Sznitman (1998). We decided to work with Brownian motions having generator Δ instead of $\frac{1}{2}\Delta$. We write $\mathbb{P}_x$ for the probability measure under which $B = B^{(1)}$ starts from $x \in \mathbb{R}^d$. The canonical (non-normalized) Brownian bridge measure on the time interval $[0,\beta]$ with initial site $x \in \mathbb{R}^d$ and terminal site $y \in \mathbb{R}^d$ is defined as $\mu^\beta_{x,y}(A) = \mathbb{P}_x(B \in A; B_\beta \in \mathrm{d}y)/\mathrm{d}y$ for $A \subset \Omega_\beta$ measurable. Hence, the Brownian bridge measure for a Brownian bridge confined to a subset $\Lambda_N \subset \mathbb{R}^d$ is defined by:

$$\mu^{\beta,N}_{x,y}(A) = \frac{\mathbb{P}_x(B \in A; B_\beta \in \mathrm{d}y, B_{[0,\beta]} \subset \Lambda_N)}{\mathrm{d}y} \qquad A \subset \Omega_\beta \text{ measurable.} \qquad (7.1)$$

If the motions are not confined to stay in Λ_N we have:

$$\mu^\beta_{x,y}(\Omega_\beta) = \frac{\mathbb{P}_x(B_\beta \in \mathrm{d}y)}{\mathrm{d}y} = (4\pi\beta)^{-d/2} e^{-\frac{1}{4\beta}|x-y|^2}.$$

The Feynman–Kac formula gives an expression for the traces of Boltzmann factor. For that we define the following interaction Hamiltonian:

$$G_{N,\beta} = \sum_{1 \leq i < j \leq N} \int_0^\beta v(|B^{(i)}_t - B^{(j)}_t|)\mathrm{d}t$$

for the N Brownian motions $B^{(1)}, \ldots, B^{(N)}$ with time horizon $[0,\beta]$.

For Dirichlet boundary conditions for the Hamilton operator (Laplace operator), i.e., the particles are enclosed in the box Λ_N, we have:

$$\mathrm{Tr}\,(e^{-\beta H_N}) = \int_{\Lambda_N} \mathrm{d}x_1 \cdots \int_{\Lambda_N} \mathrm{d}x_N \bigotimes_{i=1}^{N} \mu^{\beta,N}_{x_i,x_i}\left(e^{-G_{N,\beta}}\right).$$

This trace describes so-called Boltzmann particles, which means classical particles for which no special statistics is required. The symmetrized trace is:

$$\mathrm{Tr}_{L^2_+(\mathbb{R}^{dN})}(e^{-\beta H_N}) = \frac{1}{N!} \sum_{\sigma \in \mathfrak{S}_N} \int_{\Lambda_N} \mathrm{d}x_1 \cdots \int_{\Lambda_N} \mathrm{d}x_N \bigotimes_{i=1}^{N} \mu^{\beta,N}_{x_i,x_{\sigma(i)}}\left(e^{-G_{N,\beta}}\right). \qquad (7.2)$$

The trace formula (7.2) is the starting point for the remaining sections, it defines transformed path measures for N Brownian motions. As mentioned above there are two aspects to deal with, the interaction and the symmetrization. The interaction for dilute systems is handled in Adams and König (2008a) of these proceedings and the symmetrization is studied here. In what follows we therefore do not handle interacting motions but focus on the symmetrization. This symmetrization is the origin for the Bose–Einstein condensation and has to be understood deeply. In Section 7.2 the cycle structure of the permutations raises interesting asymptotic combinatorial questions, and we derive our first large deviations principles for the discrete shape measures under various distributions. We give an overview of the combinatorial research on permutations and integer partitions. In Section 7.3 we combine the asymptotic combinatorics with certain path empirical measures of the Brownian motions. The objective there is to gain deeper insight in a probabilistic symmetrized model which can provide information on the corresponding systems of Bosons. For the first time we derive a phase transition for mean path measure for a model with no interaction. The proof of this transition requires complete information on the cycle structure, i.e., we will use our insights from Section 7.2. We contrast the cycle structure to a spatial structure, which we analyse in Section 7.3.1. This spatial approach is new, see Adams and König (2007) and Adams and Dorlas (2007), and it gives an indirect proof for the Bose–Einstein condensation. In future, the spatial and the cycle method have to be combined to describe the transition behaviour also with interactions. This combination will enable one to prove Bose–Einstein condensation with the off-diagonal long range order behaviour criterion.

7.2 Large deviations for cycle counts

The cycle structure of permutations allows us to replace in (7.2) the sum over permutations by a sum over integer partitions. This in turn defines probability distributions on permutations and on integer partitions. We introduce some basic facts on integer partitions.

For any integer N, a partition λ of N is the collection of integers $n_1 \geq n_2 \geq \cdots \geq n_k \geq 1, k \in \{1,\ldots,N\}$, such that $\sum_{i=1}^k n_i = N$. We denote the set of all partitions of N by $\mathcal{P}_N$. Any partition $\lambda \in \mathcal{P}_N$ is determined by the sequence $\{r_k\}_{k=1}^N$ of positive integers r_k such that $\sum_{k=1}^N k r_k = N$, where we write $r_k(\lambda) = r_k$. We call the number r_k an *occupation number* or *cycle count* of the partition, and we denote the whole tuple of the cycle counts by $R_N = (r_1,\ldots,r_N)$. The multiplicity $\sharp\lambda$ of a partition is the number of cycles, i.e., $\sharp\lambda = \sum_{k=1}^N r_k$. A cycle of length k is a chain of permutations, such as 1 goes to 2, 2 goes to 3, 3 goes to 4, etc. until $k-1$ goes to k and finally k goes to 1. A permutation with exactly r_k cycles of length k is said to be of type $\{r_k\}_{k=1}^N$. Hence, each partition $\lambda \in \mathcal{P}_N$ corresponds to a conjugacy class $A(\lambda)$ of permutations, i.e., those of the same type, with exactly:

$$\sharp A(\lambda) = \frac{N!}{\prod_{k=1}^N r_k! k^{r_k}}$$

elements.

If a permutation is chosen uniformly and at random from the $N!$ possible permutations in $\mathfrak{S}_N$, then the counts r_k of cycles of length k are dependent random variables. The joint distribution of the cycle countsis given by:

$$\mathbb{P}(R_N = r) = \mathbb{1}\{\sum_{k=1}^{N} kr_k = N\} \prod_{k=1}^{N} \left(\frac{1}{k}\right)^{r_k} \frac{1}{r_k!} \qquad (7.3)$$

with $r = (r_1, \ldots, r_N) \in \mathbb{Z}_+^N$. The uniform distribution in (7.3) is called the Ewens Sampling formula with parameter $\Theta = 1$. The Ewens Sampling formula (Ewens 1972) reads:

$$\mathbb{P}(R_N = r) = \frac{N!}{\Theta(\Theta+1)\cdots(\Theta+N-1)} \mathbb{1}\{\sum_{k=1}^{N} kr_k = N\} \prod_{k=1}^{N} \left(\frac{\Theta}{k}\right)^{r_k} \frac{1}{r_k!}$$

with $r = (r_1, \ldots, r_N) \in \mathbb{Z}_+^N$. This sampling formula was analysed intensively by Kingman (1975), Kingman (1978b), and Kingman (1978a), see also Watterson (1976) for a diffusion model of the allele frequencies. There exists an extensive literature on questions related to this sampling formula and random discrete partitions. See the recent monographs Pitman (2002) and Arratia et al. (2003) for an overview and further references. These studies go back to Goncharov (1944), who studied the asymptotic behaviour of the distribution of cycle counts for the uniform (Ewens sampling with $\Theta = 1$) distribution. For permutations of single points of point process clouds in $\mathbb{R}^d$ or graphs we refer to Kolchin (1986), see further the Section 7.3.1 below.

We focus on large deviations of different distributions of the following functional of integer partitions, the so-called *discrete empirical shape measure*, or *empirical cycle count* distribution, defined as:

$$Q_N : \mathcal{P}_N \to \mathcal{M}_1(\mathbb{N}), \lambda \mapsto Q_N^\lambda(\cdot) = \frac{1}{N} \sum_{k=\cdot}^{N} r_k(\lambda), \qquad (7.4)$$

where $\mathcal{M}_1(\mathbb{N})$ is the set of probability measures on $\mathbb{N}$. We will write $Q_N^\lambda = Q_N$ in the following. The name shape measure has its roots in the two conjugate representations of integer partitions, the so-called Ferrer diagram and the Young Tableau. Define $\widehat{Q}_N(k) = Q_N(k) - Q_N(k+1)$ for any $k \in \mathbb{N}$. Then the occupation numbers are given by $r_k = N\widehat{Q}_N(k), k = 1, \ldots, N$, which define uniquely the integer partition λ. In a Ferrer diagram the partition $\{r_k\}_{k=1}^{N}$ is represented by r_k rows of k horizontal blocks. They are placed in a diagram in descending order with the longest or largest k at the top. It can also be viewed as a block diagram in the $(NQ_N(1), \ldots, NQ_N(N))$ space. Here $NQ_N(k)$ blocks are put vertically in the k-th column. The total number of rows is the multiplicity $\sharp\lambda$ of the partition and the area (the total number of blocks) of this diagram is N. A Young tableau is similar but, here, $NQ_N(k)$ blocks are put horizontally in the k-th row. One can obtain a Young tableau from a Ferrer diagram by first turning the diagram upside down and then by rotating it through 90° clockwise. The notion 'shape' comes now from the study of the asymptotic shapes of this diagrams/tableaux as $N \to \infty$ under suitable continuous scaling, see Vershik (1996) and for a recent overview Pitman (2002). Large deviations from the expected shape of the diagrams are studied in Dembo et al. (2000) for the uniform distribution.

We do not scale the discrete shape measure because we need in the limit the whole discrete cycle count distribution for our large deviations results in Section 7.3. That is, we are interested in the large N-behaviour of the discrete shape measures Q_N under different distributions of the integer partitions. Beside the uniform distribution in (7.3) and general Ewens sampling distribution our main interest is in the following distribution.

$$\nu_N^{(\text{Bose})}(\lambda) = \frac{1}{Z_N^{(\text{Bose})}(\beta)} \prod_{k=1}^{N} \left(\frac{\left(\varrho^{-1} N \right)^{r_k}}{r_k! k^{r_k}} \right) \left(\frac{1}{4\pi\beta k} \right)^{d/2 r_k} \qquad \lambda \in \mathcal{P}_N \qquad (7.5)$$

with normalization:

$$Z_N^{(\text{Bose})}(\beta) = \sum_{\lambda \in \mathcal{P}_N} \prod_{k=1}^{N} \left(\frac{\left(\varrho^{-1} N \right)^{r_k}}{r_k! k^{r_k}} \right) \left(\frac{1}{4\pi\beta k} \right)^{d/2 r_k}$$

for given $d \in \mathbb{N}$ and $\beta, \varrho > 0$. This distribution is motivated from the non-interacting Bose gas enclosed in $\Lambda_N \subset \mathbb{R}^d$ with particle density $\varrho = |\Lambda_N|/N$. We outline this in the following. Going back to trace formula (7.2) note that the conjugacy classes $A(\lambda)$ of permutations are the ones where the trace operation is constant because it is a cyclic operation. For each partition $\lambda \in \mathcal{P}_N$ we can regroup the product of the Brownian bridge measures $\mu_{x,y}^{\beta,N}$ in such a way that we concatenate r_k times k Brownian bridges to obtain r_k Brownian bridges of time horizon $[0, k\beta]$. This is possible because we integrate out the intermediate spatial points. Hence we get for the canonical partition function of the noninteracting Bose gas:

$$Z_N^{(\text{sym})}(\beta) = \sum_{\lambda \in \mathcal{P}_N} \prod_{k=1}^{N} \left(\frac{1}{r_k! k^{r_k}} \right) \bigotimes_{k=1}^{N} \left(\int_{\Lambda_N} \mathrm{d}x \mu_{x,x}^{k\beta,N} \right)^{\otimes r_k} (\Omega_\beta^{\otimes N}). \qquad (7.6)$$

The difference with (7.5) is that there we are using the free Brownian bridge measure, i.e., the motions are not confined to stay in Λ_N. However, both expressions are close and coincide in the limit $\Lambda_N \uparrow \mathbb{R}^d$ as $N \to \infty$, because of the estimation:

$$(4\pi\beta k)^{-d/2}(1 - \mathrm{e}^{-dN/4\beta}) \leq \mu_{x,x}^{k\beta,N}(\Omega_k) \leq (4\pi\beta k)^{-d/2},$$

which compares the measure with Dirichlet boundary condition for the box Λ_N with the free Brownian bridge measure. It is technically easier here to work on a torus and with periodic boundary conditions for the Laplacian.

To formulate the rate functions we need some notations. Let:
$$\mathcal{M} = \{Q \in [0,1]^{\mathbb{N}} : \sum_{l \in \mathbb{N}} Q(l) \leq 1, Q(l) \geq Q(l+1) \; \forall l \in \mathbb{N}\}$$
be the set of monotonously nonincreasing sub-probability functions on $\mathbb{N}$. For $Q \in \mathcal{M}$ define $\widehat{Q}(k) = Q(k) - Q(k+1)$ for any $k \in \mathbb{N}$. For $d \geq 1$ let:
$$\widehat{Q}^*(k) = \frac{1}{\rho(4\pi\beta)^{d/2} k^{1+d/2}}, \quad k \in \mathbb{N}, \tag{7.7}$$
be given, and define the functional:
$$\mathcal{S}^{(\mathrm{Bose})}(Q) = \sum_{k=1}^{\infty} \widehat{Q}(k) \left(\log \frac{\widehat{Q}(k)}{\widehat{Q}^*(k)} - 1 \right) \quad Q \in \mathcal{M}. \tag{7.8}$$

The corresponding functional for the uniform distribution $\nu_N^{(\mathrm{u})}$ is given as:
$$\mathcal{S}^{(\mathrm{u})}(Q) = \sum_{k=1}^{\infty} \widehat{Q}(k) \left(\log \widehat{Q}(k) k - 1 \right) \quad Q \in \mathcal{M}.$$

The uniform distribution is defined through (7.3), i.e.:
$$\nu_N^{(\mathrm{u})}(\lambda) = \frac{1}{Z_N} \prod_{k=1}^{N} \left(\frac{1}{k}\right)^{r_k} \frac{1}{r_k!} \quad \lambda \in \mathcal{P}_N,$$

with normalization $Z_N = \sum_{\lambda \in \mathcal{P}_N} \prod_{k=1}^{N} \left(\frac{1}{k}\right)^{r_k} \frac{1}{r_k!}$.

The main results follow in the next theorem.

Theorem 7.1 (Adams 2008b)

(a) Under the uniform measure $\nu_N^{(\mathrm{u})}$ the empirical discrete shape measures Q_N satisfy a large deviations principle on $\mathcal{M}$ with speed N and rate function:
$$I^{(\mathrm{u})}(Q) = \mathcal{S}^{(\mathrm{u})}(Q) - \chi \quad \text{with } \chi = \inf_{Q \in \mathcal{M}} \mathcal{S}^{(\mathrm{u})}(Q). \tag{7.9}$$

(b) Let $\varrho \in (0, \infty)$ and $\Lambda_N \subset \mathbb{R}^d$ with $\Lambda_N \uparrow \mathbb{R}^d$ and $N/|\Lambda_N| \to \varrho$ as $N \to \infty$. Under the measure $\nu_N^{(\mathrm{Bose})}$ the empirical discrete shape measures Q_N satisfy a large deviations principle on $\mathcal{M}$ with speed N and rate function:
$$I^{(\mathrm{Bose})}(Q) = \mathcal{S}^{(\mathrm{Bose})}(Q) - \chi(\beta, \varrho) \quad \text{with } \chi(\beta, \varrho) = \inf_{Q \in \mathcal{M}} \mathcal{S}^{(\mathrm{Bose})}(Q). \tag{7.10}$$

Remark 7.1 (Free energy, Adams 2007) *The variational formula (7.10) gives the specific free energy* $f(\beta, \rho) := \lim_{N \to \infty} -1/\beta|\Lambda_N| \log Z_N^{(\mathrm{sym})}(\beta)$ *for inverse temperature β and density ρ of the non-interacting Bose gas, i.e.:*
$$f(\beta, \rho) = \frac{\rho}{\beta} \inf_{Q \in \mathcal{M}} \left\{ \sum_{k=1}^{\infty} \widehat{Q}(k) \log \left(\frac{\widehat{Q}(k)}{\widehat{Q}^*(k)} - 1 \right) \right\}.$$

We analyse the variational formulae for χ and $\chi(\beta,\rho)$, and we derive an expression for the specific free energy f as a function of β and ρ. Define a dimension dependent critical density:

$$\rho_c = \begin{cases} \frac{1}{(4\pi\beta)^{d/2}}\zeta\left(\frac{d}{2}\right), & \text{for } d \geq 3, \\ +\infty, & \text{for } d = 1,2, \end{cases} \quad (7.11)$$

where ζ is the Riemann zeta function:

$$\zeta\left(\frac{d}{2}\right) = \sum_{k=1}^{\infty} k^{-\frac{d}{2}}.$$

Furthermore, denote by $g_s(\alpha)$ the so-called Bose functions (see (7.20) in Appendix 7.4)

$$g_s(\alpha) = \sum_{k=1}^{\infty} k^{-s} e^{-\alpha k} \quad \text{for all } \alpha > 0 \text{ and all } s > 0.$$

For any $\rho < \rho_c$ we denote by $\alpha = \alpha(\beta,\rho)$ the unique root of:

$$\rho = \frac{1}{(4\pi\beta)^{d/2}} \sum_{k=1}^{\infty} k^{-d/2} e^{-\alpha k}. \quad (7.12)$$

The essential difference in $d \geq 3$ and $d = 1,2$ lies in the fact that in the latter two cases the corresponding Bose functions, $g_1(\alpha)$ respectively $g_{\frac{1}{2}}(\alpha)$, diverge as $\alpha \to 0$ (see Appendix 7.4 and Gram 1925). For $d = 1,2$ there is a unique α for any density $\rho < \infty$. For $d \geq 3$ there is such an unique α given only for densities $\rho < \rho_c$. Hence, this is the mathematical origin of the so-called Bose-condensation, where for $d \geq 3$ and $\rho > \rho_c$ particles condense in the zero mode state.

Theorem 7.2 (Analysis for χ, Adams 2008b) *The functional $\mathcal{S}^{(u)}$ is convex and there is a unique minimizer Q^* for $\chi = \inf_{Q \in \mathcal{M}} \mathcal{S}^{(u)}(Q)$, and it is defined through:*

$$\widehat{Q}^*(k) = \frac{e^{-\alpha k}}{k} \quad k \in \mathbb{N} \text{ and } \alpha = \log 2.$$

The analysis for the Bose distribution gives the proof of the Bose–Einstein condensation for non-interacting Bose gas depending on the parameters d, ρ and β.

Theorem 7.3 (Analysis for $\chi(\beta,\varrho)$, Adams 2007) *For any $\rho < \infty$ in dimensions $d = 1,2$, and $\varrho < \varrho_c$ in dimensions $d \geq 3$, there is a unique minimiser $Q \in \mathcal{M}$ of the variational formula for $\chi(\beta,\varrho)$ in (7.10) with probability mass one with:*

$$\widehat{Q}(k) = \frac{e^{-\alpha k}}{\rho(4\pi\beta)^{d/2} k^{1+\frac{d}{2}}} \quad \text{for } k \in \mathbb{N},$$

whereas for dimensions $d \geq 3$ and densities $\varrho > \varrho_c$, there is no minimizer for the variational problem (7.10) with probability mass one, but the infimum is attained for any minimizing sequence $(Q_n)_{n\in\mathbb{N}}$ of $Q_n \in \mathcal{M}$ such that $Q_n \to Q^*$ as $n \to \infty$. The specific free energy for $d \geq 3$ is given by:

$$f(\beta,\varrho) = \begin{cases} -\frac{1}{(4\pi\beta)^{d/2}\beta} g_{\frac{d+2}{2}}(\alpha) - \frac{1}{\beta}\varrho\alpha\,, & \text{for } \varrho < \varrho_c \\ -\frac{1}{(4\pi\beta)^{d/2}\beta} \zeta\left(\frac{d+2}{2}\right), & \text{for } \varrho > \varrho_c, \end{cases}$$

and for $d = 1, 2$ by:

$$f(\beta,\varrho) = -\frac{1}{(4\pi\beta)^{d/2}\beta} g_{\frac{d+2}{2}}(\alpha) - \frac{\varrho\alpha}{\beta},$$

where α is the unique root of (7.12).

7.3 Large deviations for empirical path measures

In this section we present our large deviations results for the empirical path measures for N Brownian motions $B^{(1)}, \ldots, B^{(N)}$ in $\mathbb{R}^d$ with time horizon $[0,\beta]$. The empirical path measures:

$$L_N = \frac{1}{N}\sum_{i=1}^{N} \delta_{B^{(i)}}$$

are random elements in the set $\mathcal{M}_1(\Omega_\beta)$ of probability measures on the set Ω_β of continuous paths $[0,\beta] \to \mathbb{R}^d$. We analyse the large-$N$ behaviour of the distributions of L_N under different symmetrized measures in Section 7.3.1 and Section 7.3.2 respectively. In both cases we derive large deviations principles whose rate functions are given as variational problems. In Section 7.3.2 the analysis for the variational problem for the cycle structure gives the proof of a phase transition in the empirical path measure.

7.3.1 Spatial structure

We analyse the large-N asymptotic of the empirical path measure L_N under the following symmetrized probability measure:

$$\mathbb{P}^{(\text{sym})}_{m,N} = \frac{1}{N!}\sum_{\sigma\in\mathfrak{S}_N}\int_{\mathbb{R}^d}\cdots\int_{\mathbb{R}^d} m(\mathrm{d}x_1)\cdots m(\mathrm{d}x_N) \bigotimes_{i=1}^{N} \mathbb{P}^{\beta}_{x_i,x_{\sigma(i)}}, \qquad (7.13)$$

where $m \in \mathcal{M}_1(\mathbb{R}^d)$ is a probability measure and where $\mathbb{P}^{\beta}_{x,y}$ is the Brownian bridge probability measure:

$$P^{\beta}_{x,y} = \mu^{\beta}_{x,y}/\mu^{\beta}_{x,y}(\Omega_\beta) = \frac{\mu^{\beta}_{x,y}}{(4\pi\beta)^{d/2}},$$

i.e., a probability measure on Ω_β. The expectation with respect to the measure $P^{\beta}_{x,y}$ is denoted by $\mathbb{E}^{\beta}_{x,y}$. We can conceive $\mathbb{P}^{(\text{sym})}_{m,N}$ as a two-step random mechanism: first we pick uniformly a random permutation σ, then we pick N Brownian motions with initial distribution m, and the i-th motion is conditioned to terminate at the initial point of the $\sigma(i)$-th motion, for any i.

The main idea in resolving the combinatorics of the measure (7.13) is to rewrite it as a sum over pair frequencies $NQ(x,y), x, y \in \mathbb{R}^d$. Here Q is a pair probability measure with equal marginals, and $NQ(x,y)$ is the number of Brownian motions which are sent from location x to location y due to the symmetrization. We shall count the number of permutations which are admissible for a given pair probability measure Q. Furthermore, for Brownian motions we need to work with open sets of positive Lebesgue measure instead of single points. But this is a technical point and it is analysed in detail in Adams and König (2007), where one needs an additional assumption on the probability measure $m \in \mathcal{M}_1(\mathbb{R}^d)$. We will neglect these details and refer to Adams and Dorlas (2007), where symmetrized systems of random walks on graphs are analysed and applied to certain mean-field type interacting systems.

The core idea, performed in Adams and König (2007), Adams and Dorlas (2007), and Adams (2008a), is that the rewriting gives a sum over pair probability measures with two terms, one part is counting permutations for a given pair probability measure, and for any given pair probability measure the other part is a probability measure for N Brownian motions. This probability measure is now a product of not necessarily identically distributed Brownian bridge probability measures. Hence, we resolved the correlations due to the symmetrization in a two level large deviations setting (see for example Dawson and Gärtner 1994). Our rate functions consist of two parts, one deals with the combinatorics and is therefore a function of a pair probability measure and the initial measure m, the other part governs the large deviations for the empirical path measures under the corresponding probability measure.

The motivation for this novel approach is threefold. First, it is an appealing method from the mathematical point of view and originated from combinatorial methods for microcanonical ensembles in Adams (2001). Second, Bose–Einstein condensation in Onsager and Penrose (1956) is defined as an off-diagonal long range behaviour of the one-particle reduced density matrix, which measures the correlation between two spatial points. Third, we are informed by Schrödinger (1931) who considered the question of how any two spatial points are connected by random paths. The crucial observation is that this aspect of the problem can be described by pair measures. Schrödinger (1931) raised the question of the most probable behaviour of a large system of diffusing particles in thermal equilibrium. Föllmer (1988) gave a mathematical formulation of these ideas in terms of large deviations. He applied Sanov's theorem to obtain a large deviations principle for L_N when $B^{(1)}, B^{(2)}, \ldots$ are i.i.d. Brownian motions with initial distribution m and no condition at time β. The rate function is the relative entropy with respect to $\int_{\mathbb{R}^d} m(\mathrm{d}x) \mathbb{P}_x \circ B^{-1}$, where the motions start in x under $\mathbb{P}_x$. Then Schrödinger's question amounts to identifying the minimizer of that rate function under given fixed independent initial and final distributions. It turns out that the unique minimizer is of the form $\int_{\mathbb{R}^d} \int_{\mathbb{R}^d} \mathrm{d}x \mathrm{d}y \, f(x) g(y) \, \mathbb{P}^\beta_{x,y} \circ B^{-1}$, i.e. a Brownian bridge with independent initial and final distributions. The probability densities f and g are characterized by a pair of dual variational equations, which originally appeared in Schrödinger (1931) for the special case that both the initial and the final measures are Lebesgue measures.

We introduce now the rate functions for our method. With:

$$H(Q|P) = \int_{\mathbb{R}^d \times \mathbb{R}^d} Q(\mathrm{d}x) \log \frac{Q(\mathrm{d}x)}{P(\mathrm{d}x)}$$

we denote the relative entropy of the pair probability measure $Q \in \mathcal{M}_1(\mathbb{R}^d \times \mathbb{R}^d)$ with respect to $P \in \mathcal{M}_1(\mathbb{R}^d \times \mathbb{R}^d)$. Let $\mathcal{M}_1^{(\mathrm{s})}(\mathbb{R}^d \times \mathbb{R}^d)$ be the set of shift-invariant probability measures Q on $\mathbb{R}^d \times \mathbb{R}^d$, i.e., measures whose first and second marginals coincide and are both denoted by $\overline{Q}$. Note that $Q \mapsto H(Q|\overline{Q} \otimes m)$ is strictly convex.

Define the functional $I_m^{(\mathrm{sym})}$ on $\mathcal{M}_1(\Omega_\beta)$ by the following variational problem:

$$I_m^{(\mathrm{sym})}(\mu) = \inf_{Q \in \mathcal{M}_1^{(\mathrm{s})}(\mathbb{R}^d \times \mathbb{R}^d)} \left\{ H(Q|\overline{Q} \otimes m) + I^{(Q)}(\mu) \right\},$$

where:

$$I^{(Q)}(\mu) = \sup_{\Phi \in \mathcal{C}_\mathrm{b}(\Omega_\beta)} \left\{ \langle \Phi, \mu \rangle - \int_{\mathbb{R}^d} \int_{\mathbb{R}^d} Q(\mathrm{d}x, \mathrm{d}y) \log \mathbb{E}_{x,y}^\beta \left(e^{\Phi(B)} \right) \right\} \quad (7.14)$$

for $\mu \in \mathcal{M}_1(\Omega_\beta)$ and $\langle \Phi, \mu \rangle = \int_{\Omega_\beta} \Phi(\omega) \mu(\mathrm{d}\omega)$. Here $\mathcal{C}_\mathrm{b}(\Omega_\beta)$ is the space of bounded continuous functions on Ω_β. Hence, $I^{(Q)}$ is a Legendre–Fenchel transform, but *not* the one of a logarithmic moment generating function of any random variable. In particular, $I^{(Q)}$, and therefore also $I_m^{(\mathrm{sym})}$, are nonnegative, and $I^{(Q)}$ is convex as a supremum of linear functions. There seems to be no way to represent $I^{(Q)}(\mu)$ as the relative entropy of μ with respect to any measure.

Let us explore briefly the variational problem connected with the rate function $I_m^{(\mathrm{sym})}$. By $\pi_s \colon \Omega_\beta \to \mathbb{R}^d$ we denote the projection $\pi_s(\omega) = \omega_s$. The marginal measure of $\mu \in \mathcal{M}_1(\Omega_\beta)$ is denoted by $\mu_s = \mu \circ \pi_s^{-1} \in \mathcal{M}_1(\mathbb{R}^d)$; analogously we write $\mu_{0,\beta} = \mu \circ (\pi_0, \pi_\beta)^{-1} \in \mathcal{M}_1(\mathbb{R}^d \times \mathbb{R}^d)$ for the joint distribution of the initial and the terminal point of a random process with distribution μ. It is easy to see that $Q = \mu_{0,\beta}$ if $I^{(Q)}(\mu) < \infty$. Indeed, in (7.14) relax the supremum over all $\Phi \in \mathcal{C}_\mathrm{b}(\Omega_\beta)$ to all functions of the form $\omega \mapsto f(\omega_0, \omega_\beta)$ with $f \in \mathcal{C}_\mathrm{b}(\mathbb{R}^d)$. This gives that:

$$\infty > I^{(Q)}(\mu) \geq \sup_{f \in \mathcal{C}_\mathrm{b}(\mathbb{R}^d)} \left(\langle \mu_{0,\beta}, f \rangle - \langle Q, \log \mathbb{E}_{\pi_0,\pi_\beta}^\beta \left(e^{f(B_0, B_\beta)} \right) \rangle \right)$$

$$= \sup_{f \in \mathcal{C}_\mathrm{b}(\mathbb{R}^d)} \langle \mu_{0,\beta} - q, f \rangle,$$

and this implies that $\mu_{0,\beta} = Q$. In particular, the infimum in the variational problem for $I_m^{(\text{sym})}$ is uniquely attained at this Q, i.e.:

$$I_m^{(\text{sym})}(\mu) = \begin{cases} H(\mu_{0,\beta}|\mu_0 \otimes m) + \sup_{\Phi \in \mathcal{C}_b(\mathcal{C})} \left\langle \mu, \Phi - \log \mathbb{E}_{\pi_0,\pi_\beta}^\beta \left(e^{\Phi(B)} \right) \right\rangle & \text{if } \mu_0 = \mu_\beta, \\ +\infty & \text{otherwise.} \end{cases}$$

In particular, $I_m^{(\text{sym})}$ is convex.

Our main large deviations result reads as follows.

Theorem 7.4 (Large deviations for L_N) *Fix $\beta \in (0, \infty)$ and $m \in \mathcal{M}_1(\mathbb{R}^d)$. Then, as $N \to \infty$, under the symmetrized measure $\mathbb{P}_{m,N}^{(\text{sym})}$, the empirical path measures L_N satisfy a large deviations principle on $\mathcal{M}_1(\Omega_\beta)$ with speed N and rate function $I_m^{(\text{sym})}$.*

Simplifying the large deviations principle says that, as $N \to \infty$:

$$\mathbb{P}_{m,N}^{(\text{sym})}(L_N = \mu) \approx e^{-N I_m^{(\text{sym})}(\mu)}, \qquad \mu \in \mathcal{M}_1(\Omega_\beta).$$

Proof: If m has compact support the proof is in Adams and König (2007). Arbitrary initial distributions are handled in Adams (2008a). A corresponding result for symmetrized systems of random walks on graphs with applications to mean-field models is given in Adams and Dorlas (2007). □

There are also analogous results for the mean:

$$Y_N = \frac{1}{N} \sum_{i=1}^N \mu_\beta^{(i)},$$

of the N occupation measures:

$$\mu_\beta^{(i)}(dx) = \frac{1}{\beta} \int_0^\beta \delta_{B_s}(dx)\, ds, \qquad i = 1, \ldots, N.$$

We will present below these results for the very special case that m is the Lebesgue measure of finite set in $\mathbb{R}^d$. The general version can be found in Adams and König (2007) and Adams and Dorlas (2007).

Let us comment briefly on the shape of the rate functions above. The symmetrized measure $\mathbb{P}_{m,N}^{(\text{sym})}$ arises from a two-step probability mechanism. This is reflected in the representation of the rate function $I_m^{(\text{sym})}$: in a peculiar way the entropy term $H(Q|\overline{Q} \otimes m)$ describes the large deviations of the uniformly distributed random permutation σ, together with the integration over $m^{\otimes N}$. The measure Q governs a particular distribution of N independent, but not identically distributed, Brownian bridges. Under this distribution, L_N satisfies a large deviations principle with rate function $I^{(Q)}$, which also can be guessed from the Gärtner–Ellis theorem (Dembo and Zeitouni 1998, Th. 4.5.20).

Let us contrast this to the case of i.i.d. Brownian bridges $B^{(1)}, \ldots, B^{(N)}$ with starting distribution m, i.e., we replace $\mathbb{P}_{m,N}^{(\text{sym})}$ by $(\int m(\mathrm{d}x)\, \mathbb{P}_{x,x}^\beta)^{\otimes N}$. Here the empirical path measure L_N satisfies a large deviations principle with rate function:

$$I_m(\mu) = \sup_{\Phi \in \mathcal{C}_{\mathrm{b}}(\Omega_\beta)} \left\{ \langle \Phi, \mu \rangle - \log \int_{\mathbb{R}^d} m(\mathrm{d}x)\, \mathbb{E}_{x,x}^\beta\!\left(\mathrm{e}^{\Phi(B)}\right) \right\},$$

as follows from an application of Cramér's theorem (Dembo and Zeitouni 1998, Th. 6.1.3). Note that $I_m(\mu)$ is the relative entropy of μ with respect to $\int m(\mathrm{d}x)\, \mathbb{P}_{x,x}^\beta \circ B^{-1}$. Although there is apparently no reason to expect a direct comparison between the distributions of L_N under $\mathbb{P}_{m,N}^{(\text{sym})}$ and under $(\int m(\mathrm{d}x)\, \mathbb{P}_{x,x}^\beta)^{\otimes N}$, the rate functions admit a simple relation: it is easy to see that $I^{(Q)} \geq I_m$ for the measure $Q(\mathrm{d}x, \mathrm{d}y) = m(\mathrm{d}x) \delta_x(\mathrm{d}y) \in \mathcal{M}_1^{(\mathrm{s})}(\mathbb{R}^d \times \mathbb{R}^d)$, since:

$$-\int_{\mathbb{R}^d}\int_{\mathbb{R}^d} Q(\mathrm{d}x, \mathrm{d}y) \log \mathbb{E}_{x,y}^\beta\!\left(\mathrm{e}^{\Phi(B)}\right) \geq -\log \int_{\mathbb{R}^d} m(\mathrm{d}x)\, \mathbb{E}_{x,x}^\beta\!\left(\mathrm{e}^{\Phi(B)}\right).$$

In particular, $I_m^{(\text{sym})} \geq I_m$.

An interesting question is what happens if we replace in the definition of the symmetrized probability measure $\mathbb{P}_{m,N}^{(\text{sym})}$ the Brownian bridge probability measure $\mathbb{P}_{x,y}^\beta$ by $g(x,y)\mathbb{P}_{x,y}^\beta$, when $g\colon \mathbb{R}^d \times \mathbb{R}^d \to \mathbb{R}$ is a continuous function? The motivation to multiply the Brownian bridge probability measure by the spatial function g is to model (see Adams (2008a)) the spatial correlations for permutations of finitely many points of graphs or finitely many points of point process clouds in $\mathbb{R}^d$. Compare Fichtner (1991), who studied permutations of random point configurations in $\mathbb{R}^d$ and introduced the spatial weight $\mathrm{e}^{-c|x-y|^2}$ for permutations that sent the spatial point x to the spatial point y. We shall discuss no further details at this stage but formulate our general result.

Proposition 7.1 (Adams and König 2007) *Let $g\colon \mathbb{R}^d \to \mathbb{R}^d \to \mathbb{R}$ be continuous and define $\mathbb{P}_{m,N}^{(\text{sym})}$ with $\mathbb{P}_{x,y}^\beta$ replaced by $g(x,y)\mathbb{P}_{x,y}^\beta$. Then the following holds.*

(a) Theorem 7.4 remains true under the replacement. The corresponding rate function is $\mu \mapsto I_m^{(\text{sym})}(\mu) - \langle \mu_{0,\beta}, \log g \rangle$.

(b)

$$\lim_{N \to \infty} \frac{1}{N} \log \left(\frac{1}{N!} \sum_{\sigma \in \mathfrak{S}_N} \int_{(\mathbb{R}^d)^N} \prod_{i=1}^N m(\mathrm{d}x_i) \prod_{i=1}^N g(x_i, x_{\sigma(i)}) \right) \quad (7.15)$$

$$= - \inf_{Q \in \mathcal{M}_1^{(\mathrm{s})}(\mathbb{R}^d \times \mathbb{R}^d)} \left\{ H(Q | \overline{Q} \otimes m) - \langle Q, \log g \rangle \right\}.$$

(c) The unique minimizer of the rate function $\mu \mapsto I_m^{(\text{sym})}(\mu) - \langle \mu_{0,\beta}, \log g \rangle$ is given by:

$$\mu^* = \int_{\mathbb{R}^d}\int_{\mathbb{R}^d} Q^*(\mathrm{d}x, \mathrm{d}y)\, \mathbb{P}_{x,y}^\beta \circ B^{-1}, \quad (7.16)$$

where $Q^* \in \mathcal{M}_1^{(s)}$ is the unique minimizer of the formula on the right hand side of (7.15).

(d) Law of large numbers: Under the measure $g\mathbb{P}_{m,N}^{(\text{sym})}$, normalized to a probability measure, the sequence $(L_N)_{N \in \mathbb{N}}$ converges in distribution to the measure μ^* defined in (7.16).

Setting $g \equiv 1$ we derive easily the following law of large numbers for our previous case.

Corollary 7.1 *Under the measure $\mathbb{P}_{m,N}^{(\text{sym})}$, normalized to a probability measure, the sequence $(L_N)_{N \in \mathbb{N}}$ converges in distribution to the measure μ^* given by:*

$$\mu^* = \int_{\mathbb{R}^d} \int_{\mathbb{R}^d} m \otimes m(\mathrm{d}x, \mathrm{d}y)\, \mathbb{P}_{x,y}^\beta \circ B^{-1}.$$

That is, in spite of strong correlations for fixed N under $\mathbb{P}_{m,N}^{(\text{sym})}$, the initial and terminal locations $B_0^{(1)}$ and $B_\beta^{(1)}$ of the first motion become independent in the limit $N \to \infty$. One can prove this also in an elementary way, and also the fact that, for any $k \in \mathbb{N}$ and for all $i_1 < i_2 < \cdots < i_k$, the Brownian motions $B^{(i_1)}, \ldots, B^{(i_k)}$ under $\mathbb{P}_{m,N}^{(\text{sym})}$ become independent in the limit $N \to \infty$.

We finish the section with the following special case as promised above. We replace the initial distribution $m \in \mathcal{M}_1(\mathbb{R}^d)$ by the Lesbesgue measure of a set $\Lambda \subset \mathbb{R}^d$ having finite Lesbesgue measure. That is we study the non-normalized measure:

$$\mu_{\Lambda,N}^{(\text{sym})} = \frac{1}{N!} \sum_{\sigma \in \mathfrak{S}_N} \int_\Lambda \cdots \int_\Lambda \mathrm{d}x_1 \cdots \mathrm{d}x_N \bigotimes_{i=1}^N \mu_{x_i, x_{\sigma(i)}}^\beta. \qquad (7.17)$$

Apart from questions motivated from physics, this measure is also mathematically interesting. According to an analogous result of Theorem 7.4 for the mean of occupation measures, the distribution of the mean of the normalized occupation measures Y_N, under $(Z_{\Lambda,N}^{(\text{sym})})^{-1} \mu_{\Lambda,N}^{(\text{sym})}$, satisfies a large deviations principle. Here $Z_{\Lambda,N}^{(\text{sym})}$ is the normalization for the the measure (7.17). That is, we have:

$$\lim_{N \to \infty} \frac{1}{N} \log\left(\mu_{\Lambda,N}^{(\text{sym})} \circ Y_N^{-1}(\cdot)\right) = -\inf_{p \in \cdot} J_\Lambda^{(\text{sym})}(p),$$

in the weak sense on subsets of $\mathcal{M}_1(\mathbb{R}^d)$, where we introduced:

$$J_\Lambda^{(\text{sym})}(p) = \inf_{Q \in \mathcal{M}_1^{(s)}(\mathbb{R}^d \times \mathbb{R}^d)} \left\{ H(Q|\overline{Q} \otimes \text{Leb}_\Lambda) + J^{(Q)}(p) \right\} - \inf_{p \in \mathcal{M}_1(\mathbb{R}^d)} \left\{ \widetilde{J}_\Lambda^{(\text{sym})}(p) \right\}$$

with $\widetilde{J}_\Lambda^{(\text{sym})}(p) = \inf_{Q \in \mathcal{M}_1^{(s)}(\mathbb{R}^d \times \mathbb{R}^d)} \left\{ H(Q|\overline{Q} \otimes \text{Leb}_\Lambda) + J^{(Q)}(p) \right\}$ and

$$J^{(Q)}(p) = \sup_{f \in \mathcal{C}_b(\mathbb{R}^d)} \left\{ \beta \langle f, p \rangle - \int_{\mathbb{R}^d} \int_{\mathbb{R}^d} Q(\mathrm{d}x, \mathrm{d}y) \log \frac{\mathbb{E}_x\left(e^{\int_0^\beta f(B_s)\,\mathrm{d}s}; B_\beta \in \mathrm{d}y\right)}{\mathrm{d}y} \right\}.$$

The main goal is to express $J_\Lambda^{(\text{sym})}$ in much easier and more familiar terms. It turns out that $J_\Lambda^{(\text{sym})}(p)$ is identical to the energy of the square root of the

density of p, in the jargon of large deviations theory also sometimes called the
Donsker–Varadhan rate function, $I_\Lambda \colon \mathcal{M}_1(\mathbb{R}^d) \to [0, \infty]$ defined by:

$$I_\Lambda(p) = \begin{cases} \|\nabla\sqrt{\frac{\mathrm{d}p}{\mathrm{d}x}}\|_2^2, & \text{if } p \text{ has a density with square root in } H_0^1(\Lambda^\circ), \\ \infty & \text{otherwise.} \end{cases}$$

Theorem 7.5 (Adams and König 2007) *Let $\Lambda \subset \mathbb{R}^d$ be a bounded closed box. Then $\beta^{-1} J_\Lambda^{(\mathrm{sym})}(p) = I_\Lambda(p) - \inf_{p \in \mathcal{M}_1(\mathbb{R}^d)} I_\Lambda(p)$ for any $p \in \mathcal{M}_1(\mathbb{R}^d)$.*

In the theory and applications of large deviations, I_Λ plays an important role as the rate function for the normalized occupation measure of one Brownian motion (or, one Brownian bridge) in Λ, in the limit as time to tends infinity (see Gärtner 1977 and Donsker and Varadhan 1983). It is remarkable that this function turns out also to govern the large deviations for the mean of the normalized occupation measures under the symmetrized measure $\mu_{\Lambda,N}^{(\mathrm{sym})}$, in the limit of a large number of motions. Let us give an informal discussion and interpretation of this fact.

The measure $\mu_{\Lambda,N}^{(\mathrm{sym})}$ in (7.17) admits a representation which goes back to Feynman (1953) and which we want to briefly discuss. Every permutation $\sigma \in \mathfrak{S}_N$ can be written as a concatenation of cycles. Given a cycle $(i, \sigma(i), \sigma^2(i), \ldots, \sigma^{k-1}(i))$ with $\sigma^k(i) = i$ and precisely k distinct indices, the contribution coming from this cycle is independent of all the other indices. Furthermore, by the fact that $\mu_{x_i, x_{\sigma(i)}}^\beta$ is the conditional distribution given that the motion ends in $x_{\sigma(i)}$, this contribution (also executing the k integrals over $x_{\sigma^l(i)} \in \Lambda$ for $l = k-1, k-2, \ldots, 0$) turns the corresponding k Brownian bridges of length β into one Brownian bridge of length $k\beta$, starting and ending in the same point $x_i \in \Lambda$ and visiting Λ at the times $\beta, 2\beta, \ldots, (k-1)\beta$. Hence:

$$\mu_{\Lambda,N}^{(\mathrm{sym})} = \frac{1}{N!} \sum_{\sigma \in \mathfrak{S}_N} \bigotimes_{k \in \mathbb{N}} \left(\int_\Lambda \mathrm{d}y_k\, \mu_{y_k,y_k}^{k,\beta,\Lambda} \right)^{\otimes f_k(\sigma)},$$

where $f_k(\sigma)$ denotes the number of cycles in σ of length precisely equal to k, and $\mu_{x,y}^{k,\beta,\Lambda}$ is the Brownian bridge measure $\mu_{x,y}^{k\beta}$ as in (7.1), restricted to the event $\bigcap_{l=1}^k \{B_{l\beta} \in \Lambda\}$. (See de Witt and Storaeds 1970, Lemma 2.1 for related combinatorial considerations.) If $f_N(\sigma) = 1$ (i.e., if σ is a cycle), then we are considering just one Brownian bridge B of length $N\beta$, with uniform initial measure on Λ, on the event $\bigcap_{l=1}^N \{B_{l\beta} \in \Lambda\}$. Furthermore, Y_N is equal to the normalized occupation measure of this motion. For such a σ, the limit $N \to \infty$ turns into a limit for diverging time, and the corresponding large-deviation principle of Donsker and Varadhan formally applies. This reasoning applies for permutations σ having only cycles whose lengths are growing with N unboundedly. Presumably, the contribution from those permutations whose bounded cycles sum up to something of order N is strictly smaller. A thorough investigation of the large deviation properties of the cycle structure and the distribution of the cycle lengths is contained in Adams (2007) and in Section 7.3.2 below for the case of boxes $\Lambda = \Lambda_N$ having volume of order N. There, a phase transition in β for the mean path is obtained. This phase transition is absent in the present case; the fixed box Λ forces all cycle lengths to grow unbounded with N.

7.3.2 Cycle structure

We analyse the large-N behaviour of a system of N Brownian motions with time horizon $[0, \beta]$ in $\mathbb{R}^d$ confined in subsets $\Lambda_N \subset \mathbb{R}^d$, i.e., the behaviour of the system under the symmetrized measure:

$$\mathbb{P}_N^{(\text{sym})} = Z_N^{(\text{sym})}(\beta)^{-1} \frac{1}{N!} \sum_{\sigma \in \mathfrak{S}_N} \int_{\Lambda_N} dx_1 \cdots \int_{\Lambda_N} dx_N \bigotimes_{i=1}^N \mu_{x_i, x_{\sigma(i)}}^{\beta, N}, \qquad (7.18)$$

and $Z_N^{(\text{sym})}(\beta)$ is the normalization:

$$Z_N^{(\text{sym})}(\beta) = \frac{1}{N!} \sum_{\sigma \in \mathfrak{S}_N} \int_{\Lambda_N} dx_1 \cdots \int_{\Lambda_N} dx_N \bigotimes_{i=1}^N \mu_{x_i, x_{\sigma(i)}}^{\beta, N}(\Omega_\beta^N).$$

The measure in (7.18) is different from the measure (7.13) in the previous section. Here, we want to exploit our results for the discrete empirical shape measure and the formula (7.6). The core idea in formula (7.6) is to concatenate Brownian bridges to obtain Brownian bridges with larger time horizons. Therefore we study in this section large deviations of the empirical path measures for paths with unbounded time horizon. That allows us to put the Brownian bridges of time horizon $[0, k\beta]$ onto the path of unbounded time horizon. We conceive the empirical path measure as a random element in $\mathcal{M}_1(\Omega)$, hence, we need a convenient extension of any continuous path $[0, \beta] \to \mathbb{R}^d$ to a continuous path $[0, \infty) \to \mathbb{R}^d$ in the definition of the empirical path measure. For any $x \in \mathbb{R}^d$ we denote by P^x the Brownian probability measure on Ω, i.e., the canonical Wiener measure with deterministic start in $x \in \mathbb{R}^d$ (Chung and Zhao 1995). In the following we write alternatively ω_t or $\omega(t)$ for any point of a path ω. Given a path $\omega \in \Omega_\beta$ with time horizon $[0, \beta]$ define:

$$P_\omega^{(\beta)} = \delta_\omega \otimes_\beta P^{\omega_\beta(\beta)} \in \mathcal{M}_1(\Omega, \mathcal{B}),$$

where the product $\otimes_\beta$ is defined for the 'splice' of two paths, i.e., for $\omega \in \Omega_\beta$ and $\widetilde{\omega} \in \Omega$ define $\overline{\omega} \in \Omega$ by $\overline{\omega}(t) = \omega(t \wedge \beta), t \in [0, \infty)$, and $\omega \otimes_\beta \widetilde{\omega} \in \Omega$ such that $\omega \otimes_\beta \widetilde{\omega} = \overline{\omega}$ if $\widetilde{\omega}(0) \neq \omega(\beta)$ and:

$$\omega \otimes_\beta \widetilde{\omega}(t) = \begin{cases} \omega(t) & \text{for } t \in [0, \beta] \\ \widetilde{\omega}(t - \beta) & \text{for } t \in (\beta, \infty) \end{cases} \qquad (7.19)$$

if $\widetilde{\omega}(0) = \omega(\beta)$. The mapping $\omega \in \Omega_\beta \mapsto P_\omega^{(\beta)} \in \mathcal{M}_1(\Omega, \mathcal{B})$ is measurable, and the family $\{P_\omega^{(\beta)} : \omega \in \Omega_\beta\}$ satisfies the Markov property, see (Deuschel and Stroock,

2001, Lemma 4.4.21). Hence, the empirical path measure:

$$\widehat{L}_N : \Omega_\beta^N \to \mathcal{M}_1(\Omega), \omega \mapsto \widehat{L}_N(\omega) = \frac{1}{N} \sum_{i=1}^{N} \delta_{\omega^{(i)}} \otimes_\beta P^{\omega_\beta^{(i)}},$$

is $\Omega_\beta^{\otimes N}$ measurable. Here $\omega = (\omega^{(1)}, \ldots, \omega^{(N)}) \in \Omega_\beta^N$. Our main result concerns a large deviations principle for the distributions of $\widehat{L}_N$ under the symmetrized measure $\mathbb{P}_N^{(\mathrm{sym})}$. Recall that $\mathbb{P}_N^{(\mathrm{sym})}$ is a probability measure on $\Omega_\beta^{\otimes N}$.

The rate function is given by:

$$I^{(\mathrm{sym})}(\mu) = \inf_{Q \in \mathcal{M}} \left\{ S^{(\mathrm{Bose})}(Q) + I^{(Q)}(\mu) \right\} - \chi(\beta, \varrho) \qquad \mu \in \mathcal{M}_1(\Omega),$$

where:

$$I^{(Q)}(\mu) = \sup_{F \in \mathcal{C}_\mathrm{b}(\Omega)} \left\{ \langle F, \mu \rangle - \sum_{k \in \mathbb{N}} \widehat{Q}(k) \log \mathbb{E}_{0,0}^{k\beta} \left(e^{F(B)} \right) \right\} \qquad \mu \in \mathcal{M}_1(\Omega),$$

and where the function $\chi(\beta, \varrho) := \inf_{Q \in \mathcal{M}} \{ S^{(\mathrm{Bose})}(Q) \}$ is given as the negative logarithmic limit of the partition function $Z_N^{(\mathrm{sym})}(\beta)$, see Theorem 7.3, and where $\mathcal{C}_\mathrm{b}(\Omega)$ is the space of continuous bounded functions of the paths in Ω. $\mathbb{E}_{0,0}^{k\beta}$ denotes the expectation with respect to the Brownian bridge probability measure $\mathbb{P}_{0,0}^{k\beta}$ extended to a probability measure in $\mathcal{M}_1(\Omega)$. Here, $I^{(Q)}$ is a Fenchel–Legendre transform, but *not* the one of a logarithmic moment generating function of any random variable. In particular, $I^{(Q)}$, and therefore also $I^{(\mathrm{sym})}$, are nonnegative, and $I^{(Q)}$ is convex as a supremum of linear functions. There seems to be no way to represent $I^{(Q)}(\mu)$ as the relative entropy of μ with respect to any measure.

Theorem 7.6 (Large deviations for $\widehat{L}_N$, Adams 2007) *Let $\varrho \in (0, \infty)$ and $\Lambda_N \subset \mathbb{R}^d$ centred boxes with $\Lambda_N \uparrow \mathbb{R}^d$ and $N/|\Lambda_N| \to \varrho$ as $N \to \infty$.*

Under the symmetrized measure $\mathbb{P}_N^{(\mathrm{sym})}$ the empirical path measures $(\widehat{L}_N)_{N \in \mathbb{N}}$ satisfy a large deviations principle on $\mathcal{M}_1(\Omega)$ with speed N and rate function $I^{(\mathrm{sym})}$.

Remark 7.2 *To be more precise we have a large deviations principle for $\widehat{L}_N$ under the symmetrized distribution such that the initial distribution is subtracted, i.e., all motions are considered to start at the origin. This is a technical detail, and we refer to Adams (2008a) and Adams et al. (2008), where our analysis is combined with marked point processes in $\mathbb{R}^d$. However, as we focus here solely on the noninteracting case, we can relax the abstraction and let the motions start at the origin.*

We give a brief informal interpretation of the shape of the rate functions in $I^{(\mathrm{sym})}$ and $I^{(Q)}, Q \in \mathcal{M}$. As remarked earlier, the symmetrized measure $\mathbb{P}_N^{(\mathrm{sym})}$ arises from a two-step probability mechanism. This is reflected in the representation of the rate function $I^{(\mathrm{sym})}$: in a peculiar way, the term $S(Q) - \chi(\beta, \varrho)$

describes the large deviations of the discrete empirical shape measure for integer partitions. The discrete empirical shape measures Q_N governs a particular distribution of N independent, but not identically distributed, Brownian bridges. Under this distribution, $\widehat{L}_N$ satisfies a large deviations principle with rate function $I^{(Q)}$, which can also be guessed from the Gärtner–Ellis theorem (Dembo and Zeitouni 1998, Th. 4.5.20). The presence of a two-step mechanism makes it impossible to apply this theorem directly to $\mathbb{P}_N^{(\mathrm{sym})}$.

Let us contrast this to the case of i.i.d. Brownian bridges $B^{(1)}, \ldots, B^{(N)}$, starting in the origin, i.e., we replace $\mathbb{P}_N^{(\mathrm{sym})}$ by $(\mathbb{P}_{0,0}^\beta)^{\otimes N}$. Here the empirical path measure $\widehat{L}_N$ satisfies a large deviations principle with rate function:

$$I(\mu) = \sup_{F \in \mathcal{C}_{\mathrm{b}}(\Omega)} \left\{ \langle F, \mu \rangle - \log \mathbb{E}_{0,0}^\beta \big(\mathrm{e}^{F(B)} \big) \right\},$$

as follows from an application of Cramér's theorem Dembo and Zeitouni 1998, Th. 6.1.3. Note that $I(\mu)$ is the relative entropy of μ with respect to $\mathbb{P}_{0,0}^\beta \circ B^{-1}$. Although there is apparently no reason to expect a direct comparison between the distributions of L_N under $\mathbb{P}_N^{(\mathrm{sym})}$ and under $(\mathbb{P}_{0,0}^\beta)^{\otimes N}$, the rate functions admit a simple relation: it is easy to see that $I^{(Q)} \geq I$ for the measure $Q \in \mathcal{M}$ with $\widehat{Q}(k) = \delta_1$, since:

$$-\sum_{k=1}^\infty \widehat{Q}(k) \log \mathbb{E}_{0,0}^{k\beta}\big(\mathrm{e}^{F(B)}\big) \geq -\log \mathbb{E}_{0,0}^\beta\big(\mathrm{e}^{F(B)}\big).$$

In particular, $I^{(\mathrm{sym})} \geq I$.

Remark 7.3 *The techniques of the proof of Theorem 7.6 apply also to a proof of a large deviations principle under the symmetrized measure $\mathbb{P}_N^{(\mathrm{sym})}$ for the empirical path measure $\widetilde{L}_N = 1/N \sum_{i=1}^N \delta_{B^{(i)}}$, which is a random element in $\mathcal{M}_1(\Omega_\beta)$. The rate function is:*

$$\widetilde{I}^{(\mathrm{sym})}(\mu) =$$

$$\inf_{Q \in \mathcal{M}} \left\{ \mathcal{S}^{(\mathrm{Bose})}(Q) - \sup_{F \in \mathcal{C}_{\mathrm{b}}(\Omega_\beta)} \left\{ \langle F, \mu \rangle - \sum_{k=1}^\infty \widehat{Q}(k) \log \mathbb{E}_{0,0}^{k\beta} \left(\mathrm{e}^{\sum_{l=0}^{k-1} F(B_{[l\beta,(l+1)\beta]})} \right) \right\} \right\}.$$

Similar results hold for the mean Y_N of the occupation measures. However, these rate functions seem not to give enough information to derive the phase transition as in Theorem 7.7, and to obtain a probabilistic interpretation of Bose–Einstein condensation.

Our large deviations result is accompanied by an analysis of the variational formula for the rate function $I^{(\mathrm{sym})}$, i.e., the analysis for zeros of the rate function. This gives the proof of the phase transition for empirical path measures depending on the dimension and the density parameter in Theorem 7.7.

The result of Theorem 7.3 in Section 7.3.2 is an essential ingredient which leads to the analysis of the rate function $I^{(\text{sym})}$. Let:

$$A_k = \{\omega \otimes_{k\beta} \xi \colon \omega \in \Omega_k, \omega(0) = \omega(k\beta), \xi \in \Omega\} \subset \Omega, k \in \mathbb{N},$$

be the set of paths in Ω which result from the splice (7.19) of Brownian bridges paths of time horizon $[0, k\beta]$ with any path $\xi \in \Omega$.

Theorem 7.7 (Analysis of the rate function $I^{(\text{sym})}$) Adams 2007
Under the assumptions of Theorem 7.6 the following holds:

(i) $d = 1, 2$. A unique minimizer $\mu^* \in \mathcal{M}_1(\Omega)$ of the rate function $\mu \mapsto I^{(\text{sym})}(\mu)$ exists with $\sum_{k \in \mathbb{N}} k\mu^*(A_k) = 1$.

(ii) $d \geq 3$ and $\varrho < \varrho_c$. A unique minimizer $\mu^* \in \mathcal{M}_1(\Omega)$ of the rate function $\mu \mapsto I^{(\text{sym})}(\mu)$ is given with $\sum_{k \in \mathbb{N}} k\mu^*(A_k) = 1$.
For $\varrho > \varrho_c$ there is no unique minimizer given, but there exist minimizing sequences $(\mu_n)_{n \geq 1}, \mu_n \in \mathcal{M}_1(\Omega)$, with $\sum_{n=1}^{\infty} k\mu_n(A_k) = 1$ for any $n \in \mathbb{N}$ such that $\mu_n \to \mu^0 \in \mathcal{M}_1(\Omega)$ weakly as $n \to \infty$ with $\sum_{n=1}^{\infty} k\mu^0(A_k) < 1$.

Let us draw an easy corollary from this theorem.

Corollary 7.2 (Law of large numbers, Adams 2007) *Under the assumptions of Theorem 7.7 the following holds.*

(i) For $d = 1, 2$, and any density $\varrho < \infty$, there is a law of large numbers. Under the probability measure $\mathbb{P}_N^{(\text{sym})}$, the sequence $(\widehat{L}_N)_{N \in \mathbb{N}}$ converges in distribution to the measure $\mu^* \in \mathcal{M}_1(\Omega)$.

(ii) For $d \geq 3$ and $\varrho < \varrho_c$ there is a law of large numbers. Under the probability measure $\mathbb{P}_N^{(\text{sym})}$, the sequence $(\widehat{L}_N)_{N \in \mathbb{N}}$ converges in distribution to the measure $\mu^* \in \mathcal{M}_1(\Omega)$.

The main conclusion of the large deviations principle in Theorem 7.6 and Theorem 7.7 is the following phase transition for the mean empirical path measure, which gives a path measure interpretation of Bose–Einstein condensation (BEC).

Path measures and their interpretation as Bose–Einstein condensation
Let N Brownian motions with time horizon $[0, \beta]$ confined in centred sets $\Lambda_N \subset \mathbb{R}^d$ given such that $\Lambda_N \uparrow \mathbb{R}^d$ and $N/|\Lambda_N| \to \varrho \in (0, \infty)$ as $N \to \infty$. Then the following holds:

(i) For $\beta > 0$ there is a $\varrho_c = \varrho_c(\beta, d)$ such that:
no BEC: Case $\varrho < \varrho_c$ for $d \geq 3$, $\varrho > 0$ for $d = 1, 2$:
$\widehat{L}_N \to \mu^* \in \mathcal{M}_1(\Omega)$ under $\mathbb{P}_N^{(\text{sym})}$ as $N \to \infty$ with $\sum_{k=1}^{\infty} k\mu^*(A_k) = 1$
BEC: Case $\varrho < \varrho_c$ and $d \geq 3$:
$\widehat{L}_N \to \mu^0 \in \mathcal{M}_1(\Omega)$ under $\mathbb{P}_N^{(\text{sym})}$ as $N \to \infty$ with $\sum_{k=1}^{\infty} k\mu^0(A_k) < 1$.

(ii) For $\varrho \in (0, \infty)$ there exists a:

$$\beta_c = \begin{cases} \frac{1}{4\pi}\left(\frac{\varrho}{\zeta(d/2)}\right)^{2/d}, & \text{for } d \geq 3, \\ +\infty, & \text{for } d = 1, 2, \end{cases}$$

such that:
no BEC: Case $\beta < \beta_c$ for $d \geq 3$ and $\beta > 0$ for $d = 1, 2$:
$\widehat{L}_N \to \mu^* \in \mathcal{M}_1(\Omega)$ under $\mathbb{P}_N^{(\text{sym})}$ as $N \to \infty$ with $\sum_{k=1}^{\infty} k\mu^*(A_k) = 1$
BEC: Case $\beta > \beta_c$ and $d \geq 3$:
$\widehat{L}_N \to \mu^0 \in \mathcal{M}_1(\Omega)$ under $\mathbb{P}_N^{(\text{sym})}$ as $N \to \infty$ with $\sum_{k=1}^{\infty} k\mu^0(A_k) < 1$.

If $d = 1, 2$, or $\varrho < \varrho_c$ for $d \geq 3$, the mean empirical path measure has support on those paths in which one can insert, starting from time origin, a concatenation of any finite number of Brownian motions with time horizon $[0, \beta]$, i.e., for any $k \in \mathbb{N}$ one can find in paths $\omega_k \in A_k$ exactly k Brownian motions concatenated to a Brownian bridge with horizon $[0, k\beta]$. This follows from the concatenation of the Brownian motions due to the cycle structure of the permutations and due to the Lebesgue integration of any initial position in the definition of the symmetrized measure $\mathbb{P}_N^{(\text{sym})}$. If the density ϱ is high enough for $d \geq 3$, i.e., $\varrho > \varrho_c$ (or equivalently, if the inverse temperature is sufficiently large for given density, i.e., $\beta > \beta_c$, for $d \geq 3$), the mean path measure has positive weight for paths with an infinite time horizon, that is, concatenation of any finite number of Brownian motions with time horizon $[0, \beta]$, i.e., any finite cycle path in A_k, is not sufficient, because there is an excess density $(\varrho - \varrho_c)$ of Brownian motions with time horizon $[0, \beta]$. These motions concatenate to infinite long cycle, that is, these cycles grow with the system size in the thermodynamic limit. The fraction of these motions is:

$$1 - \frac{\varrho_c}{\varrho} = 1 - \left(\frac{\beta_c}{\beta}\right)^{d/2}.$$

7.4 Appendix: Bose functions

These functions are defined by:

$$g_s(\alpha) = \frac{1}{\Gamma(s)} \int_0^\infty \frac{t^{s-1}}{e^{t+\alpha} - 1} dt = \sum_{k=1}^{\infty} k^{-s} e^{-\alpha k} \quad \text{for all } \alpha > 0 \text{ and all } s > 0, \tag{7.20}$$

and also $\alpha = 0$ and $s > 1$. In the latter case:

$$g_s(0) = \sum_{k=1}^{\infty} k^{-s} = \zeta(s),$$

which is the zeta function of Riemann. The behaviour of the Bose functions about $\alpha = 0$ is given by:

$$g_s(\alpha) = \begin{cases} \Gamma(1-s)\alpha^{s-1} + \sum_{k=0}^{\infty} \zeta(s-k)\frac{(-\alpha)^k}{k!}, & s \neq 1, 2, \ldots \\ \frac{(-\alpha)^{s-1}}{(s-1)!}\left[\log\frac{1}{\alpha} + \sum_{m=1}^{s-1}\frac{1}{m}\right] + \sum_{\substack{k=0 \\ k \neq s-1}} \zeta(s-k)\frac{(-\alpha)^k}{k!}, & s = 1, 2, \ldots \end{cases}$$

At $\alpha = 0$, $g_s(\alpha)$ diverges for $s \leq 1$; indeed for all s there is some kind of singularity at $\alpha = 0$, such as a branch point. For further details see Gram (1925).

References

Adams, S. (2001). Complete equivalence of the Gibbs ensembles for one-dimensional Markov systems. *Jour. Stat. Phys.* **105**(5/6), 879–908.

Adams, S. (2007). Large deviations for empirical path measures in cycles of integer partitions. preprint arXiv:math.PR/0702052v2.

Adams, S. (2008a). Interacting Brownian bridges and probabilistic interpretation of Bose-Einstein condensation. Habilitation thesis, University Leipzig.

Adams, S. (2008b). Large deviations for shape measures of integer partitions under non-uniform distribution. Forthcoming

Adams, S. and Bru, J.-B. (2004a). Critical Analysis of the Bogoliubov Theory of Superfluidity. *Physica A* **332**, 60–78.

Adams, S. and Bru, J.-B. (2004b). Exact solution of the AVZ-Hamiltonian in the grand-canonical ensemble. *Ann. Henri Poincaré* **5**, 405–34.

Adams, S. and Bru, J.-B. (2004c). A new microscopic theory of superfluidity at all temperatures. *Ann. Henri Poincaré* **5**, 435–76.

Adams, S., Bru, J.-B. and König, W. (2006a). Large deviations for trapped interacting Brownian particles and paths. *Ann. Probab.* **34**(4), 1340–1422.

Adams, S., Bru, J.-B. and König, W. (2006b). Large systems of path-repellent Brownian motions in a trap at positive temperature. *Electronic Journal of Probability* **11**, 460–85.

Adams, S. and Dorlas, T. (2007). Asymptotic Feynman-Kac formulae for large symmetrised systems of random walks. *Annales de l'institut Henri Poincaré (B) Probabilités et Statistiques*. In press, preprint arXiv:math-ph/0610026.

Adams, S. and König, W. (2007). Large deviations for many Brownian bridges with symmetrised initial-terminal condition. *Prob. Theory Relat. Fields*. In press, pre-print arXiv:math.PR/0603702. Published online 11.09.2007 DOI 10.10007/s00440-007-0099-5.

Adams, S. and König, W. (2008). Interacting Brownian motions and the Gross-Pitaevskii formula. In: this volume. OUP.

Adams, S., König, W. and Collevecchio, A. (2008). A variational formula for the free energy of non-dilute many-particle systems. In preparation.

Arratia, R., Barbour, A. and Tavaré, S. (2003). *Logarithmic Combinatorial Structures: A Probabilistic Approach*. European Mathematical Society: Zurich.

Benfatto, G., Cassandro, M., Merola, I. and Presutti, E. (2005). Limit theorems for statistics of combinatorial partitions with applications to mean field bose gas. *Journal of Math. Phys.* **46**, 033303.

Bose, S. (1924). Plancks Gesetz und Lichtquantenhypothese. *Zeitschrift für Physik* **26**, 178–181.

Bratteli, O. and Robinson, D. (1997). *Operator Algebras and Quantum Statistical Mechanics II* (2nd ed.). Springer, Berlin.

Chung, K. and Zhao, Z. (1995). *From Brownian Motion to Schrödinger's Equation.* Springer, Berlin.

Dawson, D. and Gärtner, J. (1994). Multilevel large deviations and interacting diffusions. *Probab. Theory Relat. Fields* **98**, 423–87.

de Witt, C. and Storaeds, R. (Eds.) (1970). *Some Applications of Functional Integration in Statistical Mechanics and Field Theory.* Cargèse Lectures in Theoretical Physics: Gordon and Breach Science Publ., New York.

Dembo, A. and Zeitouni, O. (1998). *Large Deviations Techniques and Applications.* Springer, New York.

Dembo, A., Zeitouni, O. and Vershik, A. (2000). Large deviations for integer partitions. *Markov Processes and Related Fields* **6**(No.2), 147–79.

Deuschel, J. and Stroock, D. (2001). *Large Deviations.* AMS Chelsea Publishing American Mathematical Society: Providence.

Donsker, M. and Varadhan, S. (1975–1983). Asymptotic evaluation of certain Markov process expectations for large time, I–IV. *Comm. Pure Appl. Math.* **28, 29, 36**, 1–47, 279–301, 389–461, 183–212.

Dorlas, T., Martin, P. and Pulé, J. (2005). Long cycles in a perturbed mean field model of a boson gas. *Journal of. Stat. Phys.* **121**(Nos.3/4), 433–61.

Einstein, A. (1925). Quantentheorie des einatomigen idealen Gases. *Sitzber. Kgl. Preuss. Akad. Wiss.* **3**, 3–14.

Ewens, W. (1972). The sampling theory of selectively neutral alleles. *Theor. Pop. Biol.* **3**, 87–112.

Feynman, R. (1953). Atomic theory of the λ transition in Helium. *Phys. Rev.* **91**, 1291–1301.

Fichtner, K. (1991). Random permutations of countable sets. *Probab. Theory Relat. Fields* **89**, 35–60.

Föllmer, H. (1988). *Random Fields and Diffusion Processes*, Volume 1362 of *Lecture Notes in Math.* Springer, Berlin. Ecole d'Eté de Saint Flour XV–XVII.

Gärtner, J. (1977). On large deviations from the invariant measure. *Theory Probab. Appl.* **22**:1, 24–39.

Goncharov, V. (1944). On the Field of Combinatory Analysis. *Translations of the American Mathematical Society* **19**, 1–46.

Gram, J. (1925). Tafeln für die Riemannsche Zetafunktion. *Skrifter København* (**8**) 9, 311–25.

Griffin, A., Snoke, D., and Stringari, S. (1995). *Bose-Einstein Condensation.* Cambridge University Press: Cambridge.

Huang, K. (1987). *Statistical Mechanics.* Wiley: New York.

Khinchin, A. (1960). *Mathematical Foundations of Quantum Satistics.* Dover Publications: New York.

Kingman, J. (1975). Random discrete distributions. *J.R. Statist. Soc. B* **37**, 1–15.

Kingman, J. (1978a). Random partitions in population genetics. *Proc. Roy. Soc.* **361**, 1–20.

Kingman, J. (1978b). The representation of partition sructures. *J. London Math. Soc.* **18**, 374–80.

Kolchin, V. (1986). *Random Mappings.* Optimization Software, Inc., Publications Division, New York.

Landau, L. (1941). The theory of superfluidity of Helium II. *J. Phys. USSR* **5**, 71.

Lieb, E., Seiringer, R., Solovej, J. and Yngvason, J. (2005). *The Mathematics of the Bose Gas and its Condensation.* Birkhäuser Verlag, Basel.

London, F. (1938). The λ-phenomenon of liquid helium and the Bose-Einstein degeneracy. *Science* **141**, 643.

Onsager, L. and Penrose, O. (1956). Bose-Einstein condensation and liquid Helium. *Phys. Rev.* **104**, 576–84.

Penrose, O. (1951). On the quantum mechanics of Helium II. *Phil. Mag.* **42**, 1373–77.

Pitaevskii, L. and Stringari, S. (2003). *Bose-Einstein Condensation.* Clarendon Press, Oxford.

Pitman, J. (2002). *Combinatorial Stochastic Processes.* Number 1875 in Lecture Notes in Mathematics. Springer, Berlin. Ecole d'Eté de Probabilités.

Ruelle, D. (1969). *Statistical Mechanics: Rigorous Results.* Addison-Wesley: Reading, MA.

Schrödinger, E. (1931). Über die Umkehrung der Naturgesetze. *Sitzungsber. Preuß. Akad. Wiss., Phys.-Math. Kl. 1931 No.* **8/9**, 144–153.

Sznitman, A. (1998). *Brownian Motion, Obstacles and Random Media.* Springer, Berlin.

Thirring, W. (1980). *Quantenmechanik grosser Systeme.* Springer Verlag, Wien. Lehrbuch der Mathematischen Physik 4.

Vershik, A. (1996). Statistical mechanics of combinatorial partitions, and their limit shapes. *Functional Analysis and Its Applications* **30**(3), 90–105.

Watterson, G. (1976). The stationary distribution of the infinitely-many neutral alleles diffusion model. *Journal of Applied Probability* **13**(4), 639–51.

8

INTERACTING BROWNIAN MOTIONS AND THE GROSS–PITAEVSKII FORMULA

Stefan Adams and Wolfgang König

Abstract

We review probabilistic approaches to the Gross–Pitaevskii theory describing interacting dilute systems of particles. The main achievement are large deviations principles for the mean occupation measure of a large system of interacting Brownian motions in a trapping potential. The corresponding rate functions are given as variational problems whose solution provide effective descriptions of the infinite system.

8.1 Introduction

The phenomenon known as Bose–Einstein condensation (hereafter abbreviated BEC) was predicted by Einstein (1925) on the basis of ideas of the Indian physicist Bose (1924) concerning statistical description of the quanta of light: in a system of particles obeying Bose statistics and whose total number is conserved, there should be a temperature below which a finite fraction of all the particles 'condense' into the same one-particle state. Einstein's original prediction was for a non-interacting gas of particles. The predicted phase transition is associated with the condensation of atoms in the state of lowest energy and is the consequence of quantum statistical effects.

For a long time these predictions were considered a curiosity of non-interacting gases and had no practical impact. After the observation of superfluidity in liquid ^{4}He below the λ temperature (2.17 K) was made, London (1938) suggested that, despite the strong interatomic interactions, BEC indeed occurs in this system and is responsible for the superfluidity properties. A superfluid is a fluid that flows without resistance. London suggested that a large number of Helium atoms (which are Bosons) are in the translational state of lowest energy (Bose condensate), and these atoms are mixed with normal fluid consisting of atoms with higher translational energies. This interpretation has stood the test of time and is the basis of our modern understanding of the properties of the superfluid phase.

The first self-consistent theory of super-fluids was developed by Landau (1941) in terms of the spectrum of elementary excitations of the fluid. In 1947 Bogoliubov developed the first microscopic theory of interacting Bose gases, based on the concept of Bose–Einstein condensation. This initiated several theoretical studies; a recent account on the state of the art can be found in

(2004a, 2004b) and on its contribution to superfluidity theory in Adams and Bru (2004c). After Landau and Lifshitz (1951) had appeared, Penrose (1951) and Onsager and Penrose (1956) introduced the concept of the non-diagonal long-range order and discussed its relationship with BEC. An important development in the field took place with the prediction of quantized vortices by Onsager (1949) and Feynman (1955). The experimental studies on dilute atomic gases were developed much later, starting from the 1970s, benefiting from the new techniques developed in atomic physics based on magnetic and optical trapping, and advanced cooling mechanisms.

In 1995, the first experimental realizations of BEC were achieved in a system that is as different as possible from ^{4}He, namely, in dilute atomic alkali gases trapped by magnetic fields. These realizations are due to Anderson et al. (1995), Bradley et al. (1995), and Davis et al. (1995), after appropriate cooling methods had been developed. For this remarkable achievement, the Nobel prize in physics 2001 was awarded to E.A. Cornell, W. Ketterle, and C.E. Wieman. Over the last few years these systems have been the subject of an explosion of research, both experimental and theoretical. A comprehensive account on Bose–Einstein condensation is the recent monograph Pitaevskii and Stringari (2003).

Perhaps the most fascinating aspect of BEC is best illustrated by the cover of *Science* magazine of 22 December, 1995, in which the Bose condensate is declared as the 'molecule of the year'. The Bose condensate is pictured as a platoon of soldiers marching in lookstep: every atom in the condensate must behave in exactly the same way. One of the most striking consequences is that effects, which are so small that they are practically invisible at the level of a single atom, are spectacularly amplified.

Motivated by the experimental success, in a series of papers Lieb et al. (2000a), Lieb et al. (2000b), Lieb et al. (2001), and Lieb and Seiringer (2002) obtained a mathematical foundation of Bose–Einstein condensation at zero temperature. The mathematical formulation of the N-particle Boson system is in terms of an N-particle Hamilton operator, $\mathcal{H}_N$, whose ground states describe the Bosons under the influence of a trap potential and a pair potential, see Section 8.2. Lieb et al. rigorously proved that the ground state energy per particle of $\mathcal{H}_N$ (after proper rescaling of the pair potential) converges towards the energy of the well-known Gross–Pitaevskii functional. The ground state is approximated by the N-fold product of the Gross–Pitaevskii minimizer multiplied by a correlated term involving the solution of the associated scattering equation. Moreover, they also showed the convergence of the reduced density matrix, which implies the Bose–Einstein condensation. As had been generally predicted, the scattering length of the pair interaction potential plays a key role in this description.

These rigorous results are only for zero temperature, whereas the experiments show BEC at very low, but positive temperature. The mathematical understanding of BEC at positive temperature is rather incomplete yet. Its analysis represents an important challenging and ambitious research area in the field of many-particle systems. Thermodynamic equilibrium states are described by traces of $e^{-\beta \mathcal{H}_N}$, where $\beta \in (0, \infty)$ is the inverse temperature and $\mathcal{H}_N$ is the N-particle Hamilton operator. In what follows we set the Boltzmann constant $k_B = 1$. Via the Feynman–Kac formula (see e.g. Feynman 1953 and Ginibre 1970), these traces are expressed as exponential expectations of N interacting Brownian motions with time horizon $[0, \beta]$. This opens up the possibility to use probabilistic approaches for the study of these traces, in particular stochastic analysis and the theory of large deviations.

In this review we present our probabilistic approaches to dilute systems of interacting many-particle systems at positive temperature using the Gross–Pitaevskii approximation. Using the the theory of large deviations, we characterize the large-N and the large-β behaviour of various exponential expectations of N interacting Brownian motions with time horizon $[0, \beta]$ in terms of variants of the Gross–Pitaevskii variational formula. In particular we introduce and analyse a new model, which we call the Hartree model, whose ground states are the ground product states of the Hamilton operator $\mathcal{H}_N$. Their large-N behaviour is characterized in terms of the Gross–Pitaevskii formula, with the scattering length replaced by the integral of the pair interaction potential. This nice assertion is complemented by an analogous result for positive temperature. Our programme started with Adams et al. (2006a,b) which we summarize here. Further aspects are considered in Adams and Dorlas (2007a), Adams and König (2007), Adams and Dorlas (2007b), and Adams (2008a). Under current development are Adams (2007), Adams (2008b), and Adams et al. (2007) in which non-dilute systems are studied.

We give a brief introduction to the physics of dilute quantum gases and their mathematical treatment at zero temperature in Section 8.2. In particular we introduce the Gross–Pitaevskii formula and the scattering length and describe the results by Lieb et al. and our results of the ground product state. Our probabilistic models are introduced in Section 8.3. Section 8.4 is devoted to our large deviations results and the variational analysis.

8.2 Dilute quantum gases

We introduce the modelling of the Gross–Pitaevskii theory which will be the starting point for our probabilistic models in Section 8.3. Let us comment briefly on some issues of the 1995 experiments as these are the motivation for the renewed interest in the Gross–Pitaevskii theory and its analytical proof by Lieb et al.

The experimental systems are collections of individual neutral alkali-gas atoms (e.g., ^{6}Li, ^{40}K, ^{87}Rb, ^{23}Na, ^{7}Li and ^{85}Rb, ^{87}Rb, ^{133}Cs, ^{174}Yb, ^{85}Rb$_2$, and ^{6}Li$_2$), with total number N ranging from a few hundreds up to $\approx 10^{10}$, confined by magnetic and/or optical means to a relatively small region of space. Their densities range from $\approx 10^{11}$cm^{-1} to $\approx 5 \times 10^{15}$cm^{-1}, and their temperatures are typically in the range of a few tenths of nK up to $\approx 5\mu$K.

In a typical system, we are faced with several length scales. One of them is the two-body interaction energy $\hbar^2/m\alpha^2$, where m is the reduced mass of the two particles, $\hbar$ is Heisenberg's constant, and α is the scattering length (see Section 8.2.1 below), expressing the strength of the interatomic interaction. A second one is the mean interparticle spacing r_{int}, and a third one is the oscillator frequency a_{osc} of the confining trap potential. Note that the first scale does not depend on the trap geometry, whereas the oscillator frequency a_{osc}, the mean interparticle spacing, the transition temperature T_c, and the mean-field energy U_0 (to be specified later) do depend on the shape of the confining potential. Introduce the 'healing length' $\xi = (2mnU_0\hbar)^{-1/2}$ and the de Broglie wavelength λ_{DB}. Note that a_{osc} is the zero-point spread of the ground-state wave function of a free particle in the trap. The relations between these scales are as follows.

$$\alpha \ll r_{\text{int}} \approx \lambda_{\text{DB}} \leq \xi \ll a_{\text{osc}}.$$

Typical values are $\alpha \approx 50$ Å, $r_{\text{int}} \approx 2000$ Å, $\xi \approx 4000$ Å, $a_{\text{osc}} \approx 1\mu$. If one compares these numbers with those of liquid helium, one sees that the dilute gas condition $\alpha \ll r_{\text{int}}$, which is characteristic for the BEC of alkali gases, is very far from satisfied for liquid helium. As a consequence, liquid helium is a much more strongly interacting system than BEC gases, by many orders of magnitude.

We now turn to a mathematical modelling and introduce the potentials and the scattering length in Section 8.2.1 and the Gross–Pitaevskii theory in Section 8.2.2.

8.2.1 *Potentials and scattering length*

Our two fundamental ingredients are a trap potential, W, and a pair-interaction potential, v. We restrict ourselves to dimensions $d \in \{2, 3\}$. Our assumptions on W are the following.

$W: \mathbb{R}^d \to [0, \infty]$ is measurable and locally integrable on $\{W < \infty\}$ with
$$\lim_{R \to \infty} \inf_{|x|>R} W(x) = \infty. \tag{8.1}$$

In order to avoid trivialities, we assume that $\{W < \infty\}$ is either equal to $\mathbb{R}^d$ or is a bounded connected open set containing the origin. A typical choice for the trapping potentials is $W(x) = |x|^2$, see Pitaevskii and Stringari (2003).

Our assumptions on v are the following. By $B_r(x)$ we denote the open ball with radius r around $x \in \mathbb{R}^d$.

$v: [0, \infty) \to \mathbb{R} \cup \{+\infty\}$ is measurable and bounded from below,
$a := \sup\{r \geq 0 : v(r) = \infty\} \in [0, \infty), \qquad v|_{[\eta, \infty)}$ is bounded $\forall \eta > a$. $\tag{8.2}$

Note that we also admit $v(a) = +\infty$. We are mainly interested in the case where v has a singularity, i.e., either $a > 0$, or $a = 0$ and $\lim_{r \downarrow 0} v(r) = \infty$.

According to integrability properties near the origin, we distinguish two different classes as follows. We call the interaction potential v a soft-core potential if $a = 0$ and $\int_{B_1(0)} v(|x|)\,dx < +\infty$. Otherwise (i.e., if $a > 0$, or if $a = 0$ and $\int_{B_1(0)} v(|x|)\,dx = +\infty$), we call the interaction potential a hard-core potential.

We shall need the following dN-dimensional versions of the trap and the interaction potential:

$$\mathfrak{W}(x) = \sum_{i=1}^{N} W(x_i) \quad \text{and} \quad \mathfrak{v}(x) = \sum_{1 \leq i < j \leq N} v(|x_i - x_j|),$$

where $x = (x_1, \ldots, x_N) \in \mathbb{R}^{dN}$.

Let us introduce the scattering length of the pair potential, v, and its most important properties. For a detailed overview, see Lieb and Yngvason (2001). First we turn to $d \geq 3$. Let $u\colon [0, \infty) \to [0, \infty)$ be a solution of the scattering equation:

$$u'' = \frac{1}{2} uv \quad \text{on } (0, \infty), \qquad u(0) = 0. \tag{8.3}$$

Then the scattering length $\alpha(v) \in [0, \infty]$, of v is defined as:

$$\alpha(v) = \lim_{r \to \infty} \left[r - \frac{u(r)}{u'(r)} \right]. \tag{8.4}$$

If $v(0) > 0$, then $\alpha(v) > 0$, and if $\int_{a+1}^{\infty} v(r) r^{d-1}\,dr < \infty$, then $\alpha(v) < \infty$. In the pure hard-core case, i.e., $v = \infty \mathbb{1}_{[0,a)}$, we have $\alpha(v) = a$. It is easily seen from the definition that the scattering length of the rescaled potential $\xi^{-2} v(\cdot \xi^{-1})$ is equal to $\xi \alpha(v)$, for any $\xi > 0$.

There is some ambiguity of the choice of u in (8.3); positive multiples of u are also solutions, but the factor drops out in (8.4). We like to normalize u by requiring that $\lim_{R \to \infty} u'(R) = 1$. It is easily seen that (where ω_d denotes the area of the unit sphere in $\mathbb{R}^d$):

$$\int_{\mathbb{R}^d} v(|x|) \frac{u(|x|)}{|x|^{d-2}}\,dx = \omega_d \int_0^{\infty} v(r) u(r) r\,dr = 2\omega_d \int_0^{\infty} u''(r) r\,dr$$
$$= 2\omega_d \lim_{R \to \infty} \left(u'(r) r \Big|_0^R - \int_0^R u'(r)\,dr \right) \tag{8.5}$$
$$= 2\omega_d \lim_{R \to \infty} \left(u'(R) R - u(R) \right) = 2\omega_d \alpha(v).$$

As a consequence, in dimension $d = 3$, we have $\alpha(v) < \widetilde{\alpha}(v)$, where $\widetilde{\alpha}(v) = 1/8\pi \int_{\mathbb{R}^d} v(|x|)\,dx$. Indeed, u is a nonnegative convex function whose slope is always below one because of $\lim_{R \to \infty} u'(R) = 1$. By $u(0) = 0$, we have that $u(r) < r = r^{d-2}$ for any $r > 0$. With the help of (8.5) we therefore get $8\pi \alpha(v) = 2\omega_d \alpha(v) < \int_{\mathbb{R}^d} v(|x|)\,dx = 8\pi \widetilde{\alpha}(v)$.

In $d = 2$, the definition of the scattering length is slightly different. We treat first the case that $\mathrm{supp}(v) \subset [0, R_*]$ for some $R_* > 0$ and consider, for some $R > R_*$, the solution $u \colon [0, R] \to [0, \infty)$ of the scattering equation:

$$u'' = \frac{1}{2} uv \quad \text{on } [0, R], \qquad u(R) = 1, u(0) = 0.$$

Then $u(r) = \log \frac{r}{\alpha(v)} / \log \frac{R}{\alpha(v)}$ for $R_* < r < R$ for some $\alpha(v) \geq 0$, which is by definition the scattering length of v in the case that $\mathrm{supp}(v) \subset [0, R_*]$. Note that $\alpha(v)$ does not depend on R. Hence:

$$\log \alpha(v) = \frac{\log r - u(r) \log R}{1 - u(R)}, \qquad R_* < r < R.$$

For general v (i.e., not necessarily having finite support), v is approximated by compactly supported potentials, and the scattering length of v is put equal to the limit of the scattering lengths of the approximations.

The dilute gas condition ensures that the scattering length is a satisfactory measure of the interaction strength. This approximation neglects any higher energy scattering processes. We finally discuss briefly the effects of the atom-atom scattering on the properties of the many-body alkali-gas system. The fundamental result is that under some conditions the true interaction potential v of two atoms of reduced mass m may be replaced by a delta function of strength $2\pi\hbar^2 \alpha/m$. The effective interaction is:

$$v_{\mathrm{eff}}(x) = \frac{4\pi\alpha\hbar^2}{m} \delta(x), \qquad x \in \mathbb{R}^d.$$

This motivates to scale the potential in such a way that it approximates the delta function in the large N-limit. This will be done in the so-called Gross–Pitaevskii scaling in Subsection 8.2.2, which is a particular approximation of the delta function.

8.2.2 The Gross–Pitaevskii approximation

The simplest possible approximation for the wave function of a many-body system is a (correctly symmetrized) product of single-particle wave functions, i.e., the Hartree–Fock ansatz, see Thirring (1980), Fetter and Walecka (1971), or Dickhoff and Van Neck (2005). In the case of a BEC system at temperature $T = 0$, this approximation usually leads to the Gross–Pitaevskii approximation. Basically the Gross–Pitaevskii approximation suggests replacing the (time-dependent) evolution of the many-body wave functions, governed by a system of Schrödinger equations, by a one-particle non-linear Schrödinger equation (see Gross 1961, Pitaevskii 1961):

$$i\partial_t \Psi(x,t) = \Big(-\Delta + W + 4\pi\alpha |\Psi(x,t)|^2 \Big) \Psi(x,t), \qquad x \in \mathbb{R}^d, t \in \mathbb{R}_+.$$

In the stationary case, the Gross–Pitaevskii theory gives an approximation for the quantum mechanical ground state for many particles (i.e., in the limit

$N \to \infty$) as a variational problem for a single particle in an effective potential. Hence we first summarize some ground state properties for finitely many particles.

The ground-state energy per particle of the N-particle Hamilton operator:

$$\mathcal{H}_N = -\Delta + \mathfrak{W} + \mathfrak{v} \qquad \text{on } L^2(\mathbb{R}^d),$$

is given by:

$$\chi_N = \frac{1}{N} \inf_{h \in H^1(\mathbb{R}^d):\, \|h\|_2 = 1} \left\{ \|\nabla h\|_2^2 + \langle \mathfrak{W}, h^2 \rangle + \langle \mathfrak{v}, h^2 \rangle \right\}, \tag{8.6}$$

Here $H^1(\mathbb{R}^d) = \{f \in L^2(\mathbb{R}^d) \colon \nabla f \in L^2(\mathbb{R}^d)\}$ is the usual Sobolev space, and ∇ is the distributional gradient. It is standard to proof that there is a unique, continuously differentiable, minimizer $h_* \in H^1(\mathbb{R}^d)$ on the right hand side of (8.6), and that it satisfies the variational equation:

$$\Delta h_* = \mathfrak{W} h_* + \mathfrak{v} h_* - N\chi_N h_*.$$

Now we turn to the above mentioned product ansatz. Introduce the ground product state energy of $\mathcal{H}_N$, that is:

$$\chi_N^{(\otimes)} = \frac{1}{N} \inf_{h_1,\ldots,h_N \in H^1(\mathbb{R}^d):\, \|h_i\|_2 = 1\, \forall i} \langle h_1 \otimes \cdots \otimes h_N, \mathcal{H}_N h_1 \otimes \cdots \otimes h_N \rangle. \tag{8.7}$$

The replacement of the ground state energy, χ_N, by the ground product state energy, $\chi_N^{(\otimes)}$, is known as the Hartree–Fock approach (see Dickhoff and Van Neck 2005). Sometimes, the formula in (8.7) is called the Hartree formula. Obviously:

$$\chi_N^{(\otimes)} \geq \chi_N.$$

We can also write:

$$\chi_N^{(\otimes)} = \frac{1}{N} \inf_{\substack{h_1,\ldots,h_N \in H^1(\mathbb{R}^d):\\ \|h_i\|_2 = 1\, \forall i}} \left\{ \sum_{i=1}^N \left\{ \|\nabla h_i\|_2^2 + \langle W, h_i^2 \rangle \right\} + \sum_{1 \leq i < j \leq N} \langle h_i^2, V h_j^2 \rangle \right\},$$

where V denotes the integral operator with kernel $v \circ |\cdot|$, either defined for functions by $Vf(x) = \int_{\mathbb{R}^d} v(|x-y|) f(y)\, dy$ or for measures by $V\mu(x) = \int_{\mathbb{R}^d} \mu(dy)\, v(|x-y|)$. The main assertions on the formula in (8.7) and its minimizers are summarized as follows (see Adams et al. 2006a).

Lemma 8.1 (Ground product states of $\mathcal{H}_N$) *Fix $N \in \mathbb{N}$.*

(i) *There exists at least one minimizer $(h_1,\ldots,h_N)$ of the right hand side in the formula for $\chi_N^{(\otimes)}$. The set of minimizers is compact and invariant under permutation of the functions $h_1,\ldots,h_N$.*

(ii) Any minimizer $(h_1, \ldots, h_N)$ satisfies the system of differential equations:
$$\Delta h_i = -\lambda_i h_i + W h_i + h_i \sum_{j \neq i} V h_j^2, \qquad i = 1, \ldots, N,$$
with $\lambda_i = \|\nabla h_i\|_2^2 + \langle W, h_i^2 \rangle + \sum_{j \neq i} \langle h_i^2, V h_j^2 \rangle$. Furthermore, $\|h_i\|_\infty \leq C_d (\lambda_i - (N-1)\inf v)^{d/4}$ for any $i \in \{1, \ldots, N\}$, where $C_d > 0$ depends on the dimension d only.

(iii) Let v be soft-core, assume that $d \in \{2,3\}$, and let $(h_1, \ldots, h_N)$ be any minimizer. Assume that $v|_{(0,\eta)} \geq 0$ for some $\eta > 0$. In $d=3$, furthermore assume that:
$$\int_{B_1(0)} |v(|y|)|^{1+\delta} dy < \infty, \qquad \text{for some } \delta > 0.$$
Then every h_i is positive everywhere in $\mathbb{R}^d$ and continuously differentiable, and all first partial derivatives are α-Hölder continuous for any $\alpha < 1$.

(iv) Let v be hard-core, assume that $d \in \{2,3\}$, and let $(h_1, \ldots, h_N)$ be any minimizer. Then every h_i is continuously differentiable in the interior of its support, and all first partial derivatives are α-Hölder continuous for any $\alpha < 1$.

Remark 8.1

(i) Unlike for the ground states of $\mathcal{H}_N$ in (8.6), there is no convexity argument available for the formula in (8.7). This is due to the fact that a convex combination of tensor-products of functions is not tensor-product in general, and hence the domain of the infimum in (8.7) is not a convex subset of $H^1(\mathbb{R}^{dN})$. However, for $h_2, \ldots, h_N$ fixed, the minimization over h_1 enjoys the analogous convexity properties on $H^1(\mathbb{R}^d)$ as the minimization in (8.6).

(ii) If v is hard-core, it is easy to see that the distances between the supports of $h_1, \ldots, h_N$ have to be no smaller than a (see (8.2)) in order to make the value of $\langle h_1 \otimes \ldots \otimes h_N, \mathcal{H}_N h_1 \otimes \ldots \otimes h_N \rangle$ finite. The potential $\sum_{j \neq i} V h_j^2$ is equal to ∞ in the a-neighbourhood of the union of the supports of h_j with $j \neq i$, and h_i is equal to zero there (we regard $0 \cdot \infty$ as 0). In particular, minimizers of (8.7) are not of the form $(h, \ldots, h)$. In the soft-core case, this statement is not obvious at all. A partial result on this question in $d=3$ will be a by-product of Section 8.2.2 below.

We study now our main variational formulas, χ_N and $\chi_N^{(\otimes)}$, and their minimizers in the limit for diverging number N of particles. In particular, we point out some significant differences between χ_N and its product state version $\chi_N^{(\otimes)}$ in the soft-core and the hard-core case, respectively.

First we report on recent results by Lieb, Seiringer, and Yngvason on the large-N behaviour of χ_N. Let the pair functional v be as in (8.2) and assume additionally that $v \geq 0$ and $v(0) > 0$.

We shall replace v by the rescaling $v_N(\cdot) = \xi_N^{-2} v(\cdot \xi_N^{-1})$, for some appropriate ξ_N tending to zero sufficiently fast. This will provide the dilute gas condition needed. Hence, the reach of the repulsion is of order ξ_N, and its strength of order ξ_N^{-2}. Furthermore, the scattering length of v, $\alpha(v)$, is rescaled such that $\alpha(v_N) = \alpha(v)\beta_N$. If $\beta_N \downarrow 0$ sufficiently fast, this rescaling makes the system dilute, in the sense that $\alpha(v_N) \ll N^{-1/d}$. This means that the interparticle distance is much bigger than the range of the interaction potential strength. More precisely, the decay of β_N will be chosen in such a way that the pair-interaction has the same order as the kinetic term.

The mathematical description of the large-N behaviour of χ_N in this scaling, and hence the theoretical foundation of the above mentioned physical experiments, has been successfully accomplished in a recent series of papers Lieb et al. (2000a), Lieb and Yngvason (2001), Lieb et al. (2001), and Lieb and Seiringer (2002). It turned out that the well-known Gross–Pitaevskii formula adequately describes the limit of the ground states and its energy. This variational formula was first introduced in Gross (1961) and Gross (1963) and independently in Pitaevskii (1961) for the study of superfluid Helium. After its importance for the description of Bose–Einstein condensation of dilute gases in magnetic traps was realized in 1995, the interest in this formula considerably increased; see Dalfovo et al. (1999) for a summary and the monograph Pitaevskii and Stringari (2003) for a comprehensive account on Bose–Einstein condensation.

The Gross–Pitaevskii formula has a parameter $\alpha > 0$ and is defined as follows:

$$\chi_\alpha^{(GP)} = \inf_{\phi \in H^1(\mathbb{R}^d):\, \|\phi\|_2 = 1} \left\{ \|\nabla \phi\|_2^2 + \langle W, \phi^2 \rangle + 4\pi\alpha \|\phi\|_4^4 \right\}.$$

It is known by Lieb, Seiringer, and Yngvason (2000a) that $\chi_\alpha^{(GP)}$ possesses a unique minimizer $\phi_\alpha^{(GP)}$, which is positive and continuously differentiable with Hölder continuous derivatives of order one.

Since $v(0) > 0$, its scattering length $\alpha(v)$ is positive. The condition:

$$\int_{a+1}^\infty v(r) r^{d-1}\, dr < \infty$$

implies that $\alpha(v) < \infty$. Furthermore, note that the rescaled potential $\xi^{-2} v(\cdot \xi^{-1})$ has scattering length $\xi \alpha(v)$ for any $\xi > 0$.

Theorem 8.1 (Large-N asymptotic of χ_N in $d \in \{2, 3\}$) [Lieb et al. 2000a, Lieb and Yngvason 2001, Lieb et al. 2001]. *Assume that $d \in \{2,3\}$, that $v \geq 0$ with $v(0) > 0$, and $\int_{a+1}^\infty v(r) r^{d-1}\, dr < \infty$. Replace v by $v_N(\cdot) = \xi_N^{-2} v(\cdot \xi_N^{-1})$ with $\xi_N = 1/N$ in $d = 3$ and $\xi_N^2 = \alpha(v)^{-2} e^{-N/\alpha(v)} N \|\phi_{\alpha(v)}^{(GP)}\|_4^{-4}$ in $d = 2$. Let $h_N \in H^1(\mathbb{R}^{dN})$ be the unique minimizer on the right hand side of (8.6), and* define $\phi_N^2 \in H^1(\mathbb{R}^d)$ as the normalized first marginal of h_N^2, i.e.:

$$\phi_N^2(x) = \int_{\mathbb{R}^{d(N-1)}} h_N^2(x, x_2, \ldots, x_N) \, dx_2 \cdots dx_N, \qquad x \in \mathbb{R}^d.$$

Then we have:

$$\lim_{N \to \infty} \chi_N = \chi_{\alpha(v)}^{(GP)} \qquad \text{and} \qquad \phi_N^2 \to (\phi_{\alpha(v)}^{(GP)})^2 \text{ in weak } L^1(\mathbb{R}^d)\text{-sense}.$$

In particular, the proofs show that the ground state, h_N, approaches, for large N, the function:

$$(x_1, \ldots, x_N) \mapsto \prod_{i=1}^N \Big(\frac{\phi_{\alpha(v)}^{(GP)}(x_i)}{\|\phi_{\alpha(v)}^{(GP)}\|_\infty} f\big(\min\{|x_i - x_j| : j < i\}\big) \Big),$$

where $f(r) = u(r)/r$ and u is the solution of the scattering equation (8.3). In order to obtain the Gross–Pitaevskii formula as the limit of χ_N also in $d = 2$, the rescaling of v in Theorem 8.1 has to be chosen in such a way that the repulsion strength is the inverse square of the repulsion reach and such that this reach decays exponentially, which is rather unphysical.

There is an analogue of Theorem 8.1 for the Hartree model in the soft-core case, see Adams et al. (2006a). It turns out that the ground product state energy $\chi_N^{(\otimes)}$ also converges towards the Gross–Pitaevskii formula. However, in $d = 2$, it turns out that the potential v has to be rescaled differently. Furthermore, in $d \in \{2, 3\}$, the scattering length $\alpha(v)$ is replaced by the number:

$$\widetilde{\alpha}(v) := \frac{1}{8\pi} \int_{\mathbb{R}^d} v(|y|) \, dy.$$

Theorem 8.2 (Large-N asymptotic of $\chi_N^{(\otimes)}$, soft-core case) *Let $d \in \{2, 3\}$. Assume that v is a soft-core pair potential with $v \geq 0$ and $v(0) > 0$ and $\widetilde{\alpha}(v) < \infty$. In dimension $d = 3$, additionally assume that (iii) of Lemma 8.1 holds. Replace v by $v_N(\cdot) = N^{d-1} v(\cdot N)$ and let $(h_1^{(N)}, \ldots, h_N^{(N)})$ be any minimizer for the ground product state energy. Define $\phi_N^2 = \frac{1}{N} \sum_{i=1}^N (h_i^{(N)})^2$. Then we have:*

$$\lim_{N \to \infty} \chi_N^{(\otimes)} = \chi_{\widetilde{\alpha}(v)}^{(GP)} \qquad \text{and} \qquad \phi_N^2 \to \big(\phi_{\widetilde{\alpha}(v)}^{(GP)}\big)^2,$$

where the convergence of ϕ_N^2 is in the weak $L^1(\mathbb{R}^d)$-sense and weakly for the probability measures $\phi_n^2(x) \, dx$ towards the measure $(\phi_{\widetilde{\alpha}(v)}^{(GP)})^2(x) \, dx$.

Note that, in $d = 3$, the interaction potential is rescaled in the same way in Theorems 8.1 and 8.2. However, the two relevant parameters depend on different properties of the potential (the scattering length, respectively the integral) and have different values, since $\alpha(v) < \widetilde{\alpha}(v)$ (see Section 8.2.1). In particular, for N

large enough, the ground state of χ_N is *not* a product state. This implies the strictness of the inequality for the two ground state energies, for v replaced by $v_N(\cdot) = N^2 v(\cdot N)$. The phenomenon that (unrestricted) ground states are linked with the scattering length has been theoretically predicted for more general N-body problems (see Fetter and Walecka 1971, Ch. 14, Popov 1983). Indeed, Landau combined a diagrammatic method (a Born approximation of the scattering length) with Bogoliubov's approximations to almost reconstruct the scattering length from the L^1-norm of $v \circ |\cdot|$ in the (non-dilute) ground state. However, the relation between the L^1-norm and the product ground states was not rigorously known before.

In $d = 2$, a more substantial difference between the large-N behaviours of χ_N and $\chi_N^{(\otimes)}$ is apparent. Not only the asymptotic relation between the reach and the strength of the repulsion is different, but also the order of this rescaling in dependence on N. We can offer no intuitive explanation for this.

Interestingly, in the hard-core case, $\chi_N^{(\otimes)}$ shows a rather different large-N behaviour, which we want to roughly indicate in a special case. Assume that W and v are purely hard-core potentials, for definiteness we take $W = \infty \mathbb{1}_{B_1(0)^c}$ and $v = \infty \mathbb{1}_{[0,a]}$. We replace v by $v_N(\cdot) = v(\cdot/\xi_N)$ for some $\xi_N \downarrow 0$ (a pre-factor plays no role). Then $\chi_N^{(\otimes)}$ is equal to $\frac{1}{N}$ times the minimum over the sum of the principal Dirichlet eigenvalues of $-\Delta$ in N subsets of the unit ball having distance $\geq a\beta_N$ to each other, where the minimum is taken over the N sets. It is clear that the volumes of these N sets should be of order $\frac{1}{N}$, independently of the choice of ξ_N. Then their eigenvalues are at least of order $N^{2/d}$. Hence, one arrives at the statement $\liminf_{N \to \infty} N^{-2/d} \chi_N^{(\otimes)} > 0$, i.e., $\chi_N^{(\otimes)}$ tends to ∞ at least like $N^{2/d}$.

8.3 The probabilistic models

Much thermodynamic information about the Boson system is contained in the traces of the Boltzmann factor $e^{-\beta \mathcal{H}_N}$ for $\beta > 0$, like the free energy, or the pressure. Since the 1960s, interacting Brownian motions are generally used for probabilistic representations for these traces. The parameter β, which is interpreted as the inverse temperature of the system, is then the length of the time interval of the Brownian motions.

However, the traces do not contain much information about the ground state. Since the pioneering work of Donsker and Varadhan in the early 1970s it is basically known that the ground states are intimately linked with the Brownian occupation measures. This link is rigorously established via the theory of large deviations for diverging time, which corresponds to vanishing temperature.

We introduce two different models of interacting Brownian motions. These models are given in terms of transformed measures for paths of length β in terms of certain Hamiltonians. Let a family of N independent Brownian motions, $(B_t^{(1)})_{t \geq 0}, \ldots, (B_t^{(N)})_{t \geq 0}$, in $\mathbb{R}^d$ with generator $-\Delta$ be given. The Hamiltonians of both models possess a trap part and a pair-interaction part. The trap part is for both models the same, namely:

$$H_{N,\beta} = \sum_{i=1}^{N} \int_0^\beta W(B_s^{(i)})\,\mathrm{d}s. \tag{8.8}$$

The Hamiltonian of our first model consists of two parts: the trap part given in (8.8), and a pair-interaction part:

$$G_{N,\beta} = \sum_{1 \le i < j \le N} \int_0^\beta v\big(|B_s^{(i)} - B_s^{(j)}|\big)\,\mathrm{d}s.$$

We look at the distribution of the N Brownian motions under the transformed path measure:

$$\mathrm{d}\widehat{\mathbb{P}}_{N,\beta} = \frac{1}{Z_{N,\beta}} \exp(-H_{N,\beta} - G_{N,\beta})\,\mathrm{d}\mathbb{P}, \quad \text{where } Z_{N,\beta} = \mathbb{E}\big(\exp(-H_{N,\beta} - G_{N,\beta})\big).$$

Here $\mathbb{E}$ denotes the Brownian expectation for deterministic start at the origin and time horizon $[0,\beta]$. We call $\widehat{\mathbb{P}}_{N,\beta}$ the canonical ensemble model, since it is derived, via a Feynman–Kac formula, from the trace-class operator of the canonical ensemble, $\mathrm{e}^{-\beta\mathcal{H}_N}$. That is, the trace is given as:

$$\mathrm{Tr}\,(\mathrm{e}^{-\beta\mathcal{H}_N}) = \int_{\mathbb{R}^d} \mathrm{d}x_1 \cdots \int_{\mathbb{R}^d} \mathrm{d}x_N \bigotimes_{i=1}^{N} \mathbb{E}^\beta_{x_i,x_i}\big(\mathrm{e}^{-H_{N,\beta}-G_{N,\beta}}\big).$$

Here $\mathbb{E}^\beta_{x_i,x_i}$ denotes the expectation with respect to a Brownian bridge starting in x_i and terminating in x_i after time β.

However, a system of N Bosons is described by a trace of the projection to symmetric wave functions, i.e., wave functions that are invariant under permutations of the single particle indices. Hence the trace for a system of Bosons is given as:

$$\mathrm{Tr}_+(\mathrm{e}^{-\beta\mathcal{H}_N}) = \frac{1}{N!} \sum_{\sigma \in \mathfrak{S}_N} \int_{\mathbb{R}^d} \mathrm{d}x_1 \cdots \int_{\mathbb{R}^d} \mathrm{d}x_N \bigotimes_{i=1}^{N} \mathbb{E}^\beta_{x_i,x_{\sigma(i)}}\big(\mathrm{e}^{-H_{N,\beta}-G_{N,\beta}}\big),$$

where $\mathfrak{S}_N$ is the group of permutations, of N elements. These symmetrized systems are the subject of the review Adams (2008c) in these proceedings. Recent results can be found in Adams and Dorlas (2007a), Adams and König (2007), Adams (2007), and Adams (2008b).

The path measure $\mathbb{P}_{N,\beta}$ is a model for N Brownian motions in a trap W with the presence of a repellent pair interaction. We can conceive the N-tuple of the motions, $B_t = (B_t^{(1)}, \ldots, B_t^{(N)})$, as one Brownian motion in $\mathbb{R}^{dN}$. Introduce the normalized occupation measure of the dN-dimensional motion:

$$\mu_\beta(\mathrm{d}x) = \frac{1}{\beta} \int_0^\beta \delta_{B_s}(\mathrm{d}x)\,\mathrm{d}s,$$

which is a random element of the set $\mathcal{M}_1(\mathbb{R}^{dN})$ of probability measures on $\mathbb{R}^{dN}$. It measures the time spent by the tuple of N Brownian motions in a

Interacting Brownian motions and the Gross–Pitaevskii formula

given region. Note that there is only one time scale involved for all the motions, i.e., the Brownian particles interact with each other at common time units. We can write the Hamiltonians in terms of the occupation measure as:

$$H_{N,\beta} = \beta\langle \mathfrak{W}, \mu_\beta\rangle \quad \text{and} \quad G_{N,\beta} = \beta\langle \mathfrak{v}, \mu_\beta\rangle.$$

Our second Brownian model is defined in terms of another Hamiltonian. We keep the trap Hamiltonian $H_{N,\beta}$ as in (8.8), but the interaction Hamiltonian is now:

$$K_{N,\beta} = \sum_{1\leq i<j\leq N} \frac{1}{\beta}\int_0^\beta \int_0^\beta v\big(|B_s^{(i)} - B_t^{(j)}|\big)\,\mathrm{d}s\mathrm{d}t. \tag{8.9}$$

Note that the i-th Brownian motion interacts with the mean of the whole path of the j-th motion, taken over all times before β. Hence, the interaction is not a *particle* interaction, but a *path* interaction. The interaction (8.9) is related to Polaron type models Donsker and Varadhan (1983) and Bolthausen et al. (1993), where instead of several paths a single path is considered. We consider the corresponding transformed path measure:

$$\mathrm{d}\widehat{\mathbb{P}}_{N,\beta}^{(\otimes)} = \frac{1}{Z_{N,\beta}^{(\otimes)}} \exp(-H_{N,\beta} - K_{N,\beta})\,\mathrm{d}\mathbb{P}, \quad \text{where } Z_{N,\beta}^{(\otimes)} = \mathbb{E}\big(\exp(-H_{N,\beta} - K_{N,\beta})\big).$$

In Theorem 8.4 below it turns out that the large-β behaviour of $Z_{N,\beta}^{(\otimes)}$ is intimately related to the Hartree formula in (8.7). Therefore, we call this model the Hartree model. Its usefulness as a simplified model for the ground state of H_N is well-known, see the physics monograph (Dickhoff and Van Neck 2005, Ch. 12). Approximating many-body wave function by products of single-particle wave functions is known as the Hartree–Fock ansatz, usually used for wave functions of electrons (Fermions) of large atoms or molecules.

We introduce the normalized occupation measure of the i-th motion:

$$\mu_\beta^{(i)}(\mathrm{d}x) = \frac{1}{\beta}\int_0^\beta \delta_{B_s^{(i)}}(\mathrm{d}x)\,\mathrm{d}s \in \mathcal{M}_1(\mathbb{R}^d).$$

The tuple of the N occupation measures, $(\mu_\beta^{(1)}, \ldots, \mu_\beta^{(N)})$, plays a particular role in this model. We can write the Hamiltonians as:

$$H_{N,\beta} = \beta\langle \mathfrak{W}, \mu_\beta^\otimes\rangle \quad \text{and} \quad K_{N,\beta} = \beta\sum_{1\leq i<j\leq N}\langle \mu_\beta^{(i)}, V\mu_\beta^{(j)}\rangle = \beta\langle \mathfrak{v}, \mu_\beta^\otimes\rangle,$$

where we recall the operator V with kernel $v\circ|\cdot|$, and $\mu_\beta^\otimes = \mu_\beta^{(1)} \otimes \ldots \otimes \mu_\beta^{(N)}$ is the product measure.

8.4 Large deviations results

We present our main large deviations results for both the canonical ensemble and the Hartree model. In Section 8.4.1 the zero temperature (i.e., large-β) limit is considered, and in Section 8.4.2 the large-N limit, both at zero temperature and positive temperature.

8.4.1 Vanishing temperature

It turns out that the large-β behaviour of the canonical ensemble model is described by the ground state of the Hamilton operator $\mathcal{H}_N$ via a large deviations principle. The rate function I_N appearing in Theorem 8.3 is the well-known Donsker–Varadhan rate function on $\mathbb{R}^{dN}$ defined by:

$$I_N(\mu) = \begin{cases} \|\nabla\sqrt{\tfrac{\mathrm{d}\mu}{\mathrm{d}x}}\|_2^2 & \text{if } \sqrt{\tfrac{\mathrm{d}\mu}{\mathrm{d}x}} \in H^1(\mathbb{R}^{dN}) \text{ exists}, \\ \infty & \text{otherwise.} \end{cases} \qquad (8.10)$$

Note that the energy functional $\langle h, \mathcal{H}_N h\rangle$ may be rewritten $\langle h, \mathcal{H}_N h\rangle = I_N(\mu) + \langle \mathfrak{W}, \mu\rangle + \langle \mathfrak{v}, \mu\rangle$ for the probability measure $\mu(\mathrm{d}x) = h^2(x)\,\mathrm{d}x$.

Simplifying, the large deviations principle says that, as $\beta \to \infty$:

$$\mathbb{P}(\mu_\beta \approx \mu) \approx \mathrm{e}^{-N I_N(\mu)}, \qquad \mu \in \mathcal{M}_1(\mathbb{R}^{dN}).$$

This is formulated precisely in the next theorem. We refer to Appendix 8.A for the notion of what a large deviations principle is.

Weak convergence of probability measures is understood as the convergence of the integrals (expectations) against bounded continuous functions (this convergence is often called weak-$\star$ convergence).

Theorem 8.3 (Canonical ensemble model at late times) Adams et al. (2006a). *Fix* $N \in \mathbb{N}$.

(i)
$$\lim_{\beta\to\infty} \frac{1}{N\beta} \log \mathbb{E}\big(\exp(-H_{N,\beta} - G_{N,\beta})\big) = -\chi_N,$$

where χ_N is the ground-state energy per particle of the N-particle operator $\mathcal{H}_N$ given in (8.6).

(ii) As $\beta \to \infty$, the distribution of μ_β on $\mathcal{M}_1(\mathbb{R}^{dN})$ under $\widehat{\mathbb{P}}_{N,\beta}$ satisfies a principle of large deviation with speed β and rate function $\mathcal{I}_N$ given by:

$$\mathcal{I}_N(\mu) = I_N(\mu) + \langle \mathfrak{W}, \mu\rangle + \langle \mathfrak{v}, \mu\rangle - N\chi_N \text{ for } \mu \in \mathcal{M}_1(\mathbb{R}^{dN}),$$

that is:

$$\lim_{\beta\to\infty} \frac{1}{\beta} \log \widehat{\mathbb{P}}_{N,\beta}(\mu_\beta \in G) \geq -\inf_{\nu\in G} \mathcal{I}_N(\nu) \qquad G \subset \mathcal{M}_1(\mathbb{R}^d) \text{ open},$$

$$\lim_{\beta\to\infty} \frac{1}{\beta} \log \widehat{\mathbb{P}}_{N,\beta}(\mu_\beta \in F) \leq -\inf_{\nu\in F} \mathcal{I}_N(\nu) \qquad F \subset \mathcal{M}_1(\mathbb{R}^d) \text{ closed}.$$

(iii) The distribution of μ_β under $\widehat{\mathbb{P}}_{N,\beta}$ converges weakly (in the sense of probability measures) towards the measure $h_*(x)^2\,\mathrm{d}x$, where h_* is the unique minimizer in (8.6).

Remark 8.2 It is well-known (Ginibre 1970) that the bottom of the spectrum of $\mathcal{H}_N$ is related to the large-β behaviour of the trace of $e^{-\beta\mathcal{H}_N}$, more precisely:

$$\chi_N = -\lim_{\beta\to\infty} \frac{1}{N\beta} \log \operatorname{Tr}\left(e^{-\beta\mathcal{H}_N}\right).$$

Theorem 8.4 (Hartree model at late times) Adams et al. (2006a). Assume that W and v are continuous in $\{W < \infty\}$ resp. in $\{v < \infty\}$. Furthermore, assume in the soft-core case that there exists an $\varepsilon > 0$ and a decreasing function $\widetilde{v}\colon (0,\varepsilon) \to \mathbb{R}$ with $v \leq \widetilde{v}$ on $(0,\varepsilon)$, which satisfies $\int_{B_\varepsilon(0)} G(0,y)\widetilde{v}(|y|)\,dy < \infty$, where G denotes the Green's function of the free Brownian motion on $\mathbb{R}^d$. Fix $N \in \mathbb{N}$.

(i)
$$\lim_{\beta\to\infty} \frac{1}{N\beta} \log \mathbb{E}\big(\exp(-H_{N,\beta} - K_{N,\beta})\big) = -\chi_N^{(\otimes)}.$$

(ii) As $\beta \to \infty$, the distribution of the tuple $(\mu_\beta^{(1)}, \ldots, \mu_\beta^{(N)})$ of Brownian occupation measures on $\mathcal{M}_1(\mathbb{R}^d)^N$ under $\widehat{\mathbb{P}}_{N,\beta}^{(\otimes)}$ satisfies a large deviation principle with speed β and rate function:

$$I_N^{(\otimes)}(\mu_1,\ldots,\mu_N) = \sum_{i=1}^N I_1(\mu_i) + \langle \mathfrak{W}, \mu^\otimes \rangle + \langle \mathfrak{v}, \mu^\otimes \rangle - N\chi_N^{(\otimes)},$$

with $\mu_1,\ldots,\mu_N \in \mathcal{M}_1(\mathbb{R}^d)$ where I_1 is defined in (8.10), and $\mu^\otimes = \mu_1 \otimes \cdots \otimes \mu_N$ is the product measure.

(iii) The distribution of $(\mu_\beta^{(1)},\ldots,\mu_\beta^{(N)})$ under $\widehat{\mathbb{P}}_{N,\beta}^{(\otimes)}$ converges weakly (in the sense of probability measures) to the set of minimizers for ground product state energy $\chi_N^{(\otimes)}$.

8.4.2 Large systems at positive temperature

We now formulate our results on the behaviour of the Hartree model in the limit as $N \to \infty$, with $\beta > 0$ fixed. As in the zero temperature case in Theorem 8.2, we replace v by $v_N(\cdot) = N^{d-1}v(\cdot N)$; we write $K_{N,\beta}^{(N)}$ for the Hamiltonian introduced in (8.9).

First we introduce an important functional, which will play the role of a probabilistic energy functional. Define $J_\beta\colon \mathcal{M}_1(\mathbb{R}^d) \to [0,\infty]$ as the Legendre–Fenchel transform of the map $\mathcal{C}_b(\mathbb{R}^d) f \mapsto \frac{1}{\beta} \log \mathbb{E}[e^{\int_0^\beta f(B_s)\,ds}]$ on the set $\mathcal{C}_b(\mathbb{R}^d)$ of continuous bounded functions on $\mathbb{R}^d$, where $(B_s)_{s\geq 0}$ is one of the above Brownian motions. That is:

$$J_\beta(\mu) = \sup_{f\in\mathcal{C}_b(\mathbb{R}^d)} \left\{ \langle f, \mu\rangle - \frac{1}{\beta}\log\mathbb{E}\big(e^{\int_0^\beta f(B_s)\,ds}\big)\right\}, \qquad \mu \in \mathcal{M}_1(\mathbb{R}^d).$$

Here $\mathcal{M}_1(\mathbb{R}^d)$ denotes the set of probability measures on $\mathbb{R}^d$, and we write $\langle f,\mu\rangle = \int_{\mathbb{R}^d} f(x)\,\mu(dx)$ and also use the notation $\langle f,g\rangle = \int_{\mathbb{R}^d} f(x)g(x)\,dx$ for integrable functions f, g. Note that J_β depends on the initial distribution of the Brownian motion. One can show that J_β is not identical to $+\infty$. Clearly, J_β is a lower semicontinuous and convex functional on $\mathcal{M}_1(\mathbb{R}^d)$, which we endow with the topology of weak convergence induced by test integrals against continuous bounded functions. However, J_β is *not* a quadratic form coming from any linear operator. If μ possesses a Lebesgue density ϕ^2 for some L^2-normalized $\phi \in L^2$, then we also write $J_\beta(\phi^2)$ instead of $J_\beta(\mu)$. It turns out that $J_\beta(\mu) = \infty$ if μ fails to have a Lebesgue density, see Adams, Bru, and König (2006b).

In the language of the theory of large deviations, J_β is the rate function that governs a large deviations principle. The object that satisfies this principle is the mean of the N normalized occupation measures:

$$\overline{\mu}_{N,\beta} = \frac{1}{N} \sum_{i=1}^{N} \mu_\beta^{(i)}, \qquad N \in \mathbb{N}.$$

The principle follows from Cramér's theorem, see (Deuschel and Stroock 2001, Th. 3.3.11), together with the exponential tightness of the sequence $(\overline{\mu}_{N,\beta})_{N\in\mathbb{N}}$.

To apply this principle, we have to express our Hamiltonians $H_{N,\beta}$ and $K_{N,\beta}$ as functionals of $\overline{\mu}_{N,\beta}$. For the first this is an easy task and can be done for any fixed N:

$$H_{N,\beta} = N\beta \int_{\mathbb{R}^d} W(x) \frac{1}{N} \sum_{i=1}^{N} \mu_\beta^{(i)}(\mathrm{d}x) = N\beta \langle W, \overline{\mu}_{N,\beta} \rangle.$$

Now we rewrite the second Hamiltonian, which will need Brownian intersection local times and an approximation for large N. Let us first introduce the intersection local times, see Geman et al. (1984). For the following, we have to restrict to the case $d \in \{2, 3\}$.

Fix $1 \leq i < j \leq N$ and consider the process $B^{(i)} - B^{(j)}$, the so-called confluent Brownian motion of $B^{(i)}$ and $-B^{(j)}$. This two-parameter process possesses a local time process, i.e., there is a random process $(L_\beta^{(i,j)}(x))_{x \in \mathbb{R}^d}$ such that, for any bounded and measurable function $f \colon \mathbb{R}^d \to \mathbb{R}$:

$$\int_{\mathbb{R}^d} f(x) L_\beta^{(i,j)}(x)\, \mathrm{d}x = \frac{1}{\beta^2} \int_0^\beta \mathrm{d}s \int_0^\beta \mathrm{d}t\, f\big(B_s^{(i)} - B_t^{(j)}\big) \qquad (8.11)$$

$$= \int_{\mathbb{R}^d} \int_{\mathbb{R}^d} \mu_\beta^{(i,j)}(\mathrm{d}x) \mu_\beta^{(i,j)}(\mathrm{d}y) f(x-y). \qquad (8.12)$$

We scale the interaction potential, i.e., we replace v with $v_N(\cdot) = N^{d-1} v(\cdot N)$, and we denote the interaction (8.9) with this replacement by $K_{N,\beta}^{(N)}$. We rewrite it as follows:

$$K_{N,\beta}^{(N)} = \beta N^{d-1} \sum_{1 \leq i < j \leq N} \int_{\mathbb{R}^d} v(zN) L_\beta^{(i,j)}(z)\, \mathrm{d}z$$

$$= N\beta \int_{\mathbb{R}^d} v(x) \frac{1}{N^2} \sum_{1 \leq i < j \leq N} L_\beta^{(i,j)}(\tfrac{1}{N} x)\, \mathrm{d}x.$$

It is known (Geman et al. 1984, Th. 8.1) that $(L_\beta^{(i,j)}(x))_{x\in\mathbb{R}^d}$ may be chosen continuously in the space variable. Furthermore, the random variable $L_\beta^{(i,j)}(0) = \lim_{x\to 0} L_\beta^{(i,j)}(x)$ is equal to the normalized total intersection local time of the two motions $B^{(i)}$ and $B^{(j)}$ up to time β. Formally:

$$L_\beta^{(i,j)}(0) = \frac{1}{\beta^2} \int_A \mathrm{d}x \int_0^\beta \mathrm{d}s \, \frac{\mathbb{1}\{B_s^{(i)} \in \mathrm{d}x\}}{\mathrm{d}x} \int_0^\beta \mathrm{d}t \, \frac{\mathbb{1}\{B_t^{(j)} \in \mathrm{d}x\}}{\mathrm{d}x}$$
$$= \int_A \mathrm{d}x \, \frac{\mu_\beta^{(i)}(\mathrm{d}x)}{\mathrm{d}x} \frac{\mu_\beta^{(j)}(\mathrm{d}x)}{\mathrm{d}x},$$

Using the continuity of $L_\beta^{(i,j)}$, we approximate:

$$K_{N,\beta}^{(N)} \approx N\beta 4\pi\alpha(v) \frac{2}{N^2} \sum_{1\leq i<j\leq N} L_\beta^{(i,j)}(0) \approx N\beta 4\pi\alpha(v) \left\langle \frac{1}{N}\sum_{i=1}^N \mu_\beta^{(i)}, \frac{1}{N}\sum_{i=1}^N \mu_\beta^{(i)} \right\rangle$$
$$= N\beta 4\pi\alpha(v) \left\| \frac{\mathrm{d}\overline{\mu}_{N,\beta}}{\mathrm{d}x} \right\|_2^2.$$

where we conceive $\mu_\beta^{(i)}$ as densities.

Hence, using Varadhan's lemma (see (Deuschel and Stroock 2001, Th. 2.1.10) for example) and ignoring the missing continuity of the map $\mu \mapsto \|\frac{\mathrm{d}\mu}{\mathrm{d}x}\|_2^2$, this heuristic explanation is finished by:

$$\mathbb{E}\left(e^{-H_{N,\beta}-K_{N,\beta}^{(N)}} e^{N\langle f,\overline{\mu}_{N,\beta}\rangle}\right)$$
$$\approx \mathbb{E}\left(\exp\left\{-N\beta\left[\langle W-f, \overline{\mu}_{N,\beta}\rangle - 4\pi\alpha(v)\left\|\frac{\mathrm{d}\overline{\mu}_{N,\beta}}{\mathrm{d}x}\right\|_2^2\right]\right\}\right)$$
$$\approx e^{-N\beta \chi_{\alpha(v)}^{(\otimes)}(f)},$$

where:

$$\chi_\alpha^{(\otimes)}(\beta) = \inf_{\phi\in L^2(\mathbb{R}^d):\, \|\phi\|_2=1} \left\{J_\beta(\phi^2) + \langle W,\phi^2\rangle + 4\pi\alpha\|\phi\|_4^4\right\}. \tag{8.13}$$

Here we substituted $\phi^2(x)\,\mathrm{d}x = \mu(\mathrm{d}x)$, we may restrict the infimum over probability measures to the set of their Lebesgue densities ϕ^2.

Let us now give the precise formulation of our results.

Theorem 8.5 (Many-particle limit for the Hartree model) Adams et al. (2006b). *Assume that $d\in\{2,3\}$ and let in addition to the assumptions in Section 8.2.1 W be continuous in $\{W<\infty\}$ and v with $\int_{\mathbb{R}^d} v(|x|)\,\mathrm{d}x < \infty$ and $\int_{\mathbb{R}^d} v(|x|)^2\,\mathrm{d}x < \infty$. Introduce:*

$$\alpha(v) := \int_{\mathbb{R}^d} v(|y|)\,\mathrm{d}y < \infty.$$

Fix $\beta > 0$. Then, as $N \to \infty$, the mean $\overline{\mu}_{N,\beta} = \frac{1}{N}\sum_{i=1}^{N} \mu_\beta^{(i)}$ of the normalized occupation measures satisfies a large deviation principle on $\mathcal{M}_1(\mathbb{R}^d)$ under the measure with density $\mathrm{e}^{-H_{N,\beta}-K_{N,\beta}^{(N)}}$ with speed $N\beta$ and rate function:

$$I_\beta^{(\otimes)}(\mu) = \begin{cases} J_\beta(\phi^2) + \langle W, \phi^2 \rangle + \frac{1}{2}\alpha(v)\|\phi\|_4^4 & \text{if } \phi^2 = \frac{\mathrm{d}\mu}{\mathrm{d}x} \text{ exists,} \\ \infty & \text{otherwise.} \end{cases}$$

The level sets $\{\mu \in \mathcal{M}_1(\mathbb{R}^d) \colon I_\beta^{(\otimes)}(\mu) \le c\}$, $c \in \mathbb{R}$, are compact.

In order to avoid trivialities, we tacitly assume that the support of the initial distribution of the Brownian motions is contained in the set $\{W < \infty\}$.

Lemma 8.2 (Analysis of $\chi_\alpha^{(\otimes)}(\beta)$, Adams et al. 2006b) *Fix $\beta > 0$ and $\alpha > 0$.*

(i) There exists a unique L^2-normalized minimizer $\phi_* \in L^2(\mathbb{R}^d) \cap L^4(\mathbb{R}^d)$ of the right hand side of (8.13).

(ii) For any neighbourhood $\mathcal{N} \subset L^2(\mathbb{R}^d) \cap L^4(\mathbb{R}^d)$ of ϕ_*:

$$\inf_{\phi \in L^2(\mathbb{R}^d) \colon \|\phi\|_2 = 1, \phi \notin \mathcal{N}} \left\{ J_\beta(\phi^2) + \langle W, \phi^2 \rangle + 4\pi\alpha\|\phi\|_4^4 \right\} > \chi_\alpha^{(\otimes)}(\beta).$$

Here 'neighbourhood' refers to any of the three following topologies: weakly in L^2, weakly in L^4, and weakly in the sense of probability measures, if ϕ is identified with the measure $\phi(x)^2 \, \mathrm{d}x$.

Corollary 8.1 (Free energy for positive temperature) *Let the assumptions of the previous Theorem 8.5 be satisfied. Then the specific free energy per particle is:*

$$\lim_{N\to\infty} \frac{1}{-\beta N} \log \mathbb{E}\bigl(\mathrm{e}^{-H_{N,\beta}-K_{N,\beta}^{(N)}}\bigr) = \chi_{\alpha(v)}^{(\otimes)}(\beta).$$

8.A Large deviations principles

For the convenience of our reader, we repeat the notion of a large deviation principle. A family $(X_\beta)_{\beta>0}$ of random variables X_β, taking values in a topological vector space $\mathcal{X}$, satisfies the *large deviation upper bound* with speed a_β, where $a_\beta \to \infty$ for $\beta \to \infty$, and rate function $I \colon \mathcal{X} \to [0, \infty]$ if, for any closed subset F of $\mathcal{X}$:

$$\limsup_{\beta\to\infty} \frac{1}{a_\beta} \log \mathbb{P}(X_\beta \in F) \le - \inf_{x \in F} I(x),$$

and it satisfies the *large deviation lower bound* if, for any open subset G of $\mathcal{X}$:

$$\liminf_{\beta\to\infty} \frac{1}{a_\beta} \log \mathbb{P}(X_\beta \in G) \ge - \inf_{x \in G} I(x).$$

If both upper and lower bound are satisfied and, in addition, the level sets $\{I \le c\}$ are compact for any $c \in \mathbb{R}$, then one says that $(X_\beta)_\beta$ satisfies a *large deviation principle*. This notion easily extends to the situation where the distribution of X_β is not normalized, but a sub-probability distribution only.

In our large deviations results we shall rely on the following principles for the normalized Brownian occupation measures, i.e., for certain $\mathcal{M}_1(\mathbb{R}^d)$-valued random variables. For any measurable subset A of $\mathbb{R}^d$, we conceive $\mathcal{M}_1(A)$ as a closed convex subset of the space $\mathcal{M}(A)$ of all finite signed Borel measures on A, which is a topological Hausdorff vector space whose topology is induced by the set $\mathcal{C}_b(A)$ of all continuous bounded functions $A \to \mathbb{R}$. Here $\mathcal{C}_b(A)$ is the topological dual of $\mathcal{M}(A)$. The set $\mathcal{M}_1(\mathbb{R}^d)$ inherits this topology from $\mathcal{M}(A)$. When we speak of a large deviation principle of $\mathcal{M}_1(A)$-valued random variables, then we mean a principle on $\mathcal{M}(A)$ with a rate function that is tacitly extended from $\mathcal{M}_1(A)$ to $\mathcal{M}(A)$ with the value $+\infty$.

References

Adams, S. (2007). Large deviations for empirical path measures in cycles of integer partitions. preprint arXiv:math.PR/0702052v2.

Adams, S. (2008a). Interacting Brownian bridges and probabilistic interpretation of Bose-Einstein condensation. Habilitation thesis, University Leipzig.

Adams, S. (2008b). Large deviations for shape measures of integer partitions under non-uniform distribution. In preparation.

Adams, S. (2008c). Large deviations for empirical measures in cycles of integer partitions and their relation to boson symmetrisation. This volume.

Adams, S. and Bru, J.-B. (2004a). Critical analysis of the Bogoliubov Theory of superfluidity. *Physica A* **332**, 60–78.

Adams, S. and Bru, J.-B. (2004b). Exact solution of the AVZ-Hamiltonian in the grand-canonical ensemble. *Ann. Henri Poincaré* **5**, 405–34.

Adams, S. and Bru, J.-B. (2004c). A new microscopic theory of superfluidity at all temperatures. *Ann. Henri Poincaré* **5**, 435–76.

Adams, S., Bru, J.-B., and König, W. (2006a). Large deviations for trapped interacting Brownian particles and paths. *Ann. Probab.* **34**(4), 1340–1422.

Adams, S., Bru, J.-B., and König, W. (2006b). Large systems of path-repellent Brownian motions in a trap at positive temperature. *Electronic Journal of Probability* **11**, 460–85.

Adams, S., Collevecchio, A., and König, W. (2008). A variational formula for the free energy of non-dilute many-particle systems. In preparation.

Adams, S. and Dorlas, T. (2007a). Asymptotic Feynman-Kac formulae for large symmetrised systems of random walks. *Annales de l'institut Henri Poincaré (B) Probabilités et Statistiques*. In press, preprint arXiV:math-ph/0610026.

Adams, S. and Dorlas, T. (2007b). C*-algebraic approach to the Bose-Hubbard model. Journal of Math. Phys. **48** (1033404), 1–14.

Adams, S., Georgii, H., and Kotecký, R. (2008). Gibbs measures for random permutations. In preparation.

Adams, S. and König, W. (2007). Large deviations for many Brownian bridges with symmetrised initial-terminal condition. *Prob. Theory Relat. Fields*.

Accepted for publication, preprint arXiV:math.PR/0603702, published online, Doi 10.s007/0004 40-007-0099-5.

Anderson, M., Ensher, J., Matthews, M., Wieman, C., and Cornell, E. (1995). Observation of Bose-Einstein condensation in a dilute atomic vapor. *Science* **269**, 198–201.

Bolthausen, E., Deuschel, J., and Schmock, U. (1993). Convergence of path measures arising from a mean field or polaron type interaction. *Probab. Theory Related Fields* **95**, 283–310.

Bose, S. (1924). *Zeitschrift für Physik* **26**, 178.

Bradley, C., Sackett, C., Tollet, J., and Hulet, R. (1995). Evidence of Bose-Einstein condensation in an atomic gas with attractive interactions. *Phys. Rev. Lett.* **75**, 1687–90.

Dalfovo, F., Giorgini, S., Pitaevskii, L., and Stringari, S. (1999). Theory of Bose-Einstein condensation in trapped gases. *Rev. Mod. Phys.* **71:3**, 463–512.

Davis, K., Mewes, M., Andrews, M., van Druten, N., Durfee, D., Kurn, D., and Ketterle W. (1995). Bose-Einstein condensation in a gas of sodium atoms. *Phys. Rev. Lett.* **75**(22), 3969–73.

Deuschel, J. and Stroock, D. (2001). *Large Deviations.* AMS Chelsea Publishing American Mathematical Society: Providence.

Dickhoff, W. and Van Neck D. (2005). *Many-Body Theory Exposed.* World Scientific, Singapore.

Donsker, M. and Varadhan, S. (1983). Asymptotics for the polaron. *Comm. Pure Appl. Math.* **36:4**, 505–28.

Einstein, A. (1925). *Sitzber. Kgl. Preuss. Akad. Wiss.* **3**.

Fetter, A. and Walecka, J. (1971). *Quantum Theory of Many Particle Systems.* McGraw-Hill, New York.

Feynman, R. (1953). Atomic theory of the λ transition in Helium. *Phys. Rev.* **91**, 1291–1301.

Feynman, R. (1955). *Progress in Low Temperature Physics, Vol. I.*, pp. 17. Elsevier: Amsterdam.

Geman, D., Horowitz, J., and Rosen, J. (1984). A local time analysis of intersection of Brownian paths in the plane. *Ann. Probab.* **12:1**, 86–1070.

Ginibre, J. (1970). *Some Applications of Functional Integration in Statistical Mechanics and Field Theory.* Gordon and Breach Science Publ.: New York.

Gross, E. (1961). Structure of a quantized vortex in boson systems. *Nuovo Cimento* **20**, 454–77.

Gross, E. (1963). Hydrodynamics of a superfluid condensate. *J. Math. Phys.* **4:2**, 195–207.

Landau, L. (1941). The theory of superfluidity of Helium II. *J. Phys. USSR* **5**, 71.

Landau, L. and Lifshitz, E. (1951). *Statisticheskai Fizika.* Fizatgiz, Moscow.

Lieb, E. and Seiringer, R. (2002). Proof of Bose-Einstein condensation for dilute trapped gases. *Phys. Rev. Lett.* **88:17**, 170409-1-4.

Lieb, E., Seiringer, R., and Yngvason, J. (2000a). Bosons in a trap: a rigorous derivation of the Gross-Pitaevskii energy functional. *Phys. Rev. A* **61**, 043602-1-13.

Lieb, E., Seiringer, R., and Yngvason, J. (2000b). The ground state energy and density of interacting Bosons in a trap. in: Quantum Theory and Symmetries, Goslar 1999; H. D. Doebner, V. K. Dobre, J. D. Hennig and W. Luecke, eds., World Scientific, Singpore 101–110.

Lieb, E., Seiringer, R., and Yngvason, J. (2001). A rigorous derivation of the Gross-Pitaevskii energy functional for a two-dimensional Bose gas. *Comm. Math. Phys.* **224**, 17–31.

Lieb, E. and Yngvason, J. (2001). The ground state energy of a dilute two-dimensional Bose gas. *J. Statist. Phys.* **103**, 509.

London, F. (1938). The λ-phenomenon of liquid helium and the Bose-Einstein degeneracy. *Nature* **141**, 643.

Onsager, L. (1949). Statistical hydrodynamics. *Nuovo Cimento* **6**(Suppl. 2), 279–87.

Onsager, L. and Penrose, O. (1956). Bose-Einstein condensation and liquid Helium. *Phys. Rev.* **104**, 576–84.

Penrose, O. (1951). On the quantum mechanics of Helium II. *Phil. Mag.* **42**, 1373–77.

Pitaevskii, L. (1961). Vortex lines in an imperfect Bose gas. *Sov. Phys. JETP* **13**, 451–54.

Pitaevskii, L. and Stringari, S. (2003). *Bose-Einstein Condensation*. Clarendon Press, Oxford.

Popov, V. (1983). *Functional Integrals in Quantum Field Theory and Statistical Physics*. Riedel, Dordrecht.

Thirring, W. (1980). *Quantenmechanik grosser Systeme*. Springer Verlag, Wien. Lehrbuch der Mathematischen Physik 4.

9

A SHORT INTRODUCTION TO ANDERSON LOCALIZATION

Dirk Hundertmark

Abstract

We give short introduction to some aspects of the theory of Anderson localization.

9.1 Introduction

Anderson (1958) published an article where he discussed the behaviour of electrons in a dirty crystal. This is the quantum mechanical analogue of a random walk in a random environment. He considered the tight binding approximation, in which the electrons can hop from atom to atom and are subject to an external random potential modelling the random environment and gave some nonrigorous but convincing arguments that in this case such a system should lose all its conductivity properties for large enough disorder, that is, become an insulator. The electrons in such a system become trapped due to the external extensive disorder. This is in sharp contrast to the behaviour in ideal crystals which are always conductors.

But what is this supposed to mean precisely? In this note we try to shed some light on the mathematical theory of Anderson localization. Our plan can be summarized as follows:

I. Disordered Matter: the Anderson model of a quantum particle in a random environment.

II. Exponential decay of 'correlations' (fractional moments of the resolvent) as a signature for localization.

III. Finite volume criteria for the decay of correlations in a toy model (percolation): sub-harmonicity is your friend.

IV. Localization at large disorder: the self-avoiding random walk representation.

9.1.1 *The Anderson model*

The arena is given by the Hilbert space $l^2(\mathbb{Z}^d)$. On $\mathbb{Z}^d$ we usually consider the Manhattan norm $|x| = |x_1| + \cdots + |x_d|$, but for most results one can equally well use the Euclidian norm. Disordered matter is described by a *random Schrödinger operator*, often called random Hamiltonian:

$$H = H_\omega := H_0 + \lambda V_\omega$$

with H_0 the unperturbed part, coupling constant $\lambda > 0$, and the random potential V_ω which is simply a multiplication operator on $l^2(\mathbb{Z}^d)$ with matrix elements $V_\omega(x) = v_x(\omega)$, where $(v_x(\omega))_{x \in \mathbb{Z}^d}$ is a collection of random variables indexed by $\mathbb{Z}^d$. In Dirac notation:

$$V_\omega = \sum_{x \in \mathbb{Z}^d} v_x(\omega) |x\rangle\langle x|.$$

We often use Dirac's notation, since it gives a convenient way to write projections and integral kernels: $|x\rangle = \delta_x$ with δ_x the Kronecker delta function, $\delta_x(x) = 1$ and $\delta_x(y) = 0$ for $y \neq x$, and $|x\rangle\langle x|$ is the Dirac notation for the projection operator $|x\rangle\langle x| = \langle \delta_x, \cdot \rangle \delta_x$, where $\langle \cdot, \cdot \rangle$ is the usual scalar product in $l^2(\mathbb{Z}^2)$ with the convention that it is linear in the *second* component. For a bounded operator M on $l^2(\mathbb{Z}^d)$, i.e., an infinite matrix M, the expression $M(x, y) = \langle x|M|y\rangle$ denotes the x-y matrix element.

We will often consider the simplest case in which $(v_x)_{x \in \mathbb{Z}^d}$ are independent identically distributed (i.i.d) random variables with (single-site) distribution ρ. In this case the probability space is given by the product space $\Omega = \mathbb{R}^{\mathbb{Z}^d} = \{f : \mathbb{Z}^d \to \mathbb{R}\}$ and the usual product σ-algebra, and the probability measure on Ω is simply the product measure $\mathbb{P} = \otimes_{x \in \mathbb{Z}^d} \rho$. Note, however, that one does not need to consider i.i.d. random variables, the Aizenman–Molchanov approach is rather robust to correlated random potentials, as long as the assumptions on ρ are replaced by suitable uniform assumptions on conditional expectations.

The unperturbed part of the Hamiltonian is given by:

$$H_0 = T + V_0.$$

Here V_0 models, e.g., a periodic background potential, T is the kinetic energy, e.g.:

$$T\psi(x) = -\Delta\psi(x) := -\sum_{|e|=1} \psi(x+e), \text{ that is } \langle x|T|y\rangle = T(x, y) = -\delta_{|x-y|,1}$$

is the discrete Laplace, or the (negative) adjacency matrix of $\mathbb{Z}^d$ (nearest neighbour hopping). It is also possible to include a constant magnetic field, in which case the kinetic energy is modified to:

$$T(x, y) = -e^{-iA_{x,y}} \delta_{|x-y|,1}.$$

Here $A_{x,y}$ is a *phase function*, i.e., an antisymmetric function of the (oriented) bond $b = \{x, y\}$. Moreover, one can allow for more than nearest neighbour hopping in the kinetic energy T, all that is needed for most of the analysis is a sumability condition of the form:

$$\sup_{x \in \mathbb{Z}^d} \sum_{y \in \mathbb{Z}^d} e^{\mu |x-y|} |T(x,y)| < \infty,$$

for some $\mu > 0$. This decay condition can also be weakened a little bit further if one is not interested in *exponential* decay estimates but content with some polynomial decay rate.

The resolvent is given by $G(z) = G_\omega(z) = (H_\omega - z)^{-1}$ and the Green's function, the kernel of the resolvent, is given by $G(x, y; z) = \langle x|G(z)|y\rangle$.

Remark 9.1

i) We mainly focus on an i.i.d. random potential V_ω, a kinetic energy given by $T = -\Delta$ and $H_0 = -\Delta + V_0$ with V_0 periodic or at least bounded.

ii) We will take expectations of 'random variables' freely, completely ignoring measurability questions in the spirit of 'when it is interesting after integrating then it is usually measurable'. Even though sometimes one has to work hard to prove measurability one should, in this case, first be convinced that it is interesting anyway. For good reading concerning measurability and other aspects of random operators see, for example, Kirsch (1989), Kirsch and Metzger (2007), and Pastur (1980).

iii) Originally discussed by Anderson was the case $V_0 = 0$, $\rho(dv) = \frac{1}{2}\chi_{(-1,1)}(x)\,dv$, and no magnetic field, that is, $T = -\Delta$ is the negative adjacency matrix of $\mathbb{Z}^d$.

9.1.2 What is Anderson localization (for mathematicians)?

Under the conditions described above it is known, see, for example, Pastur (1980) or Kirsch (1989), that the spectrum $\sigma(H_\omega)$ is (almost surely) constant. There is a closed set $\Sigma \subset \mathbb{R}$ such that:

$$\sigma(H_\omega) = \Sigma \text{ for } \mathbb{P} \text{ almost all } \omega.$$

Moreover, for the Anderson model on $l^2(\mathbb{Z}^d)$ with an i.i.d. random potential one knows that:

$$\Sigma = \sigma(H_0) + \lambda \operatorname{supp}(\rho) \tag{9.1}$$

where $\operatorname{supp}(\rho)$ is the support of the single-site distribution ρ. But what is the (almost sure) *nature* of the spectrum?

For any self-adjoint operator H in a Hilbert space $\mathcal{H}$ there is a decomposition of the Hilbert space into invariant subspaces:

$$\mathcal{H} = \mathcal{H}_\mathrm{p} \oplus \mathcal{H}_\mathrm{sc} \oplus \mathcal{H}_\mathrm{ac} \tag{9.2}$$

with:

$\mathcal{H}_p$ = subspace corresponding to point spectrum
$\mathcal{H}_{ac}$ = subspace corresponding to absolutely continuous spectrum
$\mathcal{H}_{sc}$ = subspace corresponding to singular continuous spectrum

and a decomposition of the spectrum into not the necessarily distinct set:

$$\sigma(H) = \sigma_p(H) \cup \sigma_{ac}(H) \cup \sigma_{sc}(H). \tag{9.3}$$

Physical intuition tells us that the states in $\mathcal{H}_p$ are bound states, that is, they should stay in compact regions and the states in $\mathcal{H}_{ac}$ correspond to scattering states, corresponding to transport. This is made precise by the RAGE theorem, due to Ruelle, Amrein, Georcescu, and Enss.

Theorem 9.1 *Under some mild physically reasonable conditions on the Hamiltonian H:*

$$\varphi \in \mathcal{H}_p \iff \lim_{R \to \infty} \sup_{t \in \mathbb{R}} \|\chi_{(|x|>R)} e^{-itH} \varphi\| = 0,$$

$$\varphi \in H_c = \mathcal{H}_{ac} \oplus \mathcal{H}_{sc} \iff \lim_{T \to \infty} \frac{1}{2T} \int_{-T}^{T} \|\chi_{(|x| \leq R)} e^{-itH} \varphi\| \, dt = 0.$$

Here $\chi_{(|x| \leq R)}$ is the characteristic function of the ball $\{|x| \leq R\}$ and $\chi_{(|x|>R)}$ the characteristic function the complement of this ball.

Remark 9.2

i) The mild conditions needed in the RAGE theorem are always fulfilled for the Anderson model on $l^2(\mathbb{Z}^d)$, and are generally fulfilled for physically reasonable Schrödinger operators in the continuum $L^2(\mathbb{R}^d)$. Thus the RAGE theorem confirms the intuition that states $\varphi \in \mathcal{H}_p$ corresponds to physically bound states in the sense that up to arbitrary small errors, the time evolved state $e^{-itH}\varphi$ does not leave the compact balls $\{|x| \leq R\}$ uniformly in time.

ii) The RAGE theorem is somewhat vague for the complement of the bound states. It does not distinguish between the absolutely continuous and the singularly continuous subspaces.

iii) For a nice proof of the RAGE theorem see Hunziker and Sigal (2000).

One of the possible definition(s) of Anderson localization is as follows.

Definition 9.1 *A random Schrödinger operator H_ω of the type discussed above in Section 9.1.1 has spectral localization in an energy interval $[a,b]$ if, with probability one, the spectrum of H_ω in this interval is pure point. That is, if:*

$$\sigma(H_\omega) \cap [a,b] \subset \sigma_p(H_\omega) \text{ with probability one.}$$

The random Schrödinger operator H_ω has exponential spectral localization in $[a,b]$ if it has spectral localization in $[a,b]$ and the eigenfunctions corresponding to eigenvalues in $[a,b]$ decay exponentially.

Remark 9.3 *Thus exponential spectral localization holds in the interval $[a,b]$ if for almost all ω the random Hamiltonian H_ω has a complete set of eigenvectors $(\varphi_{\omega,n})_{n\in\mathbb{N}}$ in the energy interval $[a,b]$ obeying:*

$$|\varphi_{\omega,n}(x)| \leq C_{\omega,n} e^{-\mu|x-x_{n,\omega}|} \tag{9.4}$$

with $\mu > 0$, some finite $C_{\omega,n}$ and $x_{n,\omega}$ the centres of localization.

The physical mechanism for localization is the suppression of tunnelling over large distances due to the de-coherence effect induced by the random potential (as opposed to, say a periodic potential). Spectral localization was for quite some time the only definition used by mathematicians. From a physical point of view, one might prefer to say that Anderson localization holds if there is no transport. But what exactly does transport mean? A possible way to express this is as follows: take any initial condition φ_0 with compact support and consider:

$$\begin{aligned}\langle x^2 \rangle_{P_{H_\omega \in [a,b]}\varphi_0}(t) &:= \langle e^{-itH_\omega} P_{H_\omega \in [a,b]} \varphi_0, x^2 e^{-itH_\omega} P_{H_\omega \in [a,b]} \varphi_0 \rangle \\ &= |||x|e^{-itH_\omega} P_{H_\omega \in [a,b]} \varphi_0||^2\end{aligned} \tag{9.5}$$

here $P_{H_\omega \in [a,b]}$ is the orthogonal projection onto the spectral subspace of H_ω corresponding to energies in $[a,b]$. That is, we consider only the portion of φ_0 with energy in $[a,b]$. If the electrons with energies in $[a,b]$ move ballistically with average velocity v_{av} then roughly $x(t) \sim v_{\text{av}}t$ for large times. Hence $\langle x^2 \rangle_{P_{H_\omega \in [a,b]}\varphi_0}(t)$ will be proportional to t^2 for large t. This can certainly be interpreted as a signature for transport.

On the other hand, if the electrons in the energy interval $[a,b]$ are *localized* then it should be natural to expect that $\langle x^2 \rangle_{P_{H_\omega \in [a,b]}\varphi_0}(t)$ is bounded uniformly in t. So as long as the spectrum in $[a,b]$ is pure point or at least if exponential spectral localization holds in $[a,b]$ then, with probability one, (9.5) should be uniformly bounded in t for all suitably localized initial conditions φ_0. But this is *wrong*. It is known, see Simon (1990), that pure point spectrum implies absence of ballistic motion:

$$\lim_{t\to\infty} \frac{\langle x^2 \rangle_{P_{H_\omega \in [a,b]}\varphi_0}(t)}{t^2} = 0$$

for all compactly supported initial conditions φ_0 as soon as the spectrum in $[a,b]$ is pure point. However, del Rio, Jitomirskaya, Last, and Simon (1996) constructed examples of (nonrandom) one dimensional Schrödinger operators with pure point spectrum for all energies and exponentially localized eigenfunction for which:

$$\limsup_{t\to\pm\infty} \frac{\langle x^2 \rangle_{\varphi_0}(t)}{t^{2-\delta}} = \infty \quad \text{for all } \delta > 0, \tag{9.6}$$

for a large class of compactly supported initial conditions φ_0. That is, the mean square distance $\langle x^2 \rangle_{\varphi_0}(t)$ can grow *arbitrarily close* to the ballistic motion *even though* the operator has exponential spectral localization. Thus exponential spectral localization in the sense of Definition 9.1 is a priori not strong enough to restrict the long time dynamics of the system besides what is given by the RAGE theorem. The main failing has to do with the complete freedom of the constants $C_{\omega,n}$ in (9.4). When thinking about localization one usually thinks of all the eigenvectors being confined with some typical length scale. If the $C_{\omega,n}$ are allowed to arbitrarily grow, in n, then, in fact, the eigenvectors can be extended over arbitrarily large length scales, possibly leading to transport arbitrarily close to the ballistic motion, even though one has only pure point spectrum. This is nicely discussed in del Rio et al. (1995) with the proofs given in del Rio et al. (1996).

A physically more natural definition for localization than Definition 9.1, one which takes the dynamical properties of the physical system into account, is:

Definition 9.2 *A random Schrödinger operator has* strong dynamical localization *in an energy interval $[a,b]$ if for any $q > 0$*

$$\mathbb{E}\big[\sup_t \||X|^q e^{-itH_\omega} P_{H_\omega \in [a,b]}\varphi\|^2\big] < \infty$$

for all φ with compact support.

So strong dynamical localization holds if for any localized initial condition φ the part of φ with energy in $[a,b]$ (i.e., in the range of the spectral projection $P_{H_\omega \in [a,b]}$) has uniformly in time bounded moments of all orders. In particular, localized initial conditions stay in compact regions for all times up to arbitrary small errors. Thus, by the RAGE Theorem 9.1 strong dynamical localization in $[a,b]$ implies spectral localization in $[a,b]$, but not vice versa, as the examples in del Rio, Jitomirskaya, Last, and Simon (1996) show.

9.1.3 *Known (rigorous) results*

The following is a heavily personally biased (and fairly incomplete) list of rigorous results for Anderson localization; for example, we will completely disregard all results for continuum random operators.

1. Kunz and Souillard (1980): $d = 1$ and nice ρ: always localization, no matter how small $\lambda > 0$ is. (Extended by Carmona et al. (1987) to general ρ.)
2. Fröhlich and Spencer (1983): multi-scale analysis gives:

$$\langle x|(H_\omega - E)^{-1}|y\rangle \leq A_{\omega,x} e^{-\mu|x-y|}$$

for fixed energy $E \in \sigma(H_\omega)$ and almost all ω. This implies vanishing of the conductivity using the Kubo formula and absence of an absolutely continuous spectrum, see Fröhlich and Spencer (1983) for the former and Martinelli and Scoppola (1985) for the latter.

3. Simon and Wolff (1986): The Fröhlich–Spencer result for fixed $E \in (a,b)$ (and nice ρ) already implies exponential localization in the sense of Definition 9.1:

$$\sigma(H_\omega) \cap (a,b) \subset \sigma_p(H_\omega)$$

and the corresponding eigenfunctions decay exponentially.

4. New approach by Aizenman and Molchanov (1993): instead of proving pointwise bounds for almost all ω, try to prove bounds for *averages*:

$$\tau(x,y;z) := \mathbb{E}[|\langle x|(H_\omega - z)^{-1}|y\rangle|^s]. \tag{9.7}$$

More precisely, Aizenman and Molchanov showed that these *fractional moments* are exponentially small:

$$\tau(x,y;E+i\varepsilon) \leq A e^{-\mu|x-y|} \tag{9.8}$$

for $E \in (a,b)$, uniformly in $\varepsilon \neq 0$ and a suitable (fixed) $0 < s < 1$, in the case of large disorder or extreme energies.

Kunz and Souillard proved, in fact, what we now call strong dynamical localization by showing that $\mathbb{E}[\sup_t |\langle x|e^{itH_\omega}|y\rangle|]$ decays exponentially in the distance $|x-y|$. Their proof is similar to the proof that correlations in the one-dimensional nearest neighbour Ising model decay exponentially at all positive temperatures, see Cycon et al. (1987). One of the central themes of the approach to Anderson localization initiated by Aizenman and Molchanov is a shift in focus: instead of trying to prove almost sure (in the random potential) decay estimates for the off-diagonal Green's function $G_\omega(x,y;E) = \langle x|(H_\omega - z)^{-1}|y\rangle$ for some energy E *within* the spectrum of the Hamiltonian, one should instead study the *correlation* $\tau(x,y;z)$ given by (9.7) for $z = E + i\varepsilon$ (uniformly for small $\varepsilon > 0$). As we will see shortly, the reason to look at *fractional moments* is purely technical: to guarantee that τ is well-defined. But first we discuss some consequences of (9.8). It turns out that the Aizenman–Molchanov criterion is a *very useful* signature for localization. It implies localization in all different manifestations:

Some consequences of the localization criterion (9.8):

1. *Spectral localization:* H_ω has in (a,b) only pure point spectrum with exponentially decaying eigenfunctions.
2. *Strong dynamical localization:* wave-packages corresponding to energies in (a,b) are trapped in finite regions for all times:

$$\mathbb{E}[\sup_t |\langle x|e^{-itH_\omega} P_{H_\omega \in (a,b)}|y\rangle|] \leq \tilde{A} e^{-\tilde{\mu}|x-y|}. \tag{9.9}$$

3. *No level repulsion:* Minami (1996) showed that (9.8) implies that the local fluctuation of the energy levels of a multidimensional Anderson model in the energy range (a,b) have a Poisson statistic. The first results for $d=1$ of this type are due to Molchanov (1981).

4. *Exponential decay of the Fermi projection kernel:*

$$\mathbb{E}[|\langle x|P_{H_\omega \leq E}|y\rangle|] \leq \hat{A} e^{-\hat{\mu}|x-y|},$$

for some finite $\hat{A}$ and $\hat{\mu} > 0$, see Aizenman and Graf (1998).

Remark 9.4

i) As mentioned before, the RAGE theorem guarantees that strong dynamical localization implies spectral localization. Strong dynamical localization itself follows from the Aizenman–Molchanov criterion (9.8) and the following bound on the evolution of states in the random system: if the single site distribution has a density with respect to Lebesgue measure, $\rho(dv) = f(v)dv$, with $f \in L^p$ for some $p > 1$ then:

$$\mathbb{E}[\sup_{\|g\|_\infty \leq 1} |\langle x|g(H_\omega)P_{H_\omega \in [a,b]}|y\rangle|] \leq C \liminf_{\varepsilon \to 0} \int_a^b \mathbb{E}[|G(x,y;E+i\varepsilon)|^s]\, dE \quad (9.10)$$

for all $0 < s < (p-1)/p$. See, for example, Aizenman et al. (2001) or Hundertmark (2000) for a proof of (9.10). As soon as the Aizenman–Molchanov criterion (9.8) for localization is fulfilled, the bound (9.10) clearly implies (9.9) hence strong dynamical localization according to Definition 9.2 holds in this regime.

ii) The exponential decay of the kernel of the Fermi projection plays a central rôle in understanding the plateaus in the quantized Hall effect. As soon as for some $q > 2$ one has:

$$\xi_q = \sum_{x \in \mathbb{Z}^2} \mathbb{E}[|\langle 0|P_{H_\omega \leq E}|x\rangle|^q]^{1/q} |x| \leq C < \infty$$

for all $E \in (a,b)$, the Hall conductivity on the interval (a,b) is constant. See Aizenman and Graf (1998) and Bellissard, van Elst, and Schulz-Baldes (1994).

9.1.4 Disclaimer

Finally let us mention what we are *not discussing* in this note. Both the physics and mathematics literature on the Anderson model is huge, leaving us no chance to discuss it fully. Instead we hope that this note wets the appetite of the reader. With this in mind we want to give some, still incomplete, hints to the literature. Since we will stick to the configuration space $\mathbb{Z}^d$, there will be no mention of the very nice extension of Aizenman–Molchanov technique to certain random operators on the continuum $\mathbb{R}^d$ given in Aizenman et al. (2006). In addition, we will not discuss at all the so-called multiple-scale approach initiated by Fröhlich and Spencer (1983) nor its often very powerful extensions, see for example Germinet and Klein (2004) and the references therein. Even worse, we will not discuss the beautiful recent results for Anderson localization with Bernoulli random potentials by Bourgain and Kenig (2005) and for Poisson random potentials by Germinet et al. (2005). There are many articles on the multiscale analysis approach, for a readable book see Stollmann (2001), which, however, does not contain any of the new developments in the multiscale approach. We will also not discuss at all the results on delocalization in the Anderson model on trees instead of $\mathbb{Z}^d$, see Aizenman et al. (2006), Froese et al. (2007), and Klein (1998). For a nice and readable introduction to the physics of random Schrödinger operators, see, for example, Lifshits et al. (1988).

9.2 Why fractional moments?

The reason to consider *fractional moments* of the modulus of the Green's function is to guarantee that the expectation in (9.7) is finite. Since:

$$\tau(x, y; z) = \mathbb{E}[|G(x, y; z)|^s].$$

and G is the resolvent of a self-adjoint operator, one always has the bound $\tau(x, y; z) \leq 1/|\Im(z)|$. But can one guarantee that τ is finite for $z = E + i\varepsilon$ uniformly in $\varepsilon \neq 0$? It is here that the fractional moment plus one regularity condition on the single-site distribution ρ enters. To see this in the simplest possible case, we will explicitly show that $\tau(x, x; z)$ is finite.

Write $H = \widetilde{H} + \lambda v_x |x\rangle\langle x|$. Here $|x\rangle\langle x|$ is the Dirac notation for the rank one projection operator onto the one-dimensional subspace generated by $|x\rangle = \delta_x$. Thus $\widetilde{H}$ is the Hamiltonian H with $v_x = 0$. The resolvent formula:

$$A^{-1} - B^{-1} = A^{-1}(B-A)B^{-1} = B^{-1}(B-A)A^{-1}$$

gives:

$$(H-z)^{-1} = (\widetilde{H}-z)^{-1} - (H-z)^{-1}\lambda v_x |x\rangle\langle x|(\widetilde{H}-z)^{-1} \qquad (9.11)$$

With the resolvent $G(x, y; z) := \langle x|(H-z)^{-1}|y\rangle$ one has:

$$G(x, x; z) = \widetilde{G}(x, x; z) - \lambda v_x G(x, x; z)\widetilde{G}(x, x; z). \qquad (9.12)$$

which is a simple algebraic equation for the complex number $G(x, x; z)$. Solving for $G(x, x; z)$ gives:

$$G(x, x; z) = \frac{\widetilde{G}(x, x; z)}{1 + \lambda v_x \widetilde{G}(x, x; z)} = \frac{1}{\beta + \lambda v_x} \qquad (9.13)$$

where $\beta = (\widetilde{G}(x, x; z))^{-1}$ depends on everything but NOT on the potential at site x! In the physics literature β is often called the self-energy. The importance of formula (9.13) is that although the Green's function is usually a *very* complicated function of the potential V, its diagonal element $G(x, x; z)$ is a very simple fractional function of v_x.

Now assume that the single site distribution ρ is Hölder continuous of order $\alpha \le 1$, that is:

$$\sup_{E \in \mathbb{R}} \rho([E - \varepsilon, E + \varepsilon]) \le C_H \varepsilon^\alpha \tag{9.14}$$

for all $\varepsilon > 0$. Under this condition $\sup_z \tau(x, y; z)$ is finite for all $0 < s < \alpha$. This is easiest to see for $\tau(x, x; z)$: for all $0 < s < \alpha$

$$\sup_{z \in \mathbb{C}, x \in \mathbb{Z}^d} \tau(x, x; z) \le C_{s,\alpha} < \infty \tag{9.15}$$

with $C_{s,\alpha} \le \frac{1}{1-s/\alpha} C_H^{s/\alpha} < \infty$. The proof uses the self-energy formula (9.13) heavily. One has:

$$\mathbb{E}[|G(x, x; z)|^s | v_x] = \mathbb{E}\left[\left|\frac{1}{\beta + \lambda v_x}\right|^s \bigg| v_x\right] = \lambda^{-s} \int |\tilde\beta + v|^{-s} d\rho(v),$$

with $\tilde\beta = \beta/\lambda$, for the conditional expectation of $|G(x, x; z)|^s$ with respect to the random potential at x. Notice that one always has:

$$\int |\tilde\beta + v|^{-s} d\rho(v) = \int\int_0^{|\tilde\beta + v|^{-s}} 1 \, dt \, d\rho(v) = \int_0^\infty \int \chi_{(|\tilde\beta+v|^{-s} > t)} \, d\rho(v) dt$$
$$= \int_0^\infty \rho(|\tilde\beta + v|^{-s} > t) \, dt = \int_0^\infty \rho(|\tilde\beta + v| < t^{-1/s}) \, dt.$$

The assumption that ρ is an α-Hölder continuous probability measure and $0 < s < \alpha$ yields the bound:

$$\int_0^\infty \rho(|\tilde\beta + v| < t^{-1/s}) \, dt \le \int_0^\infty \max(1, C_H t^{-\alpha/s}) \, dt \le \int_0^r 1 dt + C_H \int_r^\infty t^{-\alpha/s} \, dt$$
$$= r + C_H(-1 + \alpha/s)^{-1} r^{1-\alpha/s} = \frac{1}{1 - s/\alpha} C_H^{s/\alpha}$$

for the optimal choice of $r = C_H^{s/\alpha}$. Note that the right hand side here does not depend on $\tilde\beta$ any more. In particular, doing first a conditional expectation with respect to the random variable v_x and then integrating over all random variables, one gets:

$$\tau(x, x; z) = \mathbb{E}\Big[\mathbb{E}[|G(x, x; z)|^s | v_x]\Big] \le \lambda^{-s} \mathbb{E}\Big[\sup_{\tilde\beta \in \mathbb{C}} \int |\tilde\beta + v|^{-s} d\rho(v)\Big] \tag{9.16}$$
$$\le (1 - s/\alpha)^{-1} C_H^{s/\alpha} \lambda^{-s}$$

which is (9.15).

Remark 9.5

i) In the model originally studied by Anderson, ρ is given by $\rho(dv) = \frac{1}{2}\chi_{[-1,1]}(v)\,dv$. In this case $C_{s,1} = C_s = (1-s)^{-1}$.

ii) The key to showing that $\tau(x,x;z)$ is bounded uniformly in $z \in \mathbb{C}$ and $x \in \mathbb{Z}^d$ was the representation (9.13). To see that also $\tau(x,y;z)$ is bounded uniformly in $z \in \mathbb{C}$ and $x,y \in \mathbb{Z}^d$ one can argue similarly using a rank-two perturbation argument. Write:

$$H = \widetilde{H} + \lambda v_x |x\rangle\langle x| + \lambda v_y |y\rangle\langle y|,$$

that is, $\widetilde{H}$ is now the H with the potential at sites x and y put to zero. The resolvent equation yields:

$$G = (H-z)^{-1} = \widetilde{G} - G\bigl(\lambda v_x |x\rangle\langle x| + \lambda v_y |y\rangle\langle y|\bigr)\widetilde{G} \tag{9.17}$$

where we put $\widetilde{G} = (\widetilde{H}-z)^{-1}$ and suppressed the dependence on z. Equation (9.17) is an equation for G on the whole Hilbert-space $l^2(\mathbb{Z}^d)$, but one can restrict it to the two-dimensional subspace spanned by the two vectors $|x\rangle$ and $|y\rangle$: defining the 2×2 matrices:

$$B = \begin{pmatrix} G(x,x) & G(x,y) \\ G(y,x) & G(y,y) \end{pmatrix} \text{ and } \widetilde{B} = \begin{pmatrix} \widetilde{G}(x,x) & \widetilde{G}(x,y) \\ \widetilde{G}(y,x) & \widetilde{G}(y,y) \end{pmatrix}$$

equation (9.17) gives:

$$B = \widetilde{B} - B \begin{pmatrix} \lambda v_x & 0 \\ 0 & \lambda v_y \end{pmatrix} \widetilde{B}. \tag{9.18}$$

Hence:

$$B\left(\mathbf{1}_{2\times 2} + \lambda \begin{pmatrix} v_x & 0 \\ 0 & v_y \end{pmatrix} \widetilde{B}\right) = \widetilde{B}$$

which can be solved as:

$$B = \widetilde{B}\left(\mathbf{1}_{2\times 2} + \lambda \begin{pmatrix} v_x & 0 \\ 0 & v_y \end{pmatrix} \widetilde{B}\right)^{-1} = \left(\Theta + \lambda \begin{pmatrix} v_x & 0 \\ 0 & v_y \end{pmatrix}\right)^{-1} \tag{9.19}$$

with $\Theta = \widetilde{B}^{-1}$. Note that here all matrix inverses are for 2×2 matrices. As in the rank-one case, the 2×2 matrix Θ depends on everything BUT the potential v_x and v_y and the off-diagonal Green's function $G(x,y;z)$ is seen to be:

$$G(x,y;z) = \left(\Theta + \lambda \begin{pmatrix} v_x & 0 \\ 0 & v_y \end{pmatrix}\right)^{-1}_{1,2} \tag{9.20}$$

that is, $G(x,y;z)$ can be identified as the off-diagonal matrix element of the 2×2 matrix $\bigl(\Theta + \lambda \begin{pmatrix} v_x & 0 \\ 0 & v_y \end{pmatrix}\bigr)^{-1}$. A similar, and not much more complicated, argument to the one leading from (9.13) to (9.16) now shows that $\sup_{x,y \in \mathbb{Z}^d, z \in \mathbb{C}\backslash\mathbb{R}} \tau(x,y;z)$ is finite as long as the single-site distribution of the random potential is Hölder continuous, for details, see Aizenman and Molchanov (1993) or Aizenman et al. (2001).

9.3 Finite-volume criteria

A main extension of the original Aizenman–Molchanov approach to localization was that one can develop *finite-volume* criteria for the exponential decay of the correlations τ, see Aizenman et al. (2001). These bounds are related to similar bounds in statistical mechanics. One restricts the random operator $H = -\Delta + V_\omega$ to a finite box $\Lambda = \{x \in \mathbb{Z}^d : |x_j| < L \text{ all } j = 1,\ldots,d\}$, and puts $G_\Lambda(z) = (H_\Lambda - z)^{-1}$ considered as an operator (or infinite matrix) in $l^2(\Lambda)$. Somewhat loosely speaking, the finite volume criteria say that as soon as:

$$\sup_{v \in \partial \Lambda} \mathbb{E}[|G_\Lambda(0, v; z)|^s] \text{ is small enough} \qquad (9.21)$$

for all $z = E + i\varepsilon$, $E \in [a, b]$, $\varepsilon > 0$, then there exist constants $A < \infty, \mu > 0$ such that:

$$\sup_{\varepsilon > 0} \tau(x, y; E + i\varepsilon) = \sup_{\varepsilon > 0} \mathbb{E}[|G(x, y; E + i\varepsilon)|^s] \leq A e^{-\mu |x - y|} \qquad (9.22)$$

for all $E \in [a, b]$. Thus in this case the correlation $\tau(x, y; E + i\varepsilon)$ decays exponentially on all of $\mathbb{Z}^d$ and the Aizenman–Molchanov criterion for localization is fulfilled, yielding strong dynamical localization etc., as discussed in the introduction. Of course, the question is what does 'small enough' in (9.21) mean precisely and, when made precise, whether this criterion is fulfilled in interesting physical situations. We are deliberately vague at this point. Let us only mention that (9.21) (in its precise formulation) is a physically very natural condition which is often quite easily seen to be true in the relevant cases. See Aizenman et al. (2001) for a more precise formulation of such a finite volume criterium for Anderson localization, and Aizenman et al. (2000) for a discussion of the physical implications of these criteria in the case of Anderson localization.

Here we would like to take the opportunity to discuss finite volume criteria for the decay of correlations in their original setting, namely statistical mechanics. The type of argument we are going to give is known as Simon–Lieb inequalities, see Simon (1980) and Lieb (1980). In fact, we will discuss the simplest possible case, namely *finite volume criteria* for percolation in the spirit of Aizenman and Newman (1984).

9.3.1 Analogy with percolation

Consider independent site percolation on $\mathbb{Z}^d$. A site $x \in \mathbb{Z}^d$ is occupied with probability p and empty with probability $1 - p$. We usually visualize this by drawing occupied sites as small solid black circles and empty sites as white circles. Recall that two sites x, y in $\mathbb{Z}^d$ are nearest neighbours if $|x - y| = \sum_{j=1}^{d} |x_j - y_j| = 1$.

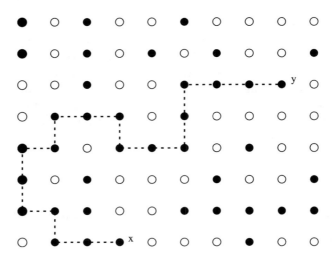

FIG. 9.1: Two points x and y connected by a path via nearest-neighbour occupied sites. The percolation with $p = 1/2$ was simulated by tossing a coin 80 times.

Definition 9.3 *Two points $x, y \in \mathbb{Z}^d$ are connected, in short $x \leftrightsquigarrow y$, if both x and y are occupied sites and one can hop from x to y via a sequence of occupied nearest neighbour sites.*

The connectivity τ is given by the probability that x and y are connected:

$$\tau(x, y) := \mathbb{P}\{x \leftrightsquigarrow y\}.$$

The situation in Definition 9.1 is sketched in Fig. 9.1. Note that, with a slight abuse of notation, we denote the connectivity with the same symbol as the correlation in the Anderson model. The point being that one can rather easily deduce finite-volume criteria for the connectivity in percolation: Let $\Lambda = \Lambda_L$ be the centred cube:

$$\Lambda_L = \{x \in \mathbb{Z}^d : |x_j| < L \text{ for all } j = 1, \ldots, d\};$$

we will always choose $L \notin \mathbb{N}$. Note that Λ has an inner and an outer boundary:

$$\partial_- \Lambda = \{x \in \Lambda : \exists v \notin \Lambda, |x - v| = 1\},$$
$$\partial_+ \Lambda = \{v \notin \Lambda : \exists x \in \Lambda, |x - v| = 1\}.$$

For a subset $B \subset \mathbb{Z}^d$ we denote by $|B|$ the 'volume' of B, that is, the number of elements in B and with $\tau(x, B) = \mathbb{P}(x \leftrightsquigarrow B)$ the probability that x is connected to some point in B. Of course, one can restrict percolation to, possibly finite, subsets A of $\mathbb{Z}^d$. In this case we denote by $\tau_A(x, y) = \mathbb{P}(x \stackrel{\text{in } A}{\leftrightsquigarrow} y)$ the probability that x and y are connected by a path in A and similarly for $\tau_A(x, B)$.

The finite volume criterion for percolation with probably the simplest proof is as follows.

Theorem 9.2 *Let Λ be a centred box in $\mathbb{Z}^d$ and $b = b_\Lambda := |\partial_+ \Lambda| \tau_\Lambda(0, \partial_- \Lambda)$. Then:*

$$\tau(x, y) \leq b^{-1} b^{|x-y|} \tag{9.23}$$

for all $x, y \in \mathbb{Z}^d$.

Note that this criterion predicts exponential decay of the connectivity as soon as $b_\Lambda < 1$. That is, strong enough finite-volume decay implies exponential decay of the connectivity τ in the infinite volume $\mathbb{Z}^d$. Since the expression b_λ is computed in a *finite volume* Λ, one can, in principle, give it to the friendly neighbourhood computational physicist in order to check on a computer if $b_\Lambda < 1$ for some maybe large box Λ.

Moreover, the criterion given in Theorem 9.2 is not only sufficient for exponential decay of τ but also *necessary*. Indeed, the volume of the boundary, $|\partial_+ \lambda|$, grows at a polynomial rate in the side-length of Λ and since the origin is in the interior of the box Λ, it is connected to $\partial_- \Lambda$ within the box Λ, if and only if they are connected, $\tau_\Lambda(0, \partial_- \Lambda) = \tau(0, \partial_- \Lambda)$. Thus, as soon as τ on $\mathbb{Z}^d$ decays with some exponential rate, one will have $b_\Lambda < 1$ for all large enough boxes Λ.

The proof of Theorem 9.2 is surprisingly simple and was the main driving force in the search of the finite volume criteria in Aizenman et al. (2001).

Proof: Let $\Lambda = \Lambda_L$ be a centred cube of side length L and $\Lambda(x) = \Lambda + x$. Assume that x and y are so far from each other that $y \notin \Lambda(x)$ and that $x \leftrightsquigarrow y$. This situation is sketched in Fig. 9.2. Since $x \leftrightsquigarrow y$, there must be a path within

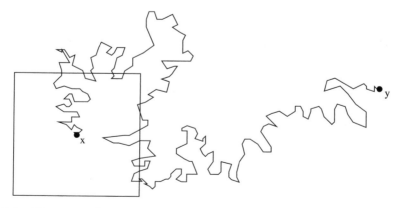

FIG. 9.2: The event that x and y are connected. Note that the path has to cross the box $\Lambda(x)$ at least once.

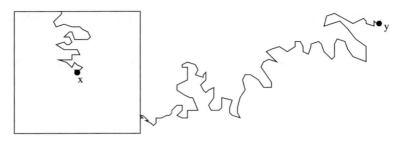

FIG. 9.3: The upper bound in equation (9.24): x is connected to the boundary of $\Lambda(x)$ within $\Lambda(x)$ and y is connected to the boundary of $\Lambda(x)$ within $\Lambda(x)^c$.

$\Lambda(x)$ from x to the (inner) boundary $\partial_-\Lambda(x)$ and a path within the complement of $\Lambda(x)$ from the (outer) boundary $\Lambda_+(x)$ to y, see Fig. 9.3.

Thus, as long as $y \notin \Lambda(x)$:

$$\tau(x,y) = \mathbb{P}(x \leftrightsquigarrow y)$$
$$\leq \mathbb{P}(x \stackrel{\text{in } \Lambda(x)}{\leftrightsquigarrow} \partial_-\Lambda(x), \partial_-\Lambda(x) \leftrightsquigarrow \partial_+\Lambda(x), \text{ and } \partial_+\Lambda(x) \stackrel{\text{in } \Lambda(x)^c}{\leftrightsquigarrow} y)$$
$$\leq \mathbb{P}(x \stackrel{\text{in } \Lambda(x)}{\leftrightsquigarrow} \partial_-\Lambda(x) \text{ and } \partial_+\Lambda(x) \stackrel{\text{in } \Lambda(x)^c}{\leftrightsquigarrow} y) \qquad (9.24)$$

by dropping the restriction that the inner boundary has to be connected to the outer boundary by some path of connected sites in $\mathbb{Z}^d$, see Fig. 9.3.

Since the events 'x is connected to $\partial_-\Lambda(x)$ within $\Lambda(x)$' and '$\partial_+\Lambda(x)$ is connected to y within the complement of $\Lambda(x)$' are *independent*, the probability in (9.24) factorizes into a product and we arrive at:

$$\tau(x,y) \leq \mathbb{P}(x \stackrel{\text{in } \Lambda(x)}{\leftrightsquigarrow} \partial_-\Lambda(x)) \mathbb{P}(\partial_+\Lambda(x) \stackrel{\text{in } \Lambda(x)^c}{\leftrightsquigarrow} y)$$
$$\leq \tau_{\Lambda(x)}(x, \partial_-\Lambda(x)) \tau_{\Lambda(x)^c}(\partial_+\Lambda(x), y)$$

By translation invariance, $\tau_{\Lambda(x)}(x, \partial_-\Lambda(x)) = \tau_\Lambda(0, \partial_-\Lambda)$. Moreover we have the simple monotonicity:

$$\tau_{\Lambda(x)^c}(\partial_+\Lambda(x), y) = \mathbb{P}(\partial_+\Lambda(x) \stackrel{\text{in } \Lambda(x)^c}{\leftrightsquigarrow} y) \leq \mathbb{P}(\partial_+\Lambda(x) \leftrightsquigarrow y)$$
$$= \tau(\partial_+\Lambda(x), y) \leq \sum_{v \in \partial_+\Lambda(x)} \tau(v, y).$$

Thus for all $y \notin \Lambda(x)$, τ obeys the bound:

$$\tau(x,y) \leq \frac{b_\Lambda}{|\partial_+\Lambda(x)|} \sum_{v \in \partial_+\Lambda(x)} \tau(v, y), \qquad (9.25)$$

which, if $b_\Lambda < 1$, says for fixed y, $\tau(x,y)$ is a *subharmonic* function of $x \notin \Lambda(y)$.

There are many ways to see that subharmonic functions have a tendency to decay exponentially. In the case at hand, possibly the easiest way is to iterate (9.25), which can be done at least $|x-y|-1$ times, and then use the a priori bound $\tau(x,y) \leq 1$. This gives:

$$\tau(x,y) \leq b_\Lambda^{|x-y|-1}$$

which is (9.23). □

Remark 9.6

i) One should notice that the above proof did not need the underlying lattice to be given by Z^d. In fact, it does not need to be a lattice at all; the proof works for percolation on arbitrary graphs $\mathcal{G}$. A similar argument as given after Theorem 9.2 then shows that the condition $b_\Lambda < 1$ is not only sufficient for the exponential decay of τ but also necessary as long as the growth of the surface volume $|\partial_+ \Lambda_L|$ of large boxes Λ_L in the graph $\mathcal{G}$ is sub-exponential.

ii) This type of idea seems to go back at least to Hammersley (1957) in the case of percolation.

iii) Using the van den Berg–Kesten inequalities for percolation one can improve on Theorem 9.2, see Aizenman and Newman (1984).

9.3.2 Some consequences from finite volume criteria

As already mentioned in Remark 9.6, the finite-volume criterion in Theorem 9.2 is a sufficient and, for a large class of graphs, also necessary condition for the exponential decay of the connectivity. For percolation on graphs it is known that there is a critical probability $0 < p_c < 1$ such that for $p < p_c$ $\tau(x,y)$ decays exponentially in the distance $|x-y|$ and for $p > p_c$ it does not. This is also related to the occurrence of an infinite connected cluster above p_c with probability one, Grimmett (1999).

The finite-volume criteria turn out to be a useful tool there. For example, they yield an algorithm to compute p_c: for $p > p_c$:

$$\liminf_{L \to \infty} b_{\Lambda_L} \geq 1$$

while for $p < p_c$ there exists at least one box Λ with $b_\Lambda < 1$. In particular, this yields lower bounds on p_c for graphs for which the precise value is not known.

The finite volume criteria can also be used to give painless proofs of the following, not necessarily obvious, facts,

1. Exponential decay of the connectivity is stable under small perturbations of parameters (for example, variation of p or slight deformations of the underlying graph) for all graphs for which the volume of boxes λ_L grows sub-exponentially in L.

2. Fast power law decay ⇒ exponential decay. (For graphs in which the surface volume growth of boxes is polynomially bounded.)
3. At critical percolation, the connectivity cannot decay too fast.

Indeed, that the exponential decay of the connectivity is stable under small perturbations of the parameters is not at all clear since one might be at a phase-transition point. That this is not the case is due to the finite-volume criteria. To show 1, assume that τ decays exponentially. Then, for some finite box Λ one must have $b_\Lambda < 1$. Since b_Λ is computed in a *finite volume*, it depends continuously on the parameters, hence wiggling them a little bit will still result in $b_\Lambda < 1$, hence the connectivity will still decay exponentially, by Theorem 9.2.

To show 2 one argues similarly, if τ decays so fast that it beats the growth of the surface volume $|\partial_+ \Lambda|$, then b_Λ will be less than 1 for all large enough boxes Λ and hence τ must decay exponentially by Theorem 9.2.

Finally, for 3, note that τ does not decay exponentially for all $p > p_c$. Hence for $p = p_c$ we must have:

$$1 \leq b_\Lambda = |\partial_+ \Lambda| \tau(0, \partial_- \Lambda)$$

for all centred boxes Λ. Otherwise, by the first fact, one would have exponential decay of the connectivity for all p slightly above p_c, which contradicts the definition of the critical probability. Thus:

$$\tau(0, \partial_- \Lambda) \geq \frac{1}{|\partial_+ \Lambda|}$$

for all centred boxes Λ.

In a similar fashion, the finite volume criteria for Anderson localization give rise to the stability results for the exponential decay of the fractional moments of the Green's function analogously to the stability results (9.1)–(9.3) for percolation above. In particular, the exponential decay of the fractional moments is stable under small perturbation of external fields, like an external (periodic) potential or an external magnetic field. For more discussions of this, see Aizenman et al. (2000).

9.4 Localization for large disorder: a simple proof

Our discussion of the finite volume criteria for Anderson localization has been, deliberately, somewhat vague. In contrast to this we would like to give a full and, we think, rather simple proof of Anderson localization which yields, in addition, very easily checkable assumptions with explicit bounds.

Theorem 9.3 *Consider the random operator $H = H_\omega = -\Delta + V_\omega$ on $l^2(\mathbb{Z}^d)$. Let the single side distribution ρ of the random potential at site 0, say, be such that:*

$$C_s = \sup_{\beta \in \mathbb{C}} \int |\beta - v|^{-s} d\rho(v) < \infty.$$

for some $0 < s < 1$. Then for all $\lambda^s > (2d-1)C_s$, the exponential bound:

$$\sup_{z \in \mathbb{C} \setminus \mathbb{R}} \mathbb{E}[|G(x,y;z)|^s] \leq A_{d,\lambda} e^{-\mu(d,\lambda)|x-y|} \tag{9.26}$$

holds. Here:

$$A_{d,\lambda} = \frac{2d(2d-1)C_s \lambda^{-s}}{(2d-1)^2[1-(2d-1)C_s \lambda^{-s}]} \tag{9.27}$$

and:

$$\mu(d,\lambda) = -\ln\left(\lambda^s/((2d-1)C_s)\right) > 0. \tag{9.28}$$

Remark 9.7

i) As the proof will show, the conclusion of the Theorem remains valid even for highly correlated random potentials $V = (v_j)_{j \in \mathbb{Z}^d}$ as long as a suitable bound of the form:

$$C_s = \sup_{\beta \in \mathbb{C}, j \in \mathbb{Z}^d} \mathbb{E}[|\beta + v_j|^{-s} | v_j] < \infty \tag{9.29}$$

for the fractional moments of the conditional expectations of the potential at site j holds.

ii) For the original Anderson model, $\rho(dv) = \frac{1}{2}\chi_{[-1,1]}(v)\,dv$. In this case $C_s = 1/(1-s)$ and one has localization at all energies as soon as:

$$\lambda > (1-s)^{-1/s}(2d-1)^{1/s}$$

for some $0 < s < 1$.

9.4.1 The self-avoiding random walk representation

The observation which leads to a simple and straightforward proof of Anderson localization for large disorder is the following self-avoiding walk (SAW) representation for the off-diagonal Green's function. That such a representation holds is not necessarily new, but that it holds for all complex energies off the real axis seems to be.

Lemma 9.1 (The self-avoiding walk representation) *Let $B \subset \mathbb{Z}^d$ be finite, $G_B(z) = (H_B - z)^{-1}$ and $G_B(x,y;z) = \langle x|G_B(z)|y\rangle$. Then:*

$$G_B(x,y;z) = \sum_{\substack{w:\text{SAW in } B \\ x \leftrightsquigarrow y}} \prod_{j=0}^{|w|} G_{B_j}(w(j),w(j);z) \tag{9.30}$$

for all $z \in \mathbb{C} \setminus \mathbb{R}$. Here w is a self-avoiding random walk connecting $x = w(0)$ and y in B, $|w|$ is the length of the walk, and the sets $B_j = B_j(w)$ are recursively defined by $B_0 = B$ and $B_{j+1} = B_j \setminus \{w(j)\}$.

Remark 9.8 *Given a self-avoiding path w in B the sets B_j are given by $B_j = B \setminus \{x, w(1), w(2), \ldots w(j-1)\}$ for $j = 1, \ldots |w|$. Thus they are a nested shrinking sequence of subsets of B depending on the self-avoiding walk only up to time-step $j - 1$. In particular, given a self-avoiding path w, the resolvent G_{B_j} does not depend any more on the potential at the previously visited places $x, w(1), \ldots, w(j-1)$. This makes the representation (9.30) very powerful.*

It is crucial for the application we have in mind that the representation (9.30) in terms of a self-avoiding random walk is valid for all $z \in \mathbb{C} \setminus \mathbb{R}$ and not only for complex z with a large enough imaginary part. Nevertheless, we will deduce Lemma 9.1 from a perturbative result which a priori is valid only for complex energies far up in the complex plane.

Lemma 9.2 (The random walk representation) *Let $B \subset \mathbb{Z}^d$ be an arbitrary subset and $G_B = (H_B - z)^{-1}$ as above. Then:*

$$G_B(x, y; z) = \sum_{\substack{w: \, RW \text{ in } B \\ x \leadsto y}} \prod_{j=0}^{|w|} \frac{1}{\lambda V(w(j)) - z} \tag{9.31}$$

for all $\Im(z)$ large enough.

Proof: Recall the resolvent formula $\frac{1}{A} - \frac{1}{B} = \frac{1}{B}(B - A)\frac{1}{A}$. Using this with the choice $A = H_B - z$ and $B = \lambda V - z$ yields:

$$\frac{1}{H_B - z} = \frac{1}{-\Delta_B + \lambda V - z} = \frac{1}{\lambda V - z} + \frac{1}{\lambda V - z} \Delta_B \frac{1}{-T_0 + \lambda V - z}$$

where Δ_B is the adjacency matrix of the graph $\mathbb{Z}^d \cap B$. Iterating the above gives:

$$\frac{1}{H_B - z} = \frac{1}{\lambda V - z} \sum_{n \geq 0} \left(\Delta \frac{1}{\lambda V - z}\right)^n \tag{9.32}$$

which is, of course, a Neumann series and converges for large enough $|\Im(z)|$. More precisely, since the operator norm $\|\Delta\| = 2d$ and V is real valued, we need $\Im(z) > 2d$ to guarantee convergence of the right hand side of (9.32).

Now we claim that (9.32) is nothing but (9.31) in disguise. Indeed, taking the x, y-matrix element of (9.32) gives:

$$G_B(x, y; z) = \langle x | \frac{1}{H_B - z} | y \rangle = \frac{1}{\lambda V(x) - z} \sum_{n \geq 0} \langle x | (\Delta_B \frac{1}{\lambda V - z})^n | y \rangle.$$

Note that:

$$\langle x | (\Delta_B \frac{1}{\lambda V - z})^n | y \rangle = \sum_{\substack{\text{paths } w \text{ of length } n \\ \text{in } B, \, x \leadsto y}} \prod_{j=1}^{n} \frac{1}{\lambda V(w(j)) - z}.$$

Setting $w(0) = x$, $w(n) = y$, and $|w| =$ the length of the path, we arrive at the random walk representation (9.31).

At first sight it might seem that the random walk representation is just a simple rewriting of a particular Neumann series for the resolvent G_B and does not necessarily deserve its own name. This is not true, however, since giving it a *new name* can drastically change the *emphasis*: the key for the proof of the self-avoiding random walk representation is the observation that every random walk leads to a self-avoiding random walk by *deleting loops*. By the random walk representation, for any set $C \subset \mathbb{Z}$, the Green function on the diagonal is given by summing over all loops of a random walk within the set C:

$$G_C(x,x;z) = \sum_{\substack{w:\,RW\text{ in }C \\ x \leftrightsquigarrow x}} \prod_{j=0}^{|w|} \frac{1}{\lambda V(w(j)) - z} \qquad (9.33)$$

$\square$

To re-sum the loops, let:

$$n_x(w) := \inf\{n : w(j) \neq x \text{ for all } j > n\},$$

that is, $n_x(w)$ is the last time the path w visited the point x.

Cut the path w from the random walk representation (9.31) into two parts, $w = (w_1, \overline{w})$, where w_1 runs from 0 up to time $n_x(w)$ ($|w_1| = n_x(w)$) and $\overline{w}$ runs from $n_x(w) + 1$ up to time $|w|$. In particular:

$$|w| = |w_1| + |\overline{w}| + 1.$$

for the lengths of the combined paths.

From (9.31) one infers:

$$G_B(x,y;z) = \sum_{\substack{w_1:\,RW\text{ in }B \\ x \leftrightsquigarrow x}} \prod_{j=0}^{|w_1|} \frac{1}{\lambda V(w_1(j)) - z} \sum_{\substack{x' \in B:\,|x'-x|=1 \\ \overline{w}:\,RW\text{ in }B,\,x' \leftrightsquigarrow y \\ \overline{w}\text{ never visits }x}} \prod_{j=0}^{|\overline{w}|} \frac{1}{\lambda V(\overline{w}(j)) - z}.$$

(9.34)

Using (9.33), the first factor is just $G_B(x,x;z)$, and appealing to the random walk representation (9.31) once more, one sees that the second factor is the resolvent of the operator $H_{B\setminus\{x\}}$, summed over the nearest neighbours of x:

$$\sum_{\substack{\overline{w}:\text{ in }B,\,x' \leftrightsquigarrow y \\ \overline{w}\text{ never visits }x}} \prod_{j=0}^{|w|} \frac{1}{\lambda V(\overline{w}(j)) - z} = \langle x'| \frac{1}{(H-z)|_{B\setminus\{x\}}} |y\rangle = G_{B\setminus\{x\}}(x',y)$$

Thus (9.34) can be rewritten as:

$$G_B(x,y;z) = G(x,x;z) \sum_{\substack{x_1 \in B \\ |x_1 - x| = 1}} G_{B\setminus\{x\}}(x_1,y). \qquad (9.35)$$

Of course, iterating (9.35) yields:

$$G_B(x,y;z) = G(x,x;z) \sum_{\substack{x_1 \in B \\ |x_1-x|=1}} G_{B \setminus \{x\}}(x_1,x_1;z) \sum_{\substack{x_2 \in B \setminus \{x\} \\ |x_2-x_1|=1}} G_{B \setminus \{x,x_1\}}(x_2,y;z)$$

$$= \ldots = \sum_{\substack{w:\text{ SAW in } B \\ x \leftrightsquigarrow y}} \prod_{j=0}^{|w|} G_{B_j}(w(j),w(j);z)$$

with $B_0 = B$ and $B_{j+1} = B_j \setminus \{w(j)\}$, which nearly finishes the proof of Lemma 9.1, except that so far we only know that (9.30) holds as long as $\Im(z)$ is large enough.

To finish the proof of Lemma (9.1), let B be a finite set. We know that the resolvent is an analytic operator-valued function for z in the complex upper half-plane. In particular, $G_B(x,y;z)$ is an analytic function on the upper half-plane and so are all the factors $G_{B_j}(w(j),w(j);z)$ in the right hand side of (9.30). The punchline is that although in a *finite* set B there are infinitely many different random walk of arbitrary length, there are only *finitely many self-avoiding* random walks. Thus for finite B, the right hand side of (9.30) is a finite sum of a finite product of analytic functions, hence it is also analytic on the upper half-plane. Since by the above both sides of (9.30) agree for z with a large enough imaginary part, by analyticity they must agree on the whole complex upper half-plane. A similar argument holds for the lower half-plane. This concludes the proof of the self-avoiding random walk representation.

9.4.2 *Proof of localization at large disorder*

In this section we use the self-avoiding random walk representation to give a straightforward proof of Anderson localization for large disorder. In some sense this proof makes precise Anderson's original heuristic argument, which uses second order perturbation theory.

Before we fully embark on the proof of Theorem 9.3, let us first note that it is enough to prove this bound for the Green function restricted to some finite set $B \subset \mathbb{Z}^d$ as long as the bounds are uniform in B. This follows from the strong resolvent convergence of H_B to H as $B \to \mathbb{Z}^d$ and Fatou's lemma, $\mathbb{E}[|G(x,y;z)|^s] \leq \liminf_{B \to \mathbb{Z}^d} \mathbb{E}[|G_B(x,y;z)|^s]$. Secondly, note that for $0 < s \leq 1$ the bound:

$$\left|\sum_{j=1}^n \alpha_j\right|^s \leq \sum_{j=1}^n |\alpha_j|^s \tag{9.36}$$

for all complex numbers $\alpha_j \in \mathbb{C}$ hold. This is one of the fundamental observations in the original Aizenman–Molchanov proof. By induction, it is enough to consider the case $n = 2$. In this case:

$$|\alpha_1 + \alpha_2|^s \leq (|\alpha_1| + |\alpha_2|)^s = \frac{|\alpha_1|}{(|\alpha_1| + |\alpha_2|)^{1-s}} + \frac{|\alpha_2|}{(|\alpha_1| + |\alpha_2|)^{1-s}}$$
$$\leq \frac{|\alpha_1|}{(|\alpha_1|)^{1-s}} + \frac{|\alpha_2|}{(|\alpha_2|)^{1-s}} = |\alpha_1|^s + |\alpha_2|^s$$

by dropping the term $|\alpha_2|$ in the first denominator, respectively $|\alpha_1|$ in the second denominator.

Now let $B \subset \mathbb{Z}^d$ be an arbitrary finite set and $0 < s < 1$. We claim that:

$$\mathbb{E}[|G_B(x,y;z)|^s] \leq \sum_{\substack{w:\text{SAW}, \\ x \leftrightsquigarrow y}} (C_s/\lambda^s)^{|w|+1} \qquad (9.37)$$

which is exponentially small in the distance $|x - y|$ (note that the right hand side does not depend on the set B anymore). Indeed, organizing the summation over the self-avoiding random walks according to their lengths one has:

$$\sum_{\substack{w:\text{SAW}, \\ x \leftrightsquigarrow y}} (C_s/\lambda^s)^{|w|+1} = \sum_{n \geq 0} (C_s \lambda^{-s})^{n+1} \sum_{\substack{w:\text{SAW}, \\ x \leftrightsquigarrow y \\ |w|=n}} 1.$$

Of course, in order to connect x with y the length n of the walk must be at least $|x - y|$. In this case, since in the first step a self avoiding walk has $2d$ of the neighbours to choose and at most $2d - 1$ from then on, one has the general bound:

$$\sum_{\substack{w:\text{SAW}, \\ x \leftrightsquigarrow y \\ |w|=n}} 1 \leq 2d(2d-1)^{n-1}.$$

This gives:

$$\sum_{\substack{w:\text{SAW}, \\ x \leftrightsquigarrow y}} (C_s/\lambda^s)^{|w|+1} \leq \sum_{n \geq |x-y|} (C_s\lambda^{-s})^{n+1} 2d(2d-1)^{n-1}$$
$$= \frac{2d}{(2d-1)^2} \frac{\left[(2d-1)C_s\lambda^{-s}\right]^{|x-y|+1}}{1-(2d-1)C_s\lambda^{-s}}$$

which is the right hand side of (9.26).

It remains to prove (9.37). Applying (9.36) to the self-avoiding random walk representation from Lemma 9.1 and taking the expectation with respect to the random potential yields the bound:

$$\mathbb{E}[|G_B(x,y;z)|^s] \leq \sum_{\substack{w:\text{SAW in } B \\ x \leftrightsquigarrow y}} \mathbb{E}\left[\prod_{j=0}^{|w|} |G_{B_j}(w(j),w(j);z)|^s\right]. \qquad (9.38)$$

We evaluate the expectation on the right hand side of (9.38) successively with the help of conditional expectations with respect to the random potential visited along the path of the self-avoiding walk w: take first the expectation with respect to $v(x) = v(w(0))$ and note that the only Green's function which depends on on $v(w(0))$ is $G_{B_0}(w(0), w(0); z)$. Thus:

$$\mathbb{E}\left[\prod_{j=0}^{|w|} |G_{B_j}(w(j), w(j); z)|^s \,\Big|\, v(w(0))\right] \qquad (9.39)$$
$$= \mathbb{E}\left[|G_{B_0}(w(0), w(0); z)|^s \,|\, v(w(0))\right] \prod_{j=1}^{|w|} |G_{B_j}(w(j), w(j); z)|^s$$

Recalling the rank-one perturbation formula (9.13) one can bound the conditional expectation on the right hand side of (9.39) simply by C_s/λ^s. Thus:

$$\mathbb{E}\left[\prod_{j=0}^{|w|} |G_{B_j}(w(j), w(j); z)|^s \,\Big|\, v(w(0))\right] \leq \frac{C_s}{\lambda^s} \prod_{j=1}^{|w|} |G_{B_j}(w(j), w(j); z)|^s \qquad (9.40)$$

Now take the conditional expectation of (9.40) with respect to $v(w(1))$ and note that all factors $G_{B_j}(w(j), w(j); z)$ with $j \geq 2$ can again be taken out of the expectation since they do not depend on $v(w(1))$. Again one uses the a priori bound $\mathbb{E}[|G_{B_1}(w(1), w(1); z)|^s | v(w(1))] \leq C_s/\lambda^s$ to see:

$$\mathbb{E}\left[\prod_{j=0}^{|w|} |G_{B_j}(w(j), w(j); z)|^s \,\Big|\, v(w(0)), v(w(1))\right] \leq \left(\frac{C_s}{\lambda^s}\right)^2 \prod_{j=2}^{|w|} |G_{B_j}(w(j), w(j); z)|^s$$
$$(9.41)$$

Iterating this procedure $|w| + 1$ times yields the bound:

$$\mathbb{E}\left[\prod_{j=0}^{|w|} |G_{B_j}(w(j), w(j); z)|^s\right] \leq \left(\frac{C_s}{\lambda^s}\right)^{|w|+1}$$

which together with (9.38) gives (9.37). This ends the proof of Theorem 9.3.

References

Aizenman, M., Elgart, A., Naboko, S., Schenker, J. H., and Stolz, G. (2006). Moment analysis for localization in random Schrödinger operators. *Invent. Math.* **163**(2), 343–413.

Aizenman, M. and Graf, G. M. (1998). Localization bounds for an electron gas. *J. Phys. A* **31**(32), 6783–806.

Aizenman, M. and Molchanov S. (1993). Localization at large disorder and at extreme energies: an elementary derivation. *Comm. Math. Phys.* **157**(2), 245–78.

Aizenman, M. and Newman C. M. (1984). Tree graph inequalities and critical behavior in percolation models. *J. Statist. Phys.* **36**(1-2), 107–43.

Aizenman, M., Schenker, J. H., Friedrich, R. M., and Hundertmark, D. (2000). Constructive fractional-moment criteria for localization in random operators. *Phys. A* **279**(1-4), 369–77. Statistical mechanics: from rigorous results to applications.

Aizenman, M., Schenker, J. H., Friedrich, R. M., and Hundertmark, D. (2001). Finite-volume fractional-moment criteria for Anderson localization. *Comm. Math. Phys.* **224**(1), 219–253. Dedicated to Joel L. Lebowitz.

Aizenman, M., Sims, R., and Warzel S. (2006). Fluctuation-based proof of the stability of ac spectra of random operators on tree graphs. In *Quantum Graphs and Their Applications*, Volume 415 of *Contemp. Math.*, pp. 1–14. Amer. Math. Soc.: Providence, RI.

Anderson, P. W. (1958, Mar). Absence of diffusion in certain random lattices. *Phys. Rev.* **109**(5), 1492–1505.

Bellissard, J., van Elst, A., and Schulz-Baldes, H. (1994). The noncommutative geometry of the quantum Hall effect. *J. Math. Phys.* **35**(10), 5373–451. Topology and physics.

Bourgain, J. and Kenig, C. E. (2005). On localization in the continuous Anderson–Bernoulli model in higher dimension. *Invent. Math.* **161**(2), 389–426.

Carmona, R., Klein, A., and Martinelli, F. (1987). Anderson localization for Bernoulli and other singular potentials. *Comm. Math. Phys.* **108**(1), 41–66.

Cycon, H. L., Froese, R. G., Kirsch, W., and Simon, B. (1987). *Schrödinger Operators with Application to Quantum Mechanics and Global Geometry* (Study ed.). Texts and Monographs in Physics Springer-Verlag: Berlin.

del Rio, R., Jitomirskaya, S., Last, Y., and Simon, B. (1995). What is localization? *Phys. Rev. Lett.* **75**(1), 117–19.

del Rio, R., Jitomirskaya, S., Last, Y., and Simon, B. (1996). Operators with singular continuous spectrum. IV. Hausdorff dimensions, rank one perturbations, and localization. *J. Anal. Math.* **69**, 153–200.

Froese, R., Hasler, D., and Spitzer, W. (2007). Absolutely continuous spectrum for the Anderson model on a tree: a geometric proof of Klein's theorem. *Comm. Math. Phys.* **269**(1), 239–57.

Fröhlich, J. and Spencer, T. (1983). Absence of diffusion in the Anderson tight binding model for large disorder or low energy. *Comm. Math. Phys.* **88**(2), 151–84.

Germinet, F., Hislop, P., and Klein, A. (2005). On localization for the Schrödinger operator with a Poisson random potential. *C. R. Math. Acad. Sci. Paris* **341**(8), 525–28.

Germinet, F. and Klein, A. (2004). A characterization of the Anderson metal-insulator transport transition. *Duke Math. J.* **124**(2), 309–50.

Grimmett, G. (1999). *Percolation* (Second ed.), Volume 321 of *Grundlehren der Mathematischen Wissenschaften [Fundamental Principles of Mathematical Sciences]*. Springer-Verlag: Berlin.

Hammersley, J. M. (1957). Percolation processes: Lower bounds for the critical probability. *Ann. Math. Statist.* **28**, 790–95.

Hundertmark, D. (2000). On the time-dependent approach to Anderson localization. *Math. Nachr.* **214**, 25–38.

Hunziker, W. and Sigal, I. M. (2000). The quantum N-body problem. *J. Math. Phys.* **41**(6), 3448–510.

Kirsch, W. (1989). Random Schrödinger operators. A course. In *Schrödinger Operators (Sønderborg, 1988)*, Volume 345 of *Lecture Notes in Phys.*, pp. 264–370. Springer: Berlin.

Kirsch, W. and Metzger, B. (2007). The integrated density of states for random Schrödinger operators. In *Spectral Theory and Mathematical Physics: A Festschrift in Honor of Barry Simon's 60th Birthday*, Volume 76 of *Proc. Sympos. Pure Math.*, pp. 649–696. Amer. Math. Soc: Providence, RI.

Klein, A. (1998). Extended states in the Anderson model on the Bethe lattice. *Adv. Math.* **133**(1), 163–84.

Kunz, H. and Souillard, B. (1980). Sur le spectre des opérateurs aux différences finies aléatoires. *Comm. Math. Phys.* **78**(2), 201–46.

Lieb, E. H. (1980). A refinement of Simon's correlation inequality. *Comm. Math. Phys.* **77**(2), 127–35.

Lifshits, I. M., Gredeskul, S. A., and Pastur, L. A. (1988). *Introduction to the Theory of Disordered Systems*. A Wiley-Interscience Publication. John Wiley & Sons Inc.: New York. Translated from the Russian by Eugene Yankovsky [E. M. Yankovskiĭ].

Martinelli, F. and Scoppola, E. (1985). Remark on the absence of absolutely continuous spectrum for d-dimensional Schrödinger operators with random potential for large disorder or low energy. *Comm. Math. Phys.* **97**(3), 465–71.

Minami, N. (1996). Local fluctuation of the spectrum of a multidimensional Anderson tight binding model. *Comm. Math. Phys.* **177**(3), 709–25.

Molchanov, S. A. (1981). The local structure of the spectrum of the one-dimensional Schrödinger operator. *Comm. Math. Phys.* **78**(3), 429–46.

Pastur, L. A. (1980). Spectral properties of disordered systems in the one-body approximation. *Comm. Math. Phys.* **75**(2), 179–96.

Simon, B. (1980). Correlation inequalities and the decay of correlations in ferromagnets. *Comm. Math. Phys.* **77**(2), 111–26.

Simon, B. (1990). Absence of ballistic motion. *Comm. Math. Phys.* **134**(1), 209–12.

Simon, B. and Wolff, T. (1986). Singular continuous spectrum under rank one perturbations and localization for random Hamiltonians. *Comm. Pure Appl. Math.* **39**(1), 75–90.

Stollmann, P. (2001). *Caught by Disorder*, Volume 20 of *Progress in Mathematical Physics*. Birkhäuser Boston Inc.: Boston, MA. Bound states in random media.

PART II

MICROSCOPIC MODELS

A
NUCLEATION AND GROWTH

10

EFFECTIVE THEORIES FOR OSTWALD RIPENING

Barbara Niethammer

10.1 Introduction

In this article we discuss the derivation and analysis of reduced models for a specific coarsening process which is known as Ostwald ripening. This phenomenon appears in the late stage of phase transitions, when—due to a change in temperature or pressure for example—the energy of the underlying system becomes nonconvex and prefers two different phases of the material. Consequently a homogeneous mixture is unstable and, in order to minimize the energy, it separates into the two stable phases. Typical examples are the condensation of liquid droplets in a supersaturated vapour and phase separation in binary alloys after rapid cooling.

With Ostwald ripening one usually denotes the case when the composition of the mixture is such that one of the two stable phases has much smaller volume fraction than the other. Then the minority phase nucleates in the form of many small droplets which first grow from a uniform background supersaturation. Once the latter is small, surface energy becomes the dominant part of the total energy and to minimize it particles start to interact via diffusional mass exchange to reduce their total surface area. As a consequence large particles grow, while smaller ones shrink and finally disappear.

Ostwald ripening is a paradigm for statistical self-similarity in coarsening systems. This means that after a transient stage the particle number density evolves in a unique self-similar fashion, which is independent of the details of the initial data. The first quantitative description of this phenomenon was given by Lifshitz and Slyozov (1961) and Wagner (1961) and is nowadays known as the classical LSW-theory. In the regime where the volume fraction of the droplets is small they derive an equation for the particle number density based on the crucial assumption that in the dilute regime the interaction between particles can be expressed solely through a common mean-field. However, it has been established by a mathematically rigorous analysis that the long-time behaviour within the LSW model is not a universal statistically self-similar one but on the contrary depends sensitively on the initial data. Hence, in order to overcome this shortcoming, one has to go beyond the mean-field assumption and take higher order effect, such as screening induced fluctuations and particle collisions, into account.

A number of different approaches to developing a corresponding theory can be found in the physics and metallurgical literature. However, the predictions based on the respective theories differ significantly and it seems that a more rigorous analysis could be helpful in resolving some of the open questions. It is the main goal of the present article to review corresponding progress on the understanding of first order corrections to the LSW theory. While we go along, we also point out some directions for future research, in particular where the combination of analytic and stochastic tools could be relevant.

For more background on results, which are not discussed in detail here, as well as for references to the applied literature, we refer to the review article (Niethammer et al. 2006).

10.2 Basic models and mean-field theories

10.2.1 *The starting point: a simplified Mullins–Sekerka evolution*

A basic model for diffusion controlled Ostwald ripening of spherical particles is a simplified Mullins–Sekerka type model which is appropriate in the case that particles have small volume fraction. In this model particles, called P_i, are distributed in a domain $\Omega \subset \mathbb{R}^3$ and are characterized by their immovable centres $X_i \in \Omega$ and their radii $R_i(t)$. Particles interact by diffusion, but in late-stage coarsening we can assume that mass exchange between particles is much faster than the growth of the interfaces. Hence we can use a quasi-steady approach, that is we assume that the potential u relaxes at each time instantaneously to equilibrium. This gives that for each time t the potential $u = u(x,t)$ solves:

$$\Delta u = 0 \quad \text{in } \Omega \backslash \bigcup_i \overline{P_i}$$
$$u = \frac{1}{R_i} \quad \text{on } \partial P_i,$$
(10.1)

where Δ and later ∇ denotes derivatives with respect to the space variable x. The second equation in (10.1) is the well-known Gibbs–Thomson law which accounts for surface tension. To define the potential uniquely, we have to couple (10.1) with suitable boundary conditions on $\partial \Omega$. In the case that Ω is bounded, a natural assumption is to consider closed systems and require:

$$\frac{\partial u}{\partial \vec{n}} = 0 \quad \text{on } \partial \Omega.$$
(10.2)

We can also consider the problem in the whole space $\Omega = \mathbb{R}^3$ in which case the appropriate boundary condition is a no-flux condition at infinity:

$$|\nabla u| \to 0 \quad \text{as } |x| \to \infty.$$
(10.3)

We easily convince ourselves that if all particles have the same size, the potential u is constant (indeed equal to the inverse radius of the particles). However, if particles have different sizes, this induces gradients in the potential and these gradients drive the system towards a state of lower energy. The Gibbs–Thomson law in (10.1) implies that u is large at small particles which have large surface area compared to their volume, and small at large particles. Hence, mass diffuses from the small to the large particles. The growth rate of a particle is simply given by the total flux towards the particle, that is:

$$\frac{d}{dt}\left(\frac{4\pi R_i^3}{3}\right) = \int_{\partial P_i} \frac{\partial u}{\partial \vec{n}} \, dS, \tag{10.4}$$

where here $\vec{n}$ denotes the outer normal to the particle.

It is not difficult to show that, if we start with a finite number of particles, which do not overlap, the problem (10.1)+(10.2) or (10.1)+(10.3) is well-posed (cf. (Dai and Pego 2005a) for the case (10.3)) and depends Lipschitz-continuously on the initial radii of the particles. As a consequence, the full time-dependent system (10.1)–(10.4) is well-posed for short times. We can extend such a local solution up to a time when a particle vanishes or when two particles touch. In the first case we just eliminate the particle and continue with the remaining ones. In this way we obtain a continuous in time, piecewise smooth solution. In the second case, where particles touch, there is no way to extend the solution in a reasonable way. In fact, the simplifying assumption that particles are spherical is not a good approximation when particles are close.

However, we are interested in the dynamics of a large set of particles with small volume fraction, and we expect that the event that particles touch is rare if it occurs at all. Hence it is plausible that it does not have an influence on the global behaviour of the system. As we shall see, the latter is true to leading order, but not if one is interested in higher order effects. We will return to this issue later in Section 10.4.

As long as the evolution is well-posed we easily verify that it preserves the total volume of the particles and decreases the surface energy. Indeed, we have:

$$\frac{d}{dt} \sum_i R_i^3 = 0 \tag{10.5}$$

and:

$$\frac{d}{dt} \sum_i R_i^2 = -\frac{\pi}{2} \int_\Omega |\nabla u|^2 \, dx. \tag{10.6}$$

In contrast to other curvature driven evolutions, such as the mean curvature flow, the Mullins–Sekerka evolution (10.1)–(10.4) is nonlocal. More precisely, the evolution of the radius of one particle depends on all the other particles in the system, since all particles interact via the potential u. A priori the interaction range between particles is large due to the slow decay of the fundamental solution of Laplace's equation. The challenge is to derive the effective growth law of a particle in a sea of surrounding particles. We will see that a key aspect in the analysis will be to establish the screening effect which identifies the effective interaction range between particles (cf. Section 10.2.5).

10.2.2 The leading order theory (LSW-theory)

Our goal is to derive from the Mullins–Sekerka model the BBGKY hierarchy for the number densities of particle radii and centres. The BBGKY hierarchy can be derived from the Liouville equation by averaging and describes the evolution of the N-particle distribution in terms of the $(N+1)$-particle distribution. To obtain a tractable system of equations one typically tries to truncate the hierarchy by a suitable closure hypothesis on the level of the one- or two particle number density. This procedure can often be justified if there is a small parameter in the system, such as in our case the volume fraction of the particles.

The formal identification of the leading order terms in the dilute regime is not difficult and goes back to the classical work by Lifshitz and Slyozov (1961) and Wagner (1961) (called nowadays the 'LSW-theory').

If the particle size is much smaller than the typical distance between the nearest neighbours one can assume that the potential u is approximately constant in space away from the particles, that is $u \approx u_\infty(t)$. In other words, each particle P_i feels the influence of the other particles only through u_∞, also called a mean-field. We then solve for particle P_i:

$$-\Delta u = 0 \quad \text{in } \mathbb{R}^3 \setminus \overline{P_i}$$
$$u = \frac{1}{R_i} \quad \text{on } \partial P_i \qquad (10.7)$$
$$u \to u_\infty \quad \text{as } |x| \to \infty,$$

whose solution is given by:

$$u(x,t) = u_\infty + \frac{1 - R_i u_\infty}{|x - X_i|}.$$

Using this solution in (10.4) we obtain the simple law:

$$\frac{d}{dt}\left(\frac{4\pi}{3} R_i^3\right) = R_i u_\infty - 1. \qquad (10.8)$$

So far, we have not specified u_∞. In the above approximation we have not yet taken into account that the evolution preserves the total volume of the particles. This constraint determines u_∞ and implies that:

$$u_\infty = \frac{\sum_{i:R_i>0} 1}{\sum_i R_i} = \frac{1}{\text{mean radius}}. \qquad (10.9)$$

We read off from (10.8)–(10.9) that the critical radius in this approximation is just the mean radius. Recall that in the coarsening picture, the critical radius typically increases, so that over time more and more particles start to shrink and finally disappear.

Based on (10.8) we can now derive an equation for the one-particle number density, that is the expected number of particles with radius R in $(R, R + dR)$, which we denote by $f_1 = f_1(R,t)$. Due to the translation invariance of the Mullins–Sekerka evolution f_1 is independent of the centres. The system (10.8)–(10.9) translates without further approximation into the following evolution law for f_1:

$$\partial_t f_1 + \partial_R \left(\frac{1}{R^2}(Ru_\infty(t) - 1) f_1 \right) = 0 \qquad (10.10)$$

with

$$u_\infty(t) = \frac{\int_0^\infty f_1(R,t)\,dR}{\int_0^\infty R f_1(R,t)\,dR}. \qquad (10.11)$$

10.2.3 Dynamic scaling and coarsening rates

Within the LSW model (10.10)–(10.11) we can investigate statistical self-similarity. In fact, we check that the equation has a scale invariance $R \sim t^{1/3}$ which is inherited from the Mullins–Sekerka evolution. It turns out that (10.10) has indeed self-similar solutions, but not only one but a one-parameter family of the form $f(R,t) = t^{-4/3} F_a(R/t)$ with $u_\infty = (at)^{1/3}$ and $a \in (0, \frac{4}{9}]$. All of the self-similar profiles have compact support, one is smooth, the other ones behave like power laws at the end of their support. LSW predict in their work that only the smooth self-similar solution is stable and is the unique scaling limit for the LSW model. As a consequence they obtain universal growth rates of the coarsening process, such as, for example, that the mean radius evolves as $\left(\frac{4}{9}t\right)^{1/3}$.

However, it has been rigorously established in (Niethammer and Pego 1999) (see also Carr and Penrose 1998 for a related model and Giron et al. 1998 for formal asymptotics) that the long-time behaviour of solutions to the LSW model is not universal, but on the contrary depends sensitively on the initial data, more precisely on the behaviour at the end of the support. Loosely speaking, if the data behave like a power law of power p, the solution converges to the self-similar solution with the same power law. The notion 'to behave like a power law' is made precise, the technical term is that the data must be regularly varying with power p at the end of their support.

Before we continue to discuss how one could overcome this weak selection problem, let us digress to discuss a related issue, which is to establish coarsening rates, that is the growth rate of typical length scales, in general. While one can often predict coarsening rates via a dimensional analysis, a rigorous treatment has only recently become available. In (Kohn and Otto, 2002) a time averaged upper bound of the coarsening rate within the Cahn–Hilliard theory has been established via a lower bound on the decay rate of the energy density. The argument uses an energy-dissipation relation and a relation between the energy and a certain appropriate length scale. This technique has been shown to be quite robust and has been applied to a large variety of other coarsening problems (Kohn and Yan 2004; Dai and Pego 2005c), in particular also to the LSW model (Dai and Pego 2005b; Pego 2007). Naturally, pointwise upper bounds are much more difficult to obtain. For the relatively simple LSW model a first result has been obtained in (Niethammer and Velàzquez, 2006a), where upper and even lower bounds on the coarsening rates have been established for data which are close to a self-similar solution. In general lower bounds cannot be expected, since there are configurations for which coarsening does not occur (e.g., all particles with equal size in the LSW model) or is extremely slow (e.g., one-dimensional coarsening in the Cahn–Hilliard equation).

It would be extremely interesting to establish lower bounds for coarsening rates using probabilistic arguments, which characterize 'typical' configurations, for which the system coarsens with the expected rate.

10.2.4 Questions around the LSW theory

We have seen that one problem in the LSW theory is the weak selection of self-similar asymptotic states, which suggests that some mechanisms are neglected in the LSW model.

Another shortcoming of the LSW model becomes apparent if one compares the predictions by LSW with experimental data. It turns out that the discrepancy is rather large: the constants in the coarsening rates are much larger and the size distributions are less narrow than predicted by the LSW theory.

It is usually argued in the applied literature that one disadvantage of the LSW theory is its mean-field nature which neglects the build up of correlations between particles, which are relevant already in the dilute regime. In other words, the LSW theory assumes that the interaction range of a particle is infinite and the contribution of all the other particles is given by a deterministic average, the mean-field. This picture however neglects screening, which implies that the interaction range of one particle is screened by its neighbours and hence finite, which leads to deviations of the effective mean-field from its average.

It is the goal of (Velàzquez, 2000) to investigate whether these discrete effects in the mean-field, and similarly in the data, change the weak-selection criterion of the LSW model over the relevant time scales, that is as long as a sufficiently large number of particles is still present. However, the analysis is restricted to a regime, in which screening effects are not relevant. It turns out that in this regime, stochastic effects do not essentially modify the effective dynamics as described by the LSW model and thus do not provide a selection mechanism.

Before we continue to give an overview of further attempts to access the effect of finite volume fraction on Ostwald ripening, we describe screening in Section 10.2.5 and review results on the rigorous derivation of the LSW model from the Mullins–Sekerka evolution in Section 10.2.6.

10.2.5 Screening

The screening effect, described above, can be most easily understood by referring to electrostatics. We briefly recall the argument which gives us the scaling of the screening length in terms of the parameters of the system.

To that aim we consider a point charge at $X_0 \in \mathbb{R}^3$ surrounded by conducting balls $P_i = B(R_i, X_i)$ which are uniformly distributed according to a number density ρ, have volume fraction $\varepsilon \ll 1$, and average radius $\langle R \rangle$. The point charge at X_0 creates an electric field and a corresponding potential G, and thus induces a negative charge on $\partial B(R_i, X_i)$. This induced charge roughly equals $-4\pi R_i G(X_i)$, where $4\pi R_i$ is the capacity of a single ball in $\mathbb{R}^3$. In a dilute system capacity is approximately additive which implies that the total negative charge density is approximately given by $-4\pi \langle R \rangle \rho G$. Hence the effective electric potential satisfies:

$$-\Delta G = \delta_{X_0} - 4\pi \langle R \rangle \rho G \quad \text{in } \mathbb{R}^3,$$

and thus:

$$G(x) = \frac{1}{4\pi |x - X_0|} e^{-\frac{|x-X_0|}{\xi}}, \tag{10.12}$$

where the 'screening length' ξ is given by:

$$\xi = \frac{1}{\sqrt{4\pi \langle R \rangle \rho}}. \tag{10.13}$$

Formula (10.12) shows that the presence of the balls has the effect that the effective range of the electric potential is limited to ξ, whereas the electric potential in a system without balls is just $\frac{1}{4\pi|x-X_0|}$ and decays slowly. Notice that the number of particles within the screening range is $\xi \rho^{1/3}$ which according to (10.13) equals $\langle R \rangle^{-1/2} \rho^{-1/6} \sim \varepsilon^{-1/6}$. Hence, in the dilute regime, the number of particles within the screening range is still large and becomes infinite as $\varepsilon \to 0$.

For further reference, we also note another relevant scaling, the ratio between typical radius and screening length, which is $\langle R \rangle / \xi \sim \varepsilon^{1/2}$.

10.2.6 Rigorous derivation of the LSW theory

The rigorous derivation of the LSW model from the Mullins–Sekerka evolution as $\varepsilon \to 0$ is by now rather complete. It is treated in a series of papers (Niethammer 1999; Niethammer and Otto 2001; Niethammer and Velàzquez 2004a,b) which deal with different assumptions on the data respectively. First, the simplest case was treated in (Niethammer, 1999), where the system size is smaller than the screening length. More precisely, one starts with $N_\varepsilon \gg 1$ well-separated particles in—say—the unit box with volume fraction $\varepsilon \ll 1$, that is $\rho = N_\varepsilon$ and $N_\varepsilon \langle R \rangle^3 = \varepsilon$. That the system size (here equal to one) is smaller than the screening length means in view of (10.13) that $\lim_{\varepsilon \to 0} \langle R \rangle N_\varepsilon \to 0$ as $\varepsilon \to 0$. In this regime it is established in (Niethammer, 1999) that the solution of the Mullins–Sekerka problem converges to the (unique) solution of the LSW model. (Well-posedness of the LSW model is established in Niethammer and Pego 2005; see also Laurençot 2002.)

In the case that the system is of the order of the screening length or larger, one obtains an inhomogeneous extension of the LSW model (Niethammer and

Otto 2001). Most interesting and natural is the case that the system size is much larger than the screening length. This implies that when rescaling the system with respect to the natural length scale, the screening length, one obtains a homogenization problem in an unbounded domain. As a consequence, energy-type estimates are not useful in the analysis. One important step in the analysis of (Niethammer and Velàzquez 2004a,b,c) is the result (Niethammer and Velàzquez 2006b) which establishes that the fundamental solution of the microscopic problem decays exponentially w.r.t. the screening length. This allows one to 'localize' the homogenization procedure in (Niethammer and Velàzquez 2004a,b. While in previous work it has been assumed that initially particles are well-separated so that they cannot touch during the evolution, Niethammer and Velàzquez (2004b) treat the case of initially randomly distributed particles. In this case particles might overlap and the evolution is defined by merging these particles into a larger one and continuing. To justify this procedure it is important to show that a very small fraction of particles can overlap and that this does not affect the macroscopic evolution law for the remaining particles. This result rules out corrections on the zero order level due to a stochastic nature of the data.

The result by Niethammer and Velàzquez (2006b) should also turn out to be useful in further related investigations. In fact, the Mullins–Sekerka evolution has not yet been considered in the setting where infinitely many particles distributed in the whole space, e.g., according to a homogeneous Poisson process. Even if one handles collision of particles in some way, global existence of a solution to this problem is not obvious, since if locally screening is very weak there could be a mass flux from infinitely far away leading to the finite time blow up of the radius of one particle. We expect, however, that if particles are initially uniformly distributed, such that there is a uniform—in a sense which has to be made precise—screening length, such a scenario does not take place and that the evolution is well-posed.

10.3 Scaling of the first order correction: a cross-over due to screening

In order to derive a perturbative theory to the LSW model which takes nonzero volume fraction into account we first have to identify the correct expansion parameter, or in other words, the scaling of the first order correction. In this chapter we review a result which rigorously establishes such a scaling. The analysis combines a variational viewpoint with elementary probability.

In the applied literature there had been a controversy about the size of the scaling of the first order correction, since numerical simulations for finite systems predicted an error of order $\varepsilon^{1/3}$, whereas theories for infinite systems predicted an error of order $\varepsilon^{1/2}$. This was first to some extent resolved by numerical simulations in (Fradkov et al. 1996), which show a cross-over from $\varepsilon^{1/3}$ to $\varepsilon^{1/2}$ when the system size becomes larger than the screening length. We will now discuss in some detail a result, which proves a refined version of this observation.

10.3.1 Set-up and assumptions

Our starting point here is the monopole approximation of (10.1), (10.3), and (10.4). In fact, it has been established in (Dai and Pego 2005a), that the monopole approximation is exact for a variant of (10.1), (10.3), and (10.4), where the Gibbs–Thomson condition is averaged, instead of the Stefan condition (10.4).

In the monopole approximation we use the ansatz $u(x,t) := -\sum_i \frac{V_i}{|X_i-x|}$ for a solution of (10.1), where $\{V_i\}_i$ are the growth rates of the particle volumes, that is $V_i := \frac{d}{dt}[\frac{4\pi}{3} R_i^3] = 4\pi R_i^2 \frac{dR_i}{dt}$. Using the Gibbs–Thomson condition in (10.1) gives to leading order the following linear system of equations:

$$\frac{1}{R_i} = u_\infty - \frac{V_i}{R_i} - \sum_{j \neq i} \frac{V_j}{d_{ij}}, \quad (10.14)$$

where $d_{ij} := |X_i - X_j|$ is the distance between particle centres and u_∞ is such that:

$$\sum_i V_i = 0. \quad (10.15)$$

We consider from now on a fixed distribution of $n \gg 1$ particles centres $\{X_i\}_i$ in a sphere of volume n (that is the number density ρ satisfies $\rho \sim 1$) which satisfies certain regularity assumptions listed below. The particle radii $\{R_i\}_i$ are identically and independently distributed according to a distribution with compact support and mean volume ε. Within this setting the screening length is given by $\xi \sim \frac{1}{\sqrt{\langle R \rangle}} \sim \varepsilon^{-1/6}$ and hence the screening length is smaller, resp. larger, than the domain size if $\xi \ll n^{1/3}$ or $\xi \gg n^{1/3}$—in other words if $\varepsilon n^2 \gg 1$ or $\varepsilon n^2 \ll 1$—respectively. We call these regimes supercritical and subcritical respectively.

In the following we estimate the deviation of the joint distribution $\{X_i, R_i, V_i\}_i$ from $\{X_i, R_i, V_i^{LSW}\}_i$, where the $\{V_i\}_i$ are determined according to (10.14) and $\{V_i^{LSW}\}_i$ are the LSW growth rates, given by the truncation of (10.14):

$$\frac{1}{R_i} = u_\infty^{LSW} - \frac{V_i^{LSW}}{R_i} \quad \text{and} \quad \sum_i V_i^{LSW} = 0. \quad (10.16)$$

Such an analysis is also called 'Snapshot' analysis, since we only estimate the difference in the rate of change of the system at a given time.

The quantity we consider in the following will be the relative deviation in the rate of change of energy, which is another convenient measure for the coarsening rate. More precisely we consider $\frac{\dot{E}^{LSW} - \dot{E}}{|\langle \dot{E}^{LSW} \rangle|}$, where E is the interfacial energy of the particles, i.e., $E = \frac{1}{2n} \sum_i R_i^2$, and its rate of change is:

$$\dot{E} = \frac{1}{n} \sum_i \frac{V_i}{R_i},$$

while:

$$\dot{E}^{LSW} = \frac{1}{n} \sum_i \frac{V_i^{LSW}}{R_i},$$

with V_i^{LSW} given by (10.16). Since the energy is decreasing, $\dot{E}$ is always negative. Likewise $\dot{E}^{LSW}$ is always negative, but we expect the difference $\dot{E}^{LSW} - \dot{E}$ to be negative for most realizations, since the LSW theory should underestimate the rate at which E is decreasing.

For the analysis we need the following regularity assumptions on the distribution of $\{X_i\}_i$. The first one ensures a certain uniformity in the distribution. We assume in the supercritical case, i.e., when the system size is much larger than the screening length, that each subdomain of size of order ξ, contains at least of the order of $\varepsilon^{-1/2}$ particles. This assumption can be shown (at least if $\varepsilon \leq \frac{1}{\ln n^5}$, cf. (Niethammer and Velàzquez 2004b)) to be satisfied with probability converging to one as $n \to \infty$.

The second assumption is less natural. We assume that the minimal distance between particles is of the order of the mean nearest neighbour distance, that is $\min_{j \neq i} d_{ij} \geq c_0 > 0$. This assumption is not satisfied with probability close to one. The number of particles which violate this assumption is small and one might expect that the inclusion would not destroy our result. It would be very interesting to establish a corresponding result rigorously, or show, on the contrary, that the above assumption is relevant.

One consequence of these two assumptions on the distribution of particle centres is that we can approximate discrete sums by the corresponding integrals, an approximation we use frequently in the proofs.

10.3.2 The result

The main result in (Hönig et al. 2005b) is that for a fixed distribution of particle centres satisfying our regularity assumptions we have with high probability (with respect to the radius distribution):

$$-\frac{\dot{E} - \dot{E}^{LSW}}{|\langle \dot{E}^{LSW} \rangle|} \sim \begin{cases} n^{-1/3} \varepsilon^{1/3} & \text{for } n \ll \varepsilon^{-1/2} \\ \varepsilon^{1/2} & \text{for } n \gg \varepsilon^{-1/2} \end{cases}. \tag{10.17}$$

Notice that this is a qualitative statement about the entire distribution, not just its expected value, which is usually considered in numerical simulations. Furthermore it makes the dependence on n precise and gives a proper crossover, that is the scalings agree in the case that $n \sim \varepsilon^{-1/3}$.

In the following $\langle \cdot \rangle$ denotes the expected value with respect to the joint probability measure P of the variables $\{R_i\}_i$.

Theorem 10.1 *(Hönig et al. 2005b, Th. 2.2) (The supercritical regime)*
If $n \ll \varepsilon^{-1/2}$ and $\varepsilon \leq \varepsilon_0$ we have with high probability that:

$$-C\varepsilon^{1/2} \leq \frac{\dot{E} - \dot{E}^{LSW}}{|\langle \dot{E}^{LSW}\rangle|} \leq -\frac{1}{C}\varepsilon^{1/2},$$

that is for all $\delta > 0$ there exists a constant $C = C(\delta)$ such that:

$$P\left(\left\{-C\varepsilon^{1/2} \leq \frac{\dot{E} - \dot{E}^{LSW}}{|\langle \dot{E}^{LSW}\rangle|} \leq -\frac{1}{C}\varepsilon^{1/2}\right\}^c\right) \leq \delta.$$

Theorem 10.2 *(Hönig et al. 2005b, Th. 2.1) (The subcritical regime) If $n \ll \varepsilon^{-1/2}$ and $\varepsilon \leq \varepsilon_0$ we have with high probability that:*

$$\frac{\dot{E} - \dot{E}^{LSW}}{|\langle \dot{E}^{LSW}\rangle|} \geq -C n^{-1/3} \varepsilon^{1/3}.$$

Furthermore:

$$\frac{\langle \dot{E} - \dot{E}^{LSW}\rangle}{|\langle \dot{E}^{LSW}\rangle|} \leq -\frac{1}{C} n^{-1/3}\varepsilon^{1/2}.$$

Remark: notice that in the subcritical regimes we only succeed to derive a lower bound, whereas we obtain an upper bound only for the expected value. It is not surprising, that subcritical systems have less good self-averaging properties than supercritical systems and, in fact, a recent rigorous result by Conti et al. (2006) shows, that for any $M > 0$ there is a finite probability $\rho_M > 0$ such that $(\dot{E} - \dot{E}^{LSW})/|\langle \dot{E}^{LSW}\rangle| > M$.

10.3.3 Sketch of proof

In the following we present the main ideas of the proof of Theorem 10.2.

We first perform the natural rescaling, by rescaling radii with respect to their typical size $\varepsilon^{1/3}$ such that (10.14) becomes:

$$\frac{1}{R_i} = u_\infty - \frac{V_i}{R_i} - \varepsilon^{1/3} \sum_{j \neq i} \frac{V_j}{d_{ij}}, \qquad (10.18)$$

where again u_∞ is such that $\sum_i V_i = 0$. Recall that the radii are distributed according to a distribution with compact support. Thus, after rescaling we can assume that $R_i \leq C_0$ for some $C_0 > 0$.

Variational formulation:

A key idea in the proof of Theorem 10.2 is that the deviation in the rate of decrease of the energy can be formulated variationally. First we observe that the solution of (10.18) can also be characterized as a minimizer of:

$$\min_{\{W_i\}_i; \sum_i W_i = 0} \left\{ \frac{1}{n} \sum_i \frac{1}{2R_i} W_i^2 + \varepsilon^{1/3} \frac{1}{n} \sum_i \sum_{j \neq i} \frac{W_i W_j}{2d_{ij}} + \frac{1}{n} \sum_i \frac{W_i}{R_i} \right\}.$$

and the solution $\{V_i\}_i$ satisfies:

$$\frac{1}{n} \sum_i \frac{1}{2R_i} V_i^2 + \varepsilon^{1/3} \frac{1}{n} \sum_i \sum_{j \neq i} \frac{V_i V_j}{2d_{ij}} + \frac{1}{n} \sum_i \frac{V_i}{R_i} = \frac{1}{n} \sum_i \frac{V_i}{2R_i} = \frac{1}{2} \dot{E}.$$

Hence:

$$\dot{E} - \dot{E}^{LSW} = \min_{\{W_i\}_i, \sum_i W_i = 0} \left\{ \frac{1}{n} \sum_i \frac{1}{R_i} W_i^2 \right.$$
$$\left. + \varepsilon^{1/3} \frac{1}{n} \sum_i \sum_{j \neq i} \frac{W_i W_j}{d_{ij}} + \frac{1}{n} \sum_i \frac{2W_i}{R_i} - \frac{1}{n} \sum_i \frac{V_i^{LSW}}{R_i} \right\}.$$

and after some elementary manipulations, recalling $V_i^{LSW} = \frac{R_i}{\overline{R}} - 1$ with $\overline{R} := \frac{1}{n} \sum_i R_i$, we find:

$$\dot{E} - \dot{E}^{LSW} = \min_{\{W_i\}_i; \sum_i W_i} \left\{ \frac{1}{n} \sum_i \frac{1}{R_i} (W_i - V_i^{LSW})^2 + \varepsilon^{1/3} \frac{1}{n} \sum_i \sum_{j \neq i} \frac{W_i W_j}{d_{ij}} \right\}. \quad (10.19)$$

Hence our goal will be to show that for any $\delta > 0$ there exists a constant $C = C(\delta)$ such that:

$$P\left(\left\{ -C \leq T \leq -\frac{1}{C} \right\}^c\right) < \delta, \quad (10.20)$$

where:

$$T := \min_{\{W_i\}_i; \sum_i W_i} \left\{ \varepsilon^{-1/2} \frac{1}{n} \sum_i \frac{1}{R_i} (W_i - V_i^{LSW})^2 + \varepsilon^{-1/6} \frac{1}{n} \sum_i \sum_{j \neq i} \frac{W_i W_j}{d_{ij}} \right\}.$$

Notice that this is exactly the statement in Theorem 10.2, since our scaling is such that $|\langle \dot{E}^{LSW} \rangle| = O(1)$.

The variational formulation has the advantage that, first, we get rid of the nonlocal term u_∞, which is not explicit, and, second, that we can obtain an upper bound by constructing a suitable test function $\{W_i\}_i$.

The upper bound:
In the supercritical case, that is the case when the system size is much larger than the screening length ξ, our intuition is that the system separates into many small subsystems of size of order ξ. With this idea in mind we divide our system into subsystems of order ξ and use the LSW construction in each subsystem j, that is $W_i := \dfrac{R_i}{\overline{R}_{[j]}} - 1$, where $\overline{R}_{[j]}$ means that we take the average over subsystem j. This construction indeed gives the desired upper bound. The computations are somewhat tedious but straightforward (see Hönig et al. 2005b for details).

The lower bound:
We now turn to the mathematically most interesting part, which is the lower bound for T. We write $T = T_0 + T_1$ with

$$T_0 := \varepsilon^{-1/2} \frac{1}{n} \sum_i \frac{1}{R_i} \left(W_i - V_i^{LSW}\right)^2, \quad T_1 := \varepsilon^{-1/6} \frac{1}{n} \sum_i \sum_{j \neq i} \frac{W_i W_j}{d_{ij}},$$

that is, T_0 is the 'good' positive part, and what we need to show is that T_1 can be split in terms which can be absorbed in T_0 and other terms which are bounded in weak-L^1, that is we aim to show that $|T_1| \leq \frac{1}{2} T_0 + \tilde{T}$, where $\tilde{T}$ is bounded in weak-L^1. (We say that T is bounded in weak-L^1 if there exists exists a constant C such that $P(|T| \geq M) \leq C/M$ for all $M > 0$.)

- Replace V_i^{LSW} by $\dfrac{R_i}{\langle R \rangle} - 1$:

  In a first step we replace in T_0 the term V_i^{LSW} by $L_i := \dfrac{R_i}{\langle R \rangle} - 1$. This has the advantage that $\langle L_i \rangle = 0$ and $\langle L_i L_j \rangle = 0$ for $i \neq j$. It is not difficult to show that the error which is made by this replacement is bounded in the supercritical regime, which ensures that $\overline{R}$ is a good approximation of $\langle R \rangle$. We omit the details here.

- Introduce cut-off length $\hat{\xi} := \delta \xi$:
  Next, we introduce a length $\hat{\xi} := \delta \xi$, where $\delta > 0$ is a small number, which will be chosen appropriately. We split the kernel:

  $$\frac{1}{d_{ij}} = \frac{e^{-d_{ij}/\hat{\xi}}}{d_{ij}} + \frac{1 - e^{-d_{ij}/\hat{\xi}}}{d_{ij}}$$

  into a far-field and near-field respectively, a splitting motivated by the screening effect and also used for example in the Ewald summation method. Accordingly we split:

  $$T_1 = \varepsilon^{-1/6} \frac{1}{n} \sum_i \sum_{j \neq i} \frac{e^{-d_{ij}/\hat{\xi}}}{d_{ij}} W_i W_j + \varepsilon^{-1/6} \frac{1}{n} \sum_i \sum_{j \neq i} \frac{1 - e^{-d_{ij}/\hat{\xi}}}{d_{ij}} W_i W_j$$
  $$=: T_{11} + T_{12}.$$

- *The 'far-field' term:*
  It turns out that the far-field term T_{12} is the simpler one to estimate. We split again:

  $$T_{12} := \varepsilon^{-1/6} \frac{1}{n} \sum_i \sum_j \frac{1 - e^{-d_{ij}/\hat{\xi}}}{d_{ij}} W_i W_j - \varepsilon^{-1/6} \frac{1}{n\hat{\xi}} \sum_i W_i^2$$
  $$=: T_{121} - T_{122}.$$

  We see that T_{121} is positive, since the kernel is even and is the Fourier transform of a positive measure and hence a function of positive type according to Bochner's theorem.

  On the other hand:

  $$T_{122} = \varepsilon^{-1/6} \frac{1}{n\hat{\xi}} \sum_i (W_i - L_i + L_i)^2$$
  $$\leq 2\varepsilon^{-1/6} \frac{1}{\hat{\xi}} \left(\frac{1}{n} \sum_i (W_i - L_i)^2 + \frac{1}{n} \sum_i L_i^2 \right)$$
  $$\leq C\varepsilon^{-1/6} \frac{1}{\hat{\xi}} \left(\varepsilon^{1/2} T_0 + \frac{1}{R^2} \right)$$

  since $R_i \leq C_0$ and since $\frac{1}{n} \sum_i L_i^2 \leq C \frac{1}{R^2}$. Recall that $\hat{\xi} \sim \varepsilon^{-1/6}$ and hence:

  $$T_{122} \leq \frac{C}{\delta} \left(\varepsilon^{1/2} T_0 + \frac{1}{R^2} \right).$$

  Using large deviation theory one can show that the expected value of all moments of $\overline{R}^{-1}$ are bounded. Hence, once we have chosen δ, we can choose e.g., $\varepsilon \leq \delta^2$ such that and T_{122} is bounded by $C\delta T_0$ plus a term which is bounded in weak-L^1.

- *The 'near-field' term:*
  It remains to estimate the near-field term T_{11}. We write:

  $$T_{11} = \varepsilon^{-1/6} \frac{1}{n} \sum_i \sum_{j \neq i} \frac{e^{-d_{ij}/\hat{\xi}}}{d_{ij}} (W_i - L_i)(W_j - L_j)$$
  $$+ 2\varepsilon^{-1/6} \frac{1}{n} \sum_i \sum_{j \neq i} \frac{e^{-d_{ij}/\hat{\xi}}}{d_{ij}} L_j (W_i - L_i) + \varepsilon^{-1/6} \frac{1}{n} \sum_i \sum_{j \neq i} \frac{e^{-d_{ij}/\hat{\xi}}}{d_{ij}} L_i L_j. \tag{10.21}$$

  The first term on the right hand side can be estimated by a kind of convolution argument and turns out to be smaller than $C\varepsilon^{1/3}(\hat{\xi})^2 T_0 \leq C\delta T_0$.

We denote the second term in (10.21) by T_{112} and have with:

$$Z_i^2 := \sum_{j \neq i} \sum_{k \neq i} \frac{e^{-\frac{d_{ij}}{\xi}} e^{-\frac{d_{ik}}{\xi}}}{d_{ij} \, d_{ik}} L_j L_k$$

that:

$$|T_{112}| \leq \varepsilon^{-1/6} \Big(\frac{1}{n} \sum_i (L_i - W_i)^2\Big)^{1/2} \Big(\frac{1}{n} \sum_i Z_i^2\Big)^{1/2}.$$

As before we argue that $\frac{1}{n}\sum_i (L_i - W_i)^2 \leq C\varepsilon^{1/2}T_0$. Furthermore, due to $\langle L_j L_k\rangle = 0$ for $j \neq k$, we have:

$$\langle Z_i^2 \rangle = \sum_{j \neq i} \frac{e^{-\frac{2d_{ij}}{\xi}}}{d_{ij}^2} \langle L_i^2 \rangle \leq C\xi,$$

where the last inequality follows from our regularity assumptions on the distribution of particle centres which allow to approximate sums by the corresponding integrals. Thus, we obtain:

$$\begin{aligned} P(|T_{112}| \geq M) &\leq \frac{1}{M} \langle |T_{112}|\rangle \\ &\leq \frac{1}{M} \varepsilon^{-1/6} \varepsilon^{1/4} \sqrt{T_0} \sqrt{\xi} \\ &= \frac{1}{M} \varepsilon^{-1/6 + 1/4 - 1/12} \sqrt{\delta T_0} \\ &\leq \frac{C}{M}(\delta T_0 + 1), \end{aligned}$$

which says that T_{112} is bounded in weak-L^1.

The third term in (10.21) can be handled similarly, we omit the proof here.

- *Summary:*
Collecting the above computations we have:

$$T \geq (1 - C\delta)T_0 + \tilde{T}_1$$

with $P(|\tilde{T}_1| \geq M) \leq \frac{C}{M}$. Choosing δ sufficiently small finishes the proof of the lower bound.

10.4 Approaches to extend the LSW model

In this section we review different approaches to derive extensions to the LSW model which take nonvanishing volume fraction into account. The theories we present now are not derived in full rigour, which due to the complexity of the problem can also not be expected. The first approach, described in Section 10.4.1, has been derived by establishing several building blocks rigorously. The model is also self-consistent for small times. However, it turns out not to be self-consistent for large times. Another approach, which overcomes this difficulty is presented in Section 10.4.2. Section 10.4.3 finally discusses an ad hoc model which takes encounters of particles into account.

10.4.1 BBGKY hierarchy to capture correlations

The first attempt to derive a corresponding theory was done in (Marqusee and Ross 1984), where an evolution of the one-point statistics under the assumption of independently and identically distributed particles is derived. However, it is obvious that the assumption of statistical independence is not preserved up to the relevant order $O(\varepsilon^{1/2})$ by the evolution and thus this theory is not self-consistent.

A more advanced theory has been developed by Marder (1987) who takes the build up of correlations into account. Let us briefly discuss why one expects a faster coarsening process due to correlation effects. Consider a system which has undergone coarsening and suppose you find a large particle. The likely reason for it being large is that it is surrounded by smaller than average particles. Because of that fact the large particle can also grow faster than predicted by the LSW mean-field theory. Equally, smaller than average particles shrink faster than predicted by the mean-field theory, and one should obtain larger coarsening rates than within the LSW model.

In order to access correlations Marder (1987) derives the evolution of the two-point statistics up to an error $o(\varepsilon^{1/2})$. Starting from the monopole approximation he generates the BBGKY hierarchy for the particle number densities, computes the growth rates which appear as coefficients in these equations, and truncates the hierarchy on the level of two-particle statistics by a closure hypothesis.

The goal of Hönig et al. (2005a) was to find a new method to identify the conditional expectations of particle growth rates under a more natural closure hypothesis than Marder's.

The assumption in Hönig et al. (2005a) is that the joint probability distribution of $\{(R_i, X_i)\}_{i\geq 1}$ satisfies a cluster expansion. More precisely, if $f_1(R_1, t)$ and $f_2(R_1, R_2, X_1, X_2, t)$ denote the one- and two-particle number densities respectively, it is assumed, with $g_2(R_1, R_2, X_1, X_2, t) := f_1(R_1)f_1(R_2) - f_2(R_1, R_2, X_1, X_2, t)$, that $\dfrac{g_2}{f_1 f_1} = O(\varepsilon^{1/2})$ and that higher order correlations are of order $o(\varepsilon^{1/2})$ and can henceforth be neglected.

Under this assumption Hönig et al. (2005a) derive that f_1, f_2 satisfy the Liouville equations:

$$\frac{\partial f_1}{\partial t} + \frac{\partial}{\partial R_1}\left(\frac{1}{R_1^2}\langle V_1\mid 1\rangle f_1\right) = 0,$$
$$\frac{\partial f_2}{\partial t} + \frac{\partial}{\partial R_1}\left(\frac{1}{R_1^2}\langle V_1\mid 1,2\rangle f_2\right) + \frac{\partial}{\partial R_2}\left(\frac{1}{R_2^2}\langle V_2\mid 1,2\rangle f_2\right) = 0, \quad (10.22)$$

where $\langle V_1\mid 1\rangle, \langle V_1\mid 1,2\rangle$ denote the expected growth rates of particles conditioned on size and position of particle (R_1, X_1) and (R_1, X_1, R_2, X_2) respectively. These are given by:

$$\langle V_1 | 1 \rangle = \left(1 + \frac{R_1}{\xi}\right)(R_1 u_\infty - 1 - \delta u_1) + o(\varepsilon^{1/2}), \tag{10.23}$$

$$\langle V_1 | 1, 2 \rangle = \left(1 + \frac{R_1}{\xi}\right)(R_1 u_\infty - 1 - (\delta u_1 + \delta u_2))$$
$$+ \frac{R_1}{d_{12}} e^{-\frac{d_{12}}{\xi}} (1 - R_2 u_\infty) + o(\varepsilon^{1/2}), \tag{10.24}$$

where for $i = 1, 2$:

$$\delta u_i = \int \frac{e^{-\frac{|y - X_1|}{\xi}}}{|y - X_1|} (1 - R u_\infty) \frac{g_2(R_i, X_i, R, y)}{f_1(R_i)} \, dR \, dy \tag{10.25}$$

and δu_i have relative size of order $O(\varepsilon^{1/2})$. The mean-field u_∞ is implicitly determined by volume conservation, which is expressed by the condition $\langle V_1 \rangle = 0$.

Notice that the terms R_1/ξ and R_1/d_{12}, etc. are terms which have typically size of order $\varepsilon^{1/2}$ due to (10.13). Hence, we recover in (10.23) to leading order the LSW theory.

We also observe that (10.24) has the expected structure. The second term on the right hand side describes how a particle (R_2, X_2) affects the growth rate of particle (R_1, X_1). If it is larger than average, the growth rate of particle (R_1, X_1) is smaller than predicted by LSW, if it is smaller than the growth rate of particle (R_1, X_1) increases. The effect is more relevant the closer particle (R_2, X_2) is to particle (R_1, X_1) and can be neglected if the distance between two particles is larger than the screening length ξ.

Nevertheless, it turned out that the model (10.22)–(10.25), despite its complexity, is still not satisfying. First, even though this is not demonstrated rigorously, it seems that the model contains no mechanism to select a unique self-similar solution. Furthermore, and most importantly, the model is not self-consistent for large times, more precisely it fails for the largest particles in the system. The argument for the latter is basically as follows. Suppose one solves (10.22)–(10.24) for uncorrelated initial data, where $f_1(R_1, 0)$ has compact support. Consequently, the support of $f_2 = f_2(R_1, R_2, X_1, X_2, 0)$ is also compact in R_1 and R_2. However, the evolution of R_1 and R_2, determined by (10.24), depends on space due to the term $e^{-d_{12}/\xi}/d_{12}$ in (10.24). Therefore, particles R_1 and R_2 which are at a distance smaller than ξ evolve differently from particles R_1 and R_2 which are at a distance much larger than ξ. As a consequence also the support of f_2 in R_1 and R_2 varies in space and we obtain regions in the variables (R_1, R_2, X_1, X_2) where f_2 identically vanished but $f_1(R_1)$ and $f_1(R_2)$ do not and consequently g_2 is of the order $f_1(R_1) f_1(R_2)$ which violates the cluster expansion.

10.4.2 Boundary layers due to fluctuations

For the reasons described in the previous section one cannot assume that correlations are small around the largest particles, and hence a uniform cluster expansion approach cannot be successful. The onset of correlations for the largest particles has instead to be described by a suitable boundary layer, that is a small region in the space of radii on which the number densities vary rapidly. A corresponding model has been derived in (Niethammer and Velàzquez 2008). The analysis is quite elaborate and the resulting model is also complicated to state in full detail. We confine ourselves here to describing the most important aspects.

The main idea in the derivation of the model is that we do not start from an expansion on the level of the number densities, but instead on the level of the trajectories of particles. This allows for a closure relation using Taylor's expansion in order to express f_2 by f_1 and $\partial_R f_1$.

One aspect is, however, very similar to the analysis of Hönig et al. (2005a). A key idea in the computation is always to describe a system of particles through the ones in a system where a particle has been removed. This is a version of Schwarz alternating method.

The resulting model has the following form:

$$\partial_t f_1 + \partial_R \left(\frac{1}{R^2}(Ru_\infty - 1)f_1 \right) = \varepsilon^{1/2} \partial_R \left(D(R) \partial_R f_1 \right) \qquad (10.26)$$

where the function $D = D(R)$ acts as a kind of diffusion coefficient and is determined via a complicated nonlocal integral equation. We refer for details to (Niethammer and Velàzquez 2008); the most relevant property of D, however, is that it is positive and has the appropriate scaling such that (10.26) has a scale invariance.

Let us emphasize again that the right hand side is seemingly a higher order term due to the factor $\varepsilon^{1/2}$. However, this is only true where f_1 is not small. For largest particles, where f_1 is small, the right hand side of (10.26) becomes of the same size as the left hand side.

By formal asymptotic expansion it is also established in (Niethammer and Velàzquez 2008) that a unique self-similar solution to (10.26) exists. This is a perturbation of the LSW self-similar solution with a Gaussian tail. Thus, the boundary layer provides a possible solution to the selection problem within the LSW theory. The induced correction to the mean particle size of order $\varepsilon^{1/4}$. Notice, that the latter does not contradict our scaling analysis in Section 10.3. For short times we have that the correction terms are of order $\varepsilon^{1/2}$. This does not say, however, what order of size of correction we can expect in a self-similar regime.

10.4.3 The LSW model with encounters

A different approach from the ones described in the last two sections has been suggested already in (Lifshitz and Slyozov 1961). As we have already mentioned, particles may collide during their evolution and merge into a larger particle. At first glance, this effect seems to be of higher order than correlations, since the number of particles per unit volume which are involved in collisions is of order $\rho\varepsilon$ and as a consequence the correction of the LSW model due to collisions should have relative size of order ε.

A model which takes this effect into account has already been suggested in (Lifshitz and Slyozov 1961). To state it, it is more convenient to change variables from radius R to volume $v := R^3$. After rescaling time by a constant, the LSW law $\dot{R} = \frac{1}{R^2}(Ru_\infty - 1)$ reads in the volume variable $\dot{v} = v^{1/3}u_\infty - 1$ and the LSW model for the density of volumes g (defined by $g(v)\,dv = f_1(R)\,dR$) is given by:

$$\partial_t g + \partial_v\big((v^{1/3}u_\infty - 1)g\big) = 0\,.$$

Introducing self-similar variables via $x := \dfrac{v}{t}$, $F(x,\tau) := t^2 g(v), \tau = \ln t$ and $\lambda = u_\infty t^{1/3}$ we obtain the equation in self-similar variables as:

$$\partial_\tau F - x\partial_x F - 2F + \partial_x\big((x^{1/3}\lambda - 1)F\big) = 0\,. \qquad (10.27)$$

To account for collisions, or 'encounters' as the phenomenon is called by Lifshitz and Slyozov, a coagulation term is added on the right hand side which is of the form:

$$\frac{1}{2}\int_0^x K(y, x-y)F(x-y)F(y)\,dy - F(x)\int_0^\infty K(x,y)F(y)\,dy\,. \qquad (10.28)$$

Since merging particles basically add their volume (this is not completely correct, since at the same time they still interact with the other particles, but sufficient for our purpose), it is assumed that K is additive and grows proportionally to $x + y$ as $x, y \to \infty$. For simplicity we set $K(x,y) := x + y$. To summarize, after normalizing to $\int xF(x)\,dx = 1$, self-similar solutions for the LSW model with encounters are given by the equation:

$$\begin{aligned}-x\partial_x F - 2F + \partial_x\big((x^{1/3}\lambda - 1)F\big)\\= \varepsilon\Big(\frac{x}{2}\int_0^x F(x-y)F(y)\,dy - xF(x)\int_0^\infty F(y)\,dy - F(x)\Big)\,.\end{aligned} \qquad (10.29)$$

Naively, one would expect that since the order of the right hand side is $O(\varepsilon)$, collisions are not as relevant as correlations and fluctuations which are of order $O(\varepsilon^{1/2})$. However, all particles can encounter other particles and thus two colliding particles of medium size produce a large particle which then dominates the long-time behaviour. Hence, for the large-time, behaviour encounters could be more relevant than fluctuations and correlations.

This conjecture is supported by an asymptotic analysis by Lifshitz and Slyozov (1961). Assuming that there exists a fast decaying solution to (10.29), they find that the correction of the growth rate of the particles is of order is of order $\dfrac{1}{|\ln\frac{1}{\varepsilon}|^2}$ and hence much larger than the correction induced by fluctuations.

It is still open, however, whether the analysis in (Lifshitz and Slyozov 1961) is correct, since is is not obvious that exponentially fast decaying solutions to (10.29) exist at all. In fact, we know from the pure coagulation equation that such a solution only exists for $\varepsilon = 1$. The situation here might be, of course, completely different. Preliminary computations by Herrmann et al. (2007) suggest, that for small ε there are both, algebraically decaying solutions as well as an exponentially decaying one. If this turns out to be correct, it is reasonable to expect, that solutions to the time dependent problem with compactly supported data converge to the self-similar solution with exponential decay and the correction to the mean radius is indeed of order $\frac{1}{|\ln\frac{1}{\varepsilon}|^2}$.

To summarize, even though the model including encounters is set up only ad hoc and is not derived from the Mullins–Sekerka evolution, which would be another challenging task, the enormous effect on the mean radius and hence the coarsening rate suggests that encounters are in fact more relevant for the long-time self-similar dynamics than fluctuations. The explanation lies in the kinetic character of the collision term, that the fraction of particles which are transported to the supercritical regime is of order ε, whereas the diffusive correction due to fluctuations in (10.26) only involves the few largest particles.

References

Carr, J. and Penrose, O. (1998). Asymptotic behaviour in a simplified Lifshitz–Slyozov equation. *Physica D* **124**, 166–76.

Conti, S., Hönig, A., Niethammer, B., and Otto, F. (2006). Non-universality in low volume fraction Ostwald Ripening. *J. Stat. Phys.* **124**, 231–59.

Dai, S. and Pego, R. L. (2005a). On the monopole approximation of the Mullins-Sekerka model. Preprint.

Dai, S. and Pego, R. L. (2005b). Universal bound on coarsening rates for mean-field models of phase transitions. *SIAM J. Math. Anal.* **37**, 2, 347–71.

Dai, S. and Pego, R. L. (2005c). An upper bound on the coarsening rate for mushy zones in a phase-field model. *Interfaces and Free Boundaries.* **7**:2,187–97.

Fradkov, V. E., Glicksman, M. E., and Marsh, S. P. (1996). Coarsening kinetics in finite clusters. *Phys. Rev. E* **53**, 3925–32.

Giron, B., Meerson, B., and Sasorov, P. V. (1998). Weak selection and stability of localized distributions in Ostwald ripening. *Phys. Rev. E* **58**, 4213–6.

Herrmann, M., Niethammer, B., and Velàzquez, J. J. L. (2007). Work in preparation.

Hönig, A., Niethammer, B., and Otto, F. (2005a). On first–order corrections to the LSW theory I: infinite systems. *J. Stat. Phys.* **119** 1/2, 61–122.

Hönig, A., Niethammer, B., and Otto, F. (2005b). On first–order corrections to the LSW theory II: finite systems. *J. Stat. Phys.* **119** 1/2, 123–64.

Kohn, R. V. and Otto, F. (2002). Upper bounds for coarsening rates. *Comm. Math. Phys.* **229**, 375–95.

Kohn, R. V. and Yan, X. (2004). Coarsening rates for models of multicomponent phase separation. *Interfaces and Free Boundaries* **6**, 135–49.

Laurençot, P. (2002). The Lifshitz–Slyozov–Wagner equation with total conserved volume. *SIAM J. Math. Anal.* **34**, 2, 257–72.

Lifshitz, I. M. and Slyozov, V. V. (1961). The kinetics of precipitation from supersaturated solid solutions. *J. Phys. Chem. Solids* **19**, 35–50.

Marder, M. (1987). Correlations and Ostwald ripening. *Phys. Rev. A* **36**, 858–74.

Marqusee, J. A. and Ross, J. (1984). Theory of Ostwald ripening: Competitive growth and its dependence on volume fraction. *J. Chem. Phys.* **80**, 536–43.

Niethammer, B. (1999). Derivation of the LSW theory for Ostwald ripening by homogenization methods. *Arch. Rat. Mech. Anal.* **147**, 2, 119–78.

Niethammer, B., Otto, F., and Velàzquez, J. J. L. (2006). On the effect of correlations, fluctuations and collisions in Ostwald ripening. In A. Mielke (Ed.), *Analysis, Modeling and Simulation of Multiscale Problems*, pp. 501–30. Springer: Berlin.

Niethammer, B. and Otto, F. (2001). Ostwald Ripening: The screening length revisited. *Calc. Var. and PDE* **13**, 1, 33–68.

Niethammer, B. and Pego, R. L. (1999). Non–self–similar behavior in the LSW theory of Ostwald ripening. *J. Stat. Phys.* **95**, 5/6, 867–902.

Niethammer, B. and Pego, R. L. (2005). Well-posedness for measure transport in a family of nonlocal domain coarsening models. *Indiana Univ. Math. J.* **54**, 2, 499–530.

Niethammer, B. and Velàzquez, J. J. L. (2004a). Homogenization in coarsening systems I: deterministic case. *Math. Meth. Mod. Appl. Sc.* **14**, 8, 1211–33.

Niethammer, B. and Velàzquez, J. J. L. (2004b). Homogenization in coarsening systems II: stochastic case. *Math. Meth. Mod. Appl. Sc.* **14**, 9, 1–24.

Niethammer, B. and Velàzquez, J. J. L. (2004c). Well–posedness for an inhomogeneous LSW–model in unbounded domains. *Math. Annalen* **328**, 3, 481–501.

Niethammer, B. and Velàzquez, J. J. L. (2006a). Global stability and bounds for coarsening rates within the LSW mean-field theory. *Comm. in PDE* **31**, 1–30.

Niethammer, B. and Velàzquez, J. J. L. (2006b). Screening in interacting particle systems. *Arch. Rat. Mech. Anal.* **180**, 3, 493–506.

Niethammer, B. and Velàzquez, J. J. L. (2008). On screening induced fluctuations in Ostwald ripening. *J. Stat. Phys.* **130**,3, 415–53.

Pego, R. L. (2007). Lectures on dynamics in models of coarsening and coagulation. Lecture Notes Series, Institute for Mathematical Sciences, National University of Singapore.

Velàzquez, J. J. L. (2000). On the effect of stochastic fluctuations in the dynamics of the Lifshitz–Slyozov–Wagner model. *J. Stat. Phys.* **99**, 57–113.

Wagner, C. (1961). Theorie der Alterung von Niederschlägen durch Umlösen. *Z. Elektrochemie* **65**, 581–94.

11

SWITCHING PATHS FOR ISING MODELS WITH LONG-RANGE INTERACTION

Nicolas Dirr

Abstract

We introduce a multi-scale model for two-phases material. The model is on the finest scale a stochastic process. The effective behaviour on larger scales is governed by deterministic nonlinear evolution equations. Due to the stochasticity on the finest scale, deviations from these limit evolution laws can happen with small probability. We describe the most likely among those deviations in two situations: first we consider the switching from one stable equilibrium of the evolution equation to another one, then we describe what happens when enforce a fast motion on a manifold of stationary solutions. This chapter is based on joint work with Giovanni Bellettini, Anna DeMasi, and Errico Presutti.

11.1 Introduction

11.1.1 *Motivation*

This note is mainly concerned with the qualitative behaviour of minimizing sequences to certain functionals, which is of course a standard topic in the Calculus of Variations and, so far, a piece of pure mathematics. The purpose of the following paragraphs is to motivate how our results are related to a multiscale description of material behaviour in the presence of thermal effects, i.e., noise.

11.1.1.1 *Gradient flow models at multiple scales* It is well known that physical models are described by a family of effective theories, such that each of them is valid in a certain time–space scale. The challenge for the mathematician is to *prove* that the effective theory on a larger scale can be derived from that on a finer scale under a suitable rescaling (and averaging out degrees of freedom). The ratio between the two (length) scales is a small parameter, and a typical mathematical result states that an 'averaged' and rescaled solution to the fine-scale equation converges to a solution of the coarse-scale equation if this parameter vanishes.

Here however, we look in a certain sense at higher-order effects. We look at deviations from the effective evolution law at a larger scale which are rare, but possible, due to a *stochastic* evolution law on a finer scale. More precisely we deal with models for the evolution of a phase boundary at three scales: at the finest scale, called *atomistic*, we deal with a stochastic process; at an intermediate scale, called *mesoscopic*, with a nonlocal reaction-diffusion equation; and at the

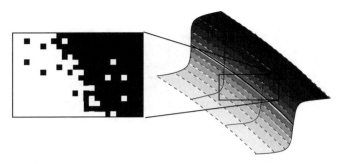

FIG. 11.1: Schematic drawing of the micro- meso- and macroscale for a model for phase boundaries.

largest scale with a geometric evolution law for the interface (mean curvature flow). (See also Fig. 11.1).

Of course the probability of *not* following the effective evolution law at a larger scale converges to zero with the ratio between the scales, but we are interested in finding among all those unlikely paths of the system that *deviate* from the effective large-scale evolution, the most likely one. This most likely deviating path may be rare, but on large samples (many independent realizations in space) or after large waiting times it will be seen. Such deviations allow in particular the switching between two local minima of the free energy of the system. (See also Olivieri and Vares (2005) and Cassandro et al. (1986).) This would be impossible for the evolution law itself, because the effective models on the meso- and macroscale have the free energy for a Lyapunov functional, i.e., the evolution can never leave an isolated local minimum.

The gradient flow dynamics in our models is a consequence of the type of stochastic process which we choose as a starting point on the finest scale (see next subsection). Gradient flow models are a good approximation in the case of strong damping/dissipation. Assume that the state of the system is described by a vector X, and the associated energy is given by $V(X)$. In the case of high friction (here $\epsilon \ll 1$) Newton's law yields (in suitable units):

$$\ddot{Y} = -\frac{1}{\epsilon}\dot{Y} - V'(Y)$$

Now let $X(t) = Y(\epsilon^{-1}t)$, then $\epsilon^2 \ddot{X} = -\dot{X} - V'(X)$ and we obtain formally as $\epsilon \to 0$ $\dot{X} = -V'(X)$, i.e., a gradient flow of the energy V.

In this note we look at two types of deviations from the gradient flow evolution, see also Fig. 11.2.

1. **Switching** The transition between two local minima of the free energy, which are separated by a potential wall. Here both minima have the same energy.

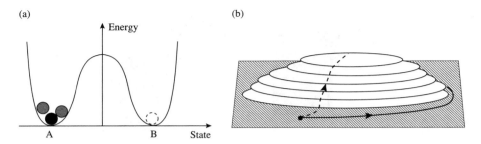

FIG. 11.2: Schematic drawing of an energy landscape for the case of 'switching' from well A to well B (left), and for the fast motion on a stationary manifold (right).

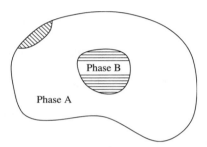

FIG. 11.3: Structure of the model: two phases, separated by a phase boundary.

2. **Motion on stationary manifold** Along a level set of the free energy the gradient vanishes, and therefore a gradient flow evolution should not move at all. We put as constraint a 'fast' motion and look for the most likely deviation. It turns out that this may be a path that leaves the level set of the energy.

In this note the two minima A, and B, correspond to the two 'phases' of a material occupying a container. The potential wall is related to the fact that the formation of a (growing) region of phase B into phase A requires a phase boundary, i.e., a region in space separating the subdomain occupied by phase A from the subdomain occupied by phase B. Near this phase boundary the system is neither close to A nor to B, hence the free energy there is higher than in any of the pure phases. (See also Fig. 11.3). The container wall is supposed to be neutral (Neumann or reflecting boundary conditions), i.e., no free energy cost is associated with the region where a subdomain occupied by one phase touches the container wall.

The manifold of stationary solutions in case 2 corresponds to the manifold of instantons (stationary solutions connecting the two phases) in a one dimensional phase field model.

11.1.2 Ising-model with Kac potential and Glauber dynamics: the three scales

Here we define more precisely our model for two-phase material. First we describe the stochastic process at the finest (atomistic) scale, then we explain the effective evolution law on the meso- and macroscale. In this note, $D = \mathbb{R}$ or $D = [-L, L]^d$, $d \in \{1, 2, 3\}$, $L \gg 1$. We state the definitions as needed later, but many of the statements hold in more general situations, see De Masi et al. (1994a).

The basic object on all scales is the *interaction kernel* $J(x, y) : D \times D \to [0, \infty)$, which we assume (for simplicity) to be smooth and which we require to have in addition the following properties:

1. $J(x, y) = J(y, x)$
2.
$$\int_D J(x, y) dy = 1$$
3. If $D = \mathbb{R}$ then $J(x, y) = \tilde{J}(|x - y|)$ for some smooth function $\tilde{J}$ with $\tilde{J}(r) = 0$ for $r > 1$.
4. If $D = [-L, L]^d$, then with $\tilde{J}$ as above
$$J(x, y) = \sum_{y' \simeq y} \tilde{J}(|x - y'|),$$
where $y' \simeq y$ means that y' is equal to y modulo reflections along the lines $\{y = \pm(2n + 1)L/2\}$ and $\{x = \pm(2n + 1)L/2\}$, $n \in \mathbb{Z}$.

Now we define the *stochastic process:* let $\Lambda = \gamma^{-1} D \cap \mathbb{Z}^d$, where $\gamma \ll 1$ and γ^{-1} is the *interaction range*.

The state space is $\{-1, 1\}^\Lambda$, an element is denoted by σ, and its value on a site X on $\mathbb{Z}^d \cap \gamma^{-1} D$ is denoted by $\sigma(X)$ and called *the spin at x*.

We can define a Markov process on this state space by defining the 'flip' rate for the spin $\sigma(X)$, i.e., loosely speaking, if $c(X, t)$ is the flip rate of the spin at X at time t, then the probability of the event that $\sigma(X)$ changes to $-\sigma(X)$ in the time interval $[t, t + dt]$ equals $c(X, t)dt$. Of course the flip rate depends in general on the state σ.

This flip rate at site X depends on the spin at X and on its neighbours (interaction) in such a way that the spin is more likely to align with its neighbours than to oppose them (ferromagnetic interaction), but the dependence is only through the average over neighbours within the interaction range (which is large, γ^{-1}.) This is called *Kac potential* or Kac interaction.

Thus the model is of local mean field type. See also the picture on the left of Fig. 11.4.

This type of dynamics, where the *total magnetization* (sum of the spins) is not conserved, is known as *Glauber dynamics*.

More precisely, with the *effective field*:
$$h_X(\sigma) := \sum_Y \gamma^d J(\gamma X, \gamma Y) \sigma(Y),$$

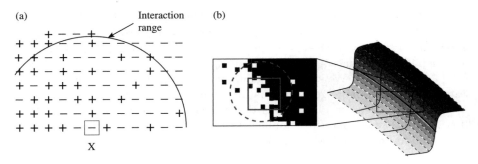

FIG. 11.4: a) interaction through average of neighbours. b) coarse graining over a box smaller than the interaction range.

felt by the spin at site X we define:

$$c(X,t) = \frac{e^{-\beta\sigma(X)h_X(\sigma(t))}}{e^{\beta h_X(\sigma(t))} + e^{-\beta h_X(\sigma(t))}},$$

where β is the inverse temperature, a quantity that controls in some sense how far the spins are from independence, i.e., from $\beta = 0$. We assume $\beta > 1$ in order to have two phases, one with magnetization close to 1 and the other one with magnetization close to -1.

As the interaction between sites depends only on averaged quantities, it is plausible to assume that correlations between spins remain small if they were small initially (propagation of chaos). If the spins are almost uncorrelated, then averaged quantities should, by some law of large numbers, be almost deterministic. This is indeed the case, for rigorous results we refer to De Masi et al. 1994a. More precisely, let us define the following *coarse graining* ('block spin transform' in De Masi et al. 1994a). Rescale space with γ, i.e., $x = \gamma X$ for $x \in D$. Then we define the piecewise constant random function $m^\gamma(x,t,\omega) \in [-1,1]$ by:

$$m^\gamma(X,t,\omega) := (\gamma^\alpha)^d \sum_{|Y_i - X_i|_\infty < \frac{1}{2}\gamma^{-\alpha}} \sigma(Y,t,\omega) \tag{11.1}$$

for $0 < \alpha < 1$, i.e., we average over many spins, but over an area which is smaller than the interaction range. It can be shown (see De Masi et al. 1994a, see also Markos et al. 1995) that on space–time domains that grow slowly in γ^{-1} (e.g., logarithmically) the random function m^γ converges in probability to a deterministic function $m(x,t)$ which solves the nonlocal nonlinear evolution equation:

$$m_t = -m + \tanh(\beta J * m), \tag{11.2}$$

which is thus the *effective evolution law* on the *mesoscopic scale*. ($*$ denotes, as usual, the convolution.)

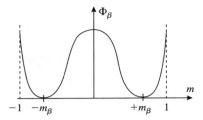

FIG. 11.5: Double-well potential with wells in $\pm m_\beta$.

This equation has some features in common with the well-known Allen–Cahn equation (see Allen and Cahn 1979):

$$\partial_t m = \Delta m - W'(m), \tag{11.3}$$

where W is a so-called *double-well potential*, i.e., it has exactly two local minima of equal depth, is strictly convex near each local minimum, and the local minima are separated by a potential barrier, see Fig. 11.5. Obviously (11.3) is the L^2-gradient flow of the energy:

$$F_{AC}(m) = \int_D (|\nabla m|^2 + W(m)) dx. \tag{11.4}$$

The mesoscopic equation (11.2) has as Lyapunov functional the *mesoscopic free energy of the Ising spins with Kac interaction*, the functional:

$$F(m) = \int_D \phi_\beta(m) \, dx + \frac{1}{4} \int_{D \times D} J(x, x')[m(x) - m(x')]^2 dx \, dx', \tag{11.5}$$

where Φ_β is a double well potential with minima (wells) at $\pm m_\beta$. m_β is defined as positive solution of $m = \tanh(\beta m)$, $\beta > 1$, and Φ_β is explicitly given as:

$$\Phi_\beta(m) = -\frac{m^2}{2} + \frac{1}{\beta} \left(\frac{1-m}{2} \log \frac{1-m}{2} + \frac{1+m}{2} \log \frac{1+m}{2} \right).$$

Φ_β is double-well potential for $\beta > 1$. (See Fig. 11.5.)

As $J \geq 0$, we see that both (11.4) and (11.5) consist of a term that penalises oscillations and a double well potential, which prefers m to be close to one of two ground states or phases. Hence the two ground states of (11.5) are the constant functions $\pm m_\beta$.

Both functionals have stationary solutions which connect the two wells and which are in some sense the lowest energy excited states (see Bellettini et al. 2005a). We call *standing wave* or *instanton* a function $\overline{m}$ which solves:

$$\overline{m} = \tanh(\beta J * \overline{m}) \tag{11.6}$$

$$\overline{m}(x) : \lim_{x \to \pm \infty} \overline{m}(x) = \pm m_\beta, \quad \overline{m}(0) = 0, \quad \overline{m}'(x) > 0. \tag{11.7}$$

FIG. 11.6: Sketch of invariant manifold in function space, arrows indicating dynamics on the manifold.

For $D = \mathbb{R}$ the equation (11.2) is invariant under translations in space, hence there exists a one-dimensional manifold of such stationary solutions, and the solution of (11.6,11.7) is unique up to translations.

If $D = [-L, L]$, then the (reflecting) boundary conditions break the translational symmetry, and of course we cannot impose any additional boundary conditions like in (11.7). The solution of (11.6) with the lowest energy will be called *finite volume instanton* $\widehat{m}_L$. (See Bellettini et al. 2005a,b). In order to make this well-defined we have to suppose that L is sufficiently large. There exist one-dimensional invariant manifolds $\mathcal{M}^{\pm}$ connecting $\widehat{m}_L$ to the ground states $m \equiv +m_\beta$ and $m \equiv -m_\beta$. (See Fig. 11.6, see also Bellettini et al. 2007.)

Now we turn to the coarsest scale, the *macroscale*. Let D denote the signed distance function and let $\overline{m}$ be the instanton on $\mathbb{R}$, i.e., the solution of (11.6,11.7). If $L = \epsilon^{-1}$, $\epsilon \to 0$, and space and time are rescaled diffusively, i.e. $T = \epsilon^2 t$, $R = \epsilon x$, then one can show by formal asymptotic expansions that both (11.3) and (11.2) admit solutions that are close to $\overline{m}(\epsilon^{-1} d(x, \Sigma(t)))$, where the hypersurface $\Sigma(t)$ evolves by mean curvature, i.e., the normal velocity equals the mean curvature multiplied by a coefficient which depends on J and β in a complicated way. This can be made rigorous, even if the mean curvature evolution admits only generalized solutions, see De Masi et al. 1994a; Markos et al. 1995.

11.1.3 Deviations from the mesoscopic equation

The fact that the random variable m^γ (see 11.1) converges to a (deterministic) solution m of (11.2) can (informally) be restated as:

$$m^\gamma(x, t, \omega) = m(x, t) + r^\gamma(x, t, \omega),$$

with r^γ small. Hence we can consider m^γ as a random perturbation of a solution to (11.2). This bears some similarities with a stochastically forced PDE, see e.g., Faris and Jona-Lasinio 1982.

This random perturbation is small, but it allows with (small) probability *deviations* from the mesoscopic equation. The probability of such deviations is exponentially small and the exponential weight of the event that m^γ is near a given function $m(x, t)$ is given (for $\gamma \to 0$) by the *action functional*, a functional:

$$I_{L;t_0,t_1}(m) : C^\infty(Q_L \times [t_0, t_1], [-1, 1]) \to [0, \infty),$$

such that, loosely speaking:

$$\mathbb{P}(m^\gamma(\omega) \in U_\delta(m)) \sim e^{-\gamma^{-2} I_{L;t_0,t_1}(m)} \quad (0 < \delta \ll 1)$$

where U_δ denotes a neighbourhood in a suitable topology (here the weak topology). Clearly $I_{L;t_0,t_1}(m) = 0$ if and only if $m(x,t)$ solves the mesoscopic limit of equation (11.2). For space reasons this discussion remains at the level of heuristics, for precise definitions of large deviation principles and action functionals in a similar context we refer to the books (Olivieri and Vares 2005; Freidlin and Wentzell 1998).

Hence if we are interested in the most likely deviation satisfying certain constraints, e.g. the constraint to start in one local minimum and to end up in another one (switching), then we have to minimize the action functional under these constraints.

The action functional for the Ising model with Kac potential and Glauber dynamics on the mesoscopic scale has been rigorously derived by F. Comets in the case of periodic boundary conditions and on finite time intervals, (Comets 1987). The extension of the derivation to large (logarithmical in γ) time intervals that would allow for a diffusive rescaling (macroscale) is far from trivial and is a work in progress.

For simplicity we replace the Comets rate function by an 'easier functional'. The extension to the true Comets functional and then to the Ising system may still require a nontrivial work, but we believe that the main physical features of the actual tunnelling excursion are already captured by our results.

First we replace (11.2) by the L^2 gradient flow of (11.5) i.e.:

$$m_t = -\frac{1}{\beta}\tanh^{-1}(m) + J * m. \tag{11.8}$$

As tanh is invertible, being a stationary solution of (11.8) is equivalent to (11.6).

The simplified action functional is given by the L^2- norm of the right hand side $r(x,t)$ which makes the given path $m(x,t)$ a solution of $\partial_t m = -\frac{1}{\beta}\tanh^{-1}(m) + J * m + r(x,t)$, i.e.:

$$I_{L;T}(m) = \frac{1}{4}\int_0^T \int_{[-L,L]^d} \left[m_t + \frac{\delta F(m)}{\delta m}\right]^2 dx\, dt. \tag{11.9}$$

(If $D = \mathbb{R}$, then we simply write I_T.)

The two different types of constraints we impose are as follows:

1. *'Switching' constraint.* ([4]) If $D = [-L,L]^D$, $d \in \{2,3\}$ then we consider $\inf_T \min_{m \in \mathcal{U}_{L,T}} I_{L;T}(m)$, where

$$\mathcal{U}_{L,T} = \left\{m \in C^\infty([-L,L]^d \times [0,T]) : m(r,0) = -m_\beta,\ m(r,T) = m_\beta\right\}, \tag{11.10}$$

i.e., there is *no time constraint*,

2. *Fast motion on invariant manifold.* (Masi et al. 2006) If $D = \mathbb{R}$, i.e. $L = \infty$, then we minimize $I_T(m)$ over the set

$$\mathcal{U}_{T,R} = \{m \in C^\infty(\mathbb{R}),\ m(r,0) = \overline{m}(r),\ m(r,\epsilon^{-2}T) = \overline{m}(r - \epsilon^{-1}R)\} \tag{11.11}$$

11.2 Results

11.2.1 'Switching' ($d=2,3$)

Let the space dimension be 2 or 3, let J be as in Section 11.1.2, 11.4., and let $I_{L,T}$ and $\mathcal{U}_{L,T}$ be as in (11.9) and (11.10), respectively. Define the *cost of switching* as:

$$P_L := \inf_{T>0} \inf_{\mathcal{U}_{L,T}} I_{L;T}(m) \tag{11.12}$$

Let $\widehat{m}$ be the finite volume instanton, and $\widehat{m}^e$ its planar extension to higher dimensions, i.e., $\widehat{m}^e(x) = \widehat{m}(x \cdot e_1)$, where e_1 is the first unit vector in $\mathbb{R}^d$. Moreover let the functions $v^\pm : D \times \mathbb{R} \to \mathbb{R}$ describe the dynamics on the invariant manifolds $\mathcal{M}^\pm$, i.e. $v^\pm(x,t)$ solve (11.8) and:

$$\lim_{s \to -\infty} \|v^\pm(\cdot,s) - \widehat{m}^e(\cdot)\|_{L^2(D)} = 0, \quad \lim_{s \to +\infty} \|v^\pm(\cdot,s) - \pm m_\beta\|_{L^2(D)} = 0.$$

(See Fig. 11.6). Then we have (Theorems 2.3 and 2.4 in Bellettini et al. 2007, for the extension to $d=3$ see the appendix in Bellettini et al. 2007):

Theorem 11.1   1. *For L large enough*

$$P_L = L^{d-1} F^{(1)}(\widehat{m}_L),$$

where $F^{(1)}$ denotes the free energy (see 11.5) in one dimension, i.e., for $D = [-L, L]$.

2. *For all L large enough, if $\{T_n, u_n\}$ is a minimising sequence for 11.12, then $\lim_{n \to +\infty} T_n = +\infty$ and, for any $\epsilon > 0$ there exists a positive integer n_ϵ such that for any $n \geq n_\epsilon$, u_n (or its image under a the symmetry group of a square/cube) has the following properties. There is $s \in (0, T_n)$ so that $\|u_n(\cdot, s) - \widehat{m}^e\|_2 \leq \epsilon$ and there are τ' and τ'' positive so that:*

$$\|u_n(\cdot,t) - v^{(-)}(\cdot, \tau' - t)\|_{L^2(D)} \leq \epsilon, \quad t \in [0, s] \tag{11.13}$$

$$\|u_n(\cdot,t) - v^{(+)}(\cdot, -\tau'' + (t-s))\|_{L^2(D)} \leq \epsilon, \quad t \in [s, T_n]. \tag{11.14}$$

Theorem 11.1 proves that the lowest cost of switching is obtained by paths which have (approximately) planar level sets, and which (approximately) follow the one-dimensional manifolds connecting the saddle $\widehat{m}^e$ and the wells $\pm m_\beta$ (stable local minima). First the path 'climbs up' in the *time reverse direction* and then, after crossing the saddle, follows the forward time direction.

We see that (in the limit) the free energy jumps at time 0 to a value which then remains (up to exponentially small corrections) constant: in the limit the whole penalty is paid at time 0^+. Thus the pattern b) in Fig. 11.7 (pay everything at once) is more favourable than pattern a) (pay 'continuously').

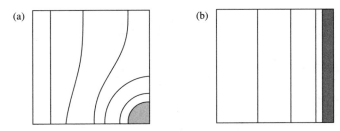

FIG. 11.7: In a) and b) we depict two possible paths on the macroscale. In a) a small droplet of the + phase (dark region) nucleates at a vertex of the square. It then invades the square as time increases, gradually changing its interface, and eventually becomes a rectangle. In b) we have initially a nucleation of a flat interface (dark rectangular region), which smoothly invades the square. b) is optimal, a) is not.

It is important to note that the optimal pattern differs initially from the Wulff shape, i.e., the shape that minimizes the free energy given the volume. For small volumes the Wulff shape is a droplet, only for large volumes a plane.

Our results use information from the macroscale (e.g., about Wulff shapes) but are essentially mesoscopic: we control the L^2-norm in $D = [-L, L]^d$ *without rescaling* (i.e., without division by L^d). Thus we obtain for the case treated here a much stronger result than the ones one could obtain by Γ-convergence methods for action functionals (e.g., as in Ros 2005).

11.2.2 *Optimal displacement of instanton on diffusive scale*

Let $D = R$, let J be as in Section 11.1.2, 11.1.4., let $\overline{m}$ solve (11.6–11.7), and let $I_T := I_{\infty,T}$ and $\mathcal{U}_{T,R}$ be as in (11.9) and (11.11), respectively. We shorthand $\overline{m}_r(x) = \overline{m}(x - r)$. In this case we will move to the macroscale, i.e., rescale space and time diffusively with $\epsilon > 0$. Therefore define the macrocost of displacing an instanton by R macrounits in macrotime T as:

$$W(R, T) = \lim_{\epsilon \to 0} \inf_{m:\ m(0) = \overline{m}, m(T) = \overline{m}_{\epsilon^{-1}R}} I_{\epsilon^{-2}T}(m_t, m)$$

When solving this variational problem, we are not forced to stay on the instanton manifold, we are allowed to leave it (e.g. by nucleating more interface) and return to it only at the end of the time interval. (See Fig. 11.8) The cost of a nucleation is $2\mathcal{F}(\overline{m})$, (see Bellettini et al. 2005a,b) and the optimal cost of n nucleations and subsequent uniform displacement of the resulting $2n+1$ interfaces until they 'collide' is:

$$w_n(R, T) := n2\mathcal{F}(\overline{m}) + (2n+1)\left\{\frac{1}{\mu}\left(\frac{V}{2n+1}\right)^2 T\right\}$$

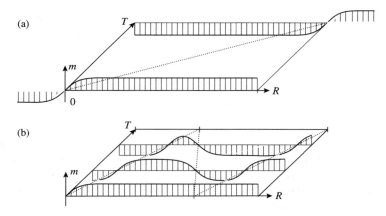

FIG. 11.8: We depict two possible paths on the macroscale. On the top a) an instanton is displaced with uniform velocity, and on the bottom b) new instantons are nucleated and subsequently displacement until they 'collide'. The top picture is optimal for 'slow' constraints ($V = R/T$ small) and the bottom picture is better for 'faster' constraints ($V = R/T$ large).

The following theorem (Theorem 2.3 in De Masi et al. 2006) holds:

Theorem 11.2 *For all $R > 0$ and $T > 0$ and $V = R/T$*

$$W(R,T) = \min_{n \in \mathbb{N}_0} w_n(R,T),$$

and the minimum is attained at $n(V)$ if

$$\mathcal{F}(\overline{m})[(2n(V))^2 - 1] \leq \frac{V^2 T}{\mu} \leq \mathcal{F}(\overline{m})\left([2(n(V)+1)]^2 - 1\right)$$

Hence 'the faster, the more nucleations'.

Let us mention that for the Allen–Cahn equation there are rigorous results on Γ-limits for action functionals, see e.g., Kohn et al. 2006, Röger and Mugnai 2007, but the methods do not seem to work for the nonlocal mesoscopic equation treated here.

11.3 Elements of the proofs

11.3.1 Common ideas

The proofs of both theorems have three important ingredients in common:

1. **The linear stability of the 1-d instanton:** the linearization of both (11.2) and (11.8) around $\overline{m}$ yields:

$$\partial_t v = -\mathcal{L}^{(1)} v$$

where $\mathcal{L}^{(1)}$ has zero as smallest eigenvalue with eigenfunction $\overline{m}'(x) > 0$ which is separated by a spectral gap from the rest of the spectrum. This fact (see De Masi et al. 1994a,b) yields important information even in the higher dimensional case.

2. **Coarse graining** with two parameters $\ell^- \ll \ell^+$: a function m is averaged over cubes of sidelength ℓ^-, and a phase indicator $\eta^{(\zeta,\ell_-)}$, is associated with m : $\eta^{(\zeta,\ell_-)} = +1$ (-1) on a cube if the average over the cube differs from $+m_\beta$ $(-m_\beta)$ by less than the *accuracy* ζ. If the average is close neither to $+m_\beta$ nor to $-m_\beta$, then $\eta^{(\zeta,\ell_-)} = 0$ on that cube. Because of the nonlocality of the interaction, $\eta^{(\zeta,\ell_-)} \neq 0$ does not guarantee that the cube contributes little to the free energy. This is only the case if it is surrounded by a 'safety zone' of 'like-minded' cubes.

$\Theta^{(\zeta,\ell_-,\ell_+)}(m;r) = \pm 1$ in a ℓ^+-cube, if $\eta^{(\zeta,\ell_-)}(m;\cdot) = \pm 1$ in all ℓ^--cubes in that ℓ^+-cube *and in all neighbouring ℓ^+-cubes*. See also Fig. 11.9. Loosely speaking we call 'contours' the regions where $\Theta = 0$, and these contours contribute to the free energy proportional to their volume. For more details we refer to Bellettini et al. 2007 or Presutti 2008.

3. **Reversibility:** this important property of the action functional is not a feature of our simplified action functional, but already present in the Comets action functional (Comet 1987) and related to the way the microscopic stochastic dynamics is constructed out of a microscopic energy (Hamiltonian).

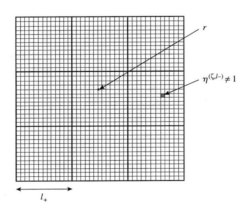

FIG. 11.9: Nine large squares of sidelength ℓ_+. The small squares are have sidelength of ℓ_-. Even if $\eta^{(\zeta,\ell_-)}(m;\cdot) = 1$ in all small squares except the one in grey, nonetheless $\Theta^{(\zeta,\ell_-,\ell_+)}(m;r) = 0$.

It connects the action of a path $m(t)$ on $[t_0, t_1]$, the action of the time reversal of that path $m^{rev}(t) = m(t_1 - t)$ and the free energy by the formula:

$$I_{L;t_0,t_1}(m) = \underbrace{I_{L;t_0,t_1}(m^{rev})}_{\geq 0} + [F(m(t_1)) - F(m(t_0))] \qquad (11.15)$$

We immediately see:
- the action of the path is at least as big as the largest free energy difference the path encounters
- if this largest free energy difference is encountered at a saddle point which is dynamically connected to $m(t_0)$ and $m(t_1)$ then the free energy difference equals the action.

11.3.2 Switching

We turn to the elements of the proof of Theorem 11.1.

1. Upper bound using the invariant manifolds $\mathcal{M}^\pm$ (see Section 11.1.2 and Fig. 11.6) and the reversibility (Section 11.3.1, 3.) we construct a for T sufficiently large paths $m_{\epsilon,T}$ with $I_{L,T}(m_{\epsilon,T}) = L^{d-1}F^{(1)}(\widehat{m}_L) + \epsilon$, hence any 'candidate' for minimization must be at least as cheap:

$$I_{L,T_n}(m_n) \leq L^{d-1}F^{(1)}(\widehat{m}_L) + \epsilon \qquad (11.16)$$

For the construction of these invariant manifolds we refer to Bellettini et al. 2007 and Buttà et al. 2003.

2. Reversibility: by reversibility and (11.16), m_n can never reach a point of free energy higher than the r.h.s. of (11.16), i.e., $\sup_{t \in [0,T]} F_L(m_n(\cdot, t)) \leq LF^{(1)}(\widehat{m}_L) + \epsilon$

On the other hand, we know that $\int_D m_n(x,t)\mathrm{d}x$ is continuous in time, so there must be a time t_0 with $\int_D m_n(x,t)\mathrm{d}x = 0$. Intuitively, the idea is to proceed as follows: the minimizer of the free energy under the constraint $\int_D m_n(x,t)\mathrm{d}x = 0$ is $\widehat{m}^e$, and the energy landscape is not degenerate around this minimizer, hence $m_n(t_0)$ must be close (on the mesoscopic scale) to $\widehat{m}^e$. While this intuitive idea is correct, the rigorous proof is not straightforward. First we show that the constraint minimizer of the free energy must be close to a plane in *macroscopic* coordinates, then we improve this estimate, so that we get closeness in mesoscopic coordinates.

3. Wulff shape in $[-1,1]^d$: we use the following auxiliary result from geometric analysis: if $d = 2, 3$ and $|\theta - 1/2| \ll 1$, then:

$$\inf_{A \subseteq Q_1,\ |A| = \theta,\ \text{Neumann b.c.}} \mathrm{Per}(A, Q_1) = 1,$$

which is achieved only by a hyperplane.

In $d = 2$ this is almost obvious: minimizers have constant mean curvature, hence are segments of planes or segments of circles. Now the minimization can be done explicitly. In higher dimensions, however, there are many hypersurfaces with constant mean curvature (including all minimal surfaces). For minimal hypersurfaces and their properties we refer to Massari and Miranda 1984. In $d = 3$ the question can be settled (see Ros 2005), in $d \geq 4$ it is, to our knowledge, an open problem. This auxiliary result is the only reason for the restriction $d \in \{2,3\}$.

4. Γ-convergence: the (macroscale) result from 3. is related to constraint minimizers of the free (mesoscopic) energy by the following Γ-convergence result, [1]: Let $G_L(v) := \frac{F(v(L \cdot))}{L^{d-1}}$ (acting on functions on the unit cube) then $\Gamma - \lim_{L \to \infty} G_L(v) = c_\beta \mathrm{Per}(\{m = m_\beta\}, [-1,1]^d)$. An argument by contradiction using 3. and 4. yields: for any $\delta > 0$ there exists $L(\delta)$ such that for $L > L(\delta)$:

$$\|m_n(t) - \overline{m}_\xi\|_{L^2([-L,L]^d)} \leq \delta L^d, \text{ if } L^{-2}\left|\int_{[-L,L]^d} m_n\right| \ll 1.$$

Here $\overline{m}_\xi(x) = \overline{m}(x \cdot e_1 - \xi)$ (up to symmetries of the cube/square), and ξ is chosen such that $\int_{[-L,L]^d} m_n = \int_{[-L,L]^d} \overline{m}_\xi$.

This is not yet the bound we need, because on the large (mesoscopic) square/cube, the error grows with L, and we need to increase L several times in the course of the proof. (Recall that the theorem holds for L sufficiently large only.)

5. Improved bound: we have to show that m_n also on mesoscale close to planar shape:

$$\|m_n(t) - \overline{m}_\xi\|_{L^2([-L,L]^d)} \leq L^{-100}, \text{ if } L^{-2}\left|\int_{[-L,L]^d} m\right| \ll 1.$$

(The exponent 100 is chosen arbitrarily)

The idea is to use the spectral gap of the planar instanton ($\overline{m}(x \cdot e_1)$) on the channel $\mathbb{R} \times [-L, L]^{d-1}$. The spectral gap result itself is not an immediate consequence of the 1-d result because we need to quantify the dependence of the spectral gap on L, see 7. below.

In order to transfer our problem to the channel we use the coarse graining introduced in Section 11.3.1, 11.3.2. We know from 4. that most of the cubes are +-cubes close to one face of the cube, and most of them are − close to the other face. If there is a positive (negative) 'connection' (see Fig. 11.10) then there is a configuration in the channel which equals the given configuration between the two connections and has a free energy which is not much higher than that of the given configuration on the cube.

If there is no such connection, then we are able to show (by estimating the number of cubes that are neither + not − in a way motivated by Bodineau and Ioffe 2004) that the free energy must be higher than the upper bound from 1, hence this case can be excluded. The *spectral gap* for the linearization of (11.8) on

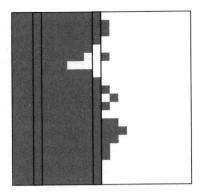

FIG. 11.10: The left strip is a connection of grey (+) cubes, the right strip is not.

the channel then implies that the energy landscape orthogonal to the instantons looks like a parabola, i.e., there is C_L s.t.

$$F_L(u) - F_L(\overline{m}) \geq \frac{1}{C_L} \|u - \overline{m}\|^2_{L^2(\mathbb{R} \times [-L,L]^{d-1})},$$

in a small but finite L^2-neighbourhood of the instanton manifold. From this the desired mesoscopic bound follows: if the energy at t_0 is close to that of $\widehat{m}^e$, then the function $m(\cdot t_0)$ must be close to $\widehat{m}^e$ in $L^2([-L, L]^d)$.

6. Attraction by invariant manifold: step 5. implies that m_n gets very close to $\widehat{m}^e$. Again stability arguments are used to show that it gets 'attracted' by $\mathcal{W}^\pm$ and stays close to the manifolds all the time. Care has to be taken because we need to show that the *time-reversal* of the path stays close to $\mathcal{W}^+$. Using reversibility and the bound on the action, we can show that this time-reversal satisfies the mesoscopic evolution equation (11.8) with a *forcing term* $b(x, t)$ which is small in space–time L^2. If this forcing is small enough, then the solution stays close to the evolution on the invariant manifold (which solves 11.8 exactly).

7. Linear stability: in many of the previous arguments we used the fact that the linearized mesoscopic evolution in a square, cube, or channel has a spectral gap. This is not an obvious consequence of the 1-d result for infinite volume in (De Masi et al. 1994b,c) because the translational symmetry is broken, and it does not follow immediately from the results in (De Masi et al. 1998), because we need the scaling of the spectral gap in L. We use a *probabilistic proof:* motivated by the analysis in (De Masi et al. 1998) we associate a *Markov jump process* with the linear operator. The invariant measure of this process is related to the principal eigenfunction, and the loss of memory for this process is related to the decay rate of the semigroup generated by the linear operator on the orthogonal complement of the span of the principal eigenfunction. Note that the smallest eigenvalue is not zero on bounded domains, but exponentially small in L.

The auxiliary results for the Markov jump process are proved by a coupling argument. The idea goes roughly as follows: a function which is orthogonal to the principal eigenfunction has expectation zero under the invariant measure. Hence we have to measure the distance of the law of the Markov process at time t from the invariant measure. In order to do so, we could consider two random walkers, one starting from the actual initial position, the other one starting from the invariant measure. A measure for the distance between the law of these two walkers (Wasserstein distance) is given by constructing an optimal coupling, i.e., a measure on the product of the two state spaces such that the marginals equal the laws of the two walkers and such that the expectation of the distance of the two walkers is as small as possible. This distance—at a given time—can be estimated by the probability that the two walkers have met by this time. This leads to the task of 'coupling' two processes (instead of two laws at a fixed time) in such a way that they are likely to meet soon. As the x_1-coordinates of both walkers have a drift towards the origin, we can couple them first and then couple the remaining coordinates, which are essentially independent jump processes without drift. Therefore we obtain the scaling in L from basic probabilistic facts.

Finally we would like to point out that the step from a 1-d spectral gap to a result in a cube/channel is immediate for the Allen–Cahn equation: the linearization around a function with planar symmetry has the form:

$$-\Delta m + V((x_1))m = \mathcal{L}_x^{(1)} m(x_1, x_2) - \partial^2_{x_2 x_2} m(x_1, x_2),$$

and the Laplacian in x_2 commutes with the '1-d operator' $\mathcal{L}_x^{(1)}$. Such a splitting does not exist for our nonlocal evolution operator.

11.3.3 Displacement of an instanton

1. Upper bound by construction: we construct explicitly a path with the claimed cost by moving, nucleating, and 'crashing' instantons explicitly. (See also Bellettini et al. 2005a,b.)

2. Good/bad intervals (contours in time): in analogy to the course graining in space introduced in Section 11.3.1, 1., we define a coarse graining in time. For the spatial coarse graining a cube was 'good' (not a contour) if it contributed little to the *free energy*. Now we call a time interval good if that interval (and its left neighbour) contribute little to the *action*. Note that we need the 'safety zone' only towards the past. More precisely, define a partition of time axis: $\{S[j, j+1), j \in \mathbb{N}\}$ and let $\phi^{(\delta, S)}(u; t) = 1$, if:

$$\int_{jS}^{(j+1)S} \|u_t - f(u)\|_2^2 < \delta,$$

FIG. 11.11: An element of the three-instanton manifold $\mathcal{M}^{(3)}$.

and 0 otherwise. Here and later we shorthand $f(u)$ for the right hand side of (11.8).

Let $G_{\text{tot}} = \{t \leq \epsilon^{-2}T : \Phi^{(\delta,S)}(u;t) = 1\}$ and call the sub-intervals of G_{tot} 'good intervals'. As the total action is bounded, there are only finitely many bad intervals and (by reversibility) only finitely many contours.

3. Multi-instanton manifold: for $\bar{\xi} = (\xi_1, \ldots, \xi_k) \in \mathbb{R}^k$, $\xi_1 < \ldots < \xi_k$ $\xi_{i+1} - \xi_i \gg 1$, $\xi_0 := -\infty$, $\xi_{k+1} := \infty$:

$$\overline{m}_{\bar{\xi}}(x) := \begin{cases} \overline{m}(x - \xi_j), & x \in \left[\frac{\xi_{j-1} + \xi_j}{2}, \frac{\xi_{j+1} + \xi_j}{2}\right], j \text{ odd}, \\ \overline{m}(\xi_j - x), & x \in \left[\frac{\xi_{j-1} + \xi_j}{2}, \frac{\xi_{j+1} + \xi_j}{2}\right], j \text{ even}. \end{cases}$$

$\mathcal{M}^{(k)} = \{\overline{m}_{\bar{\xi}}(x) : \bar{\xi} \in \mathbb{R}^k\}$, $\mathcal{M} = \bigcup_{k \geq 1} \mathcal{M}^{(k)}$ See also Fig. 11.11. This multi-instanton manifold plays an important role, because the path has to stay close to it during the good intervals. This follows from the next step:

4. Permanence away from multi-instanton manifold: by compactness arguments and the uniqueness of the infinite volume instanton (up to translations) we can show the following:

For any $\vartheta > 0$ there is $\rho > 0$ such that the following holds:
Let $m \in L^\infty(\mathbb{R}; (-1, 1))$ have an odd number p of mixed contours, let $\mathcal{F}(m) \leq P$ and let $d_\mathcal{M}(m)^2 \geq \vartheta$.($d$ denotes the L^2-distance) Then:

$$\int_\mathbb{R} f(m)^2 \geq \rho.$$

As a consequence of this estimate (and the fact that by reversibility and the upper bound the oscillation in time of the free energy is uniformly bounded for optimal paths) is that if S (length of time intervals) is chosen sufficiently large depending on ρ and the bound on the action, then u has to get close to $\mathcal{M}$ before the good interval begins. (The action was already small in the 'safety interval' before the good interval.)

5. Crashing/enlarging droplets: we can use linearization techniques only as long as the path is sufficiently close to the multi-instanton manifold. Bad intervals have to be controlled in a different way. After having passed through a bad interval and/or having been away from $\mathcal{M}$ on an interval (e.g., because the centres ξ_i were not separated from each other sufficiently), there are three possibilities:

1. The number of mixed contours (separating $\Theta = 1$ from $\Theta = -1$) is the same as in the last good interval before the bad one: then the bad interval

is 'useless'. As the displacement during that interval is negligible in comparison to $\epsilon^{-1}R$, we can ignore that interval for the purpose of finding a lower bound.

2. The number of mixed contours has decreased, i.e. two contours have collided: we replace the path by one with similar energy where we collide the contours explicitly, similarly as in the construction of the upper bound. (See also Bellettini et al. 2005a,b.) In this step the *comparison principle* for (11.8) comes into play: we construct solutions which 'overshoot', i.e., $m(\epsilon^{-2}T, x) \leq \overline{m}_{\epsilon^{-1}R}$ instead of having equality. Nevertheless, we can derive a lower bound.

3. The number of mixed contours has increased, which means that a nucleation of a droplet (or pair of instantons) must have taken place. We replace the path by one with an 'optimal' nucleation. (Time-reversal of the 'crashing' of a droplet, see also Bellettini et al. 2005a,b.)

6. Linearization: the tangent space to $\mathcal{M}$ is generated by shifts of centres. A typical element is obtained by patching together functions of the type $\overline{m}'_{\xi_j}(x) = \overline{m}'(x - \xi_j)$. ($\overline{m}'(x) = \frac{d}{dx}\overline{m}(x)$.) The linearized evolution around $\overline{m}_{\underline{\xi}}$ is denoted by $L_{\overline{m}_{\underline{\xi}}}$ and has the property that $L_{\overline{m}_{\underline{\xi}}}(\overline{m}'_{\xi_j}) \sim 0$, (vanishing as $\epsilon \to 0$). Moreover, as a consequence of the spectral gap of the linearization around a *single* instanton (De Masi et al. 1994b; De Masi et al. 1995) we get:

$$(Lu, u)_{L^2} \leq \omega \|u\|_2^2, \text{ if } (u, \overline{m}'_{\xi_i}) = 0, \text{ for all } i.$$

Now denote by u the difference of the path m from its projection on the multi-instanton manifold and let t_0 be the time when the path has reached a small neighbourhood of $\mathcal{M}$, after having been attracted towards it during the safety interval before the good interval. Moreover let $\sigma_i \in \{\pm 1\}$, for left/right motions of the centres. Let b be the forcing required to make the path m a solution of (11.8) with that forcing, i.e.:

$$I_T = \int_0^T \int_{\mathbb{R}} b^2 dx dt.$$

Then we compute:

$$\|u(\cdot, t)\|_2^2 \leq e^{-(t-t_0)\omega/2}\|u(\cdot, t_0)\|_2^2 + cSU_j^2$$

$$\sigma_i[\xi_i(t) - \xi_i(t_0)] \leq -\frac{1}{\|\overline{m}'\|_2^2}\int_{t_0}^t (b, \overline{m}'_{\xi_i(t)}) + c[\|u(\cdot, t_0)\|_2^2 + SU_j^2] \quad (11.17)$$

$$U_j^2 = \int_{(j-1)S}^{(j+1)S} \|b(\cdot, s)\|_2^2 + \text{higher order}$$

(t_0 is the time when the path has reached a small neighbourhood of $\mathcal{M}$, after having been attracted towards it during the safety interval before the good interval, and $\sigma_i \in \{\pm 1\}$).

First note that u stays small (order 1 in ϵ). Then note that any part of the forcing b which is orthogonal to $\overline{m}'_{\xi}$ contributes to the action, but not to the displacement. Hence it costs without helping to satisfy the constraint and is therefore useless. So the optimal choice is:

$$b(x,t) = V_i(t)\overline{m}'_{\xi_i(t)}(x).$$

where V_i is the velocity of the $i-th$ centre ξ_i.

Then we sum up over all good intervals. The total displacement (the sum of the $\xi_i(t) - \xi_i(t_{\mathrm{in}})$ is of order ϵ^{-1}, while the sum of the U_J-terms is bounded by the total action and hence of order 1 in ϵ and therefore negligible.

7. Conclusion: expanding the action in the good intervals and counting the cost of nucleations in the bad intervals, we obtain:

$$\frac{1}{4}\int_0^{\epsilon^{-2}T} \|b(t)\|_2^2 \geq \int_{G_{\mathrm{tot}}} \sum_i \frac{V_i(t)^2}{\mu} + 2n\mathcal{F}(\overline{m}) + o(1)$$

Equating terms of highest order in (11.17) and using the constraint yields:

$$\sum_{i=1}^{n^*} \int |v_i^0(t)| \geq \epsilon^{-1} R + \text{lower orders in } \epsilon.$$

The claimed result follows by solving:

$$\min_{t_i, v_i: \sum V_i t_i = R\epsilon^{-1}} \sum V_i^2 t_i.$$

Acknowledgement

Most of the results which are explained in this paper have been obtained by collaboration of the author with Giovanni Bellettini, Anna DeMasi, and Errico Presutti. The author would like to thank Giovanni Bellettini for reading a preliminary version of this note and for suggesting many improvements.

References

Alberti, G., Bellettini, G., Cassandro, M., and Presutti, E. (1996) Surface tension in Ising systems with Kac potentials. *J. Statist. Phys.*, **82** (3-4):743–96.

Allen S. and Cahn, J. (1979) A microscopic theory for antiphase motion and its application to antiphase domain coarsening. *Acta Metall.*, **27**: 1084–95.

Bellettini, G., De Masi, A., Dirr, N., and Presutti, E. (2007a) Stability of invariant manifolds in one and two dimensions. *Nonlinearity*, **20**(3): 537–82.

Bellettini, G., De Masi, A., Dirr, N., and Presutti, E. (2007b) Tunneling in two dimensions. *Comm. Math. Phys.*, **269**(3):715–63.

Bellettini, G., De Masi, A., and Presutti, E. (2005a) Energy levels of a nonlocal functional. *J. Math. Phys.*, **46**(8):083302, 31.

Bellettini, G., De Masi, A., and Presutti, E. (2005b) Tunnelling in nonlocal evolution equations. *J. Nonlinear Math. Phys.*, **12**(suppl. 1): 50–63.

Bodineau, T. and Ioffe, D. (2004) Stability of interfaces and stochastic dynamics in the regime of partial wetting. *Ann. Henri Poincaré*, **5**(5):871–914.

Buttà, P., De Masi, A., and Rosatelli, E. (2003) Slow motion and metastability for a nonlocal evolution equation. *J. Statist. Phys.*, **112**(3-4):709–64.

Cassandro, M., Olivieri, E., and Picco, P. (1986) Small random perturbations of infinite-dimensional dynamical systems and nucleation theory. *Ann. Inst. H. Poincaré Phys. Théor.*, **44**(4):343–96.

Comets, F. (1987) Nucleation for a long range magnetic model. *Ann. Inst. H. Poincaré Probab. Statist.*, **23**(2):135–78.

De Masi, A., Gobron, T., and Presutti, E. (1995) Travelling fronts in non-local evolution equations. *Arch. Rational Mech. Anal.*, **132**(2):143–205.

De Masi, A., Olivieri, E., and Presutti, E. (1998) Spectral properties of integral operators in problems of interface dynamics and metastability. *Markov Process. Related Fields*, **4**(1):27–112.

De Masi, A., Orlandi, E., Presutti, E., and Triolo, L. (1994a) Glauber evolution with the Kac potentials. I. Mesoscopic and macroscopic limits, interface dynamics. *Nonlinearity*, **7**(3):633–96.

De Masi, A., Orlandi, E., Presutti, E., and Triolo, L. (1994b) Stability of the interface in a model of phase separation. *Proc. Roy. Soc. Edinburgh Sect. A*, **124**(5):1013–22.

De Masi, A., Orlandi, E., Presutti, E., and Triolo, L. (1994c) Uniqueness and global stability of the instanton in nonlocal evolution equations. *Rend. Mat. Appl. (7)*, **14**(4):693–723.

De Masi, A., Dirr, N., and Presutti, E. (2006) Interface instability under forced displacements. *Ann. Henri Poincaré*, **7**(3):471–511.

Faris, W. G. and Jona-Lasinio, G. (1982) Large fluctuations for a nonlinear heat equation with noise. *J. Phys. A*, **15**(10):3025–55.

Freidlin, M. I. and Wentzell, A. D. (1998) *Random perturbations of dynamical systems*, volume 260 of *Grundlehren der Mathematischen Wissenschaften [Fundamental Principles of Mathematical Sciences]*. Springer-Verlag, New York, second edition. (Translated from the 1979 Russian original by Joseph Szücs.)

Katsoulakis, M. A. and Souganidis, P. E. (1995) Generalized motion by mean curvature as a macroscopic limit of stochastic Ising models with long range interactions and Glauber dynamics. *Comm. Math. Phys.*, **169**(1): 61–97.

Kohn, R. V., Reznikoff, M. G., and Tonegawa, Y. (2006) Sharp-interface limit of the Allen-Cahn action functional in one space dimension. *Calc. Var. Partial Differential Equations*, **25**(4):503–34.

Massari, U. and Miranda, M. (1984) *Minimal surfaces of codimension one*, volume 91 of *North-Holland Mathematics Studies*. North-Holland Publishing Co., Amsterdam., Notas de Matemática [Mathematical Notes], 95.

Olivieri, E. and Vares, M. (2005) *Large deviations and metastability*, volume 100 of *Encyclopedia of Mathematics and its Applications*. Cambridge University Press, Cambridge.

Presutti, E. (2008) *Scaling Limits in Statistical Mechanics and Microstructures in Continuum Mechanics*. Theoretical and Mathematical Physics. Springer.

Röger, M. and Mugnai, L. (2007) The Allen-Cahn action functional in higher dimensions. Technical report, Preprint series of the Max-Planck-Institute for Mathematics in the Sciences.

Ros, A. (2005) The isoperimetric problem. In *Global theory of minimal surfaces*, volume 2 of *Clay Math. Proc.*, pp. 175–209. Amer. Math. Soc., Providence, RI.

12

NUCLEATION AND DROPLET GROWTH AS A STOCHASTIC PROCESS

Oliver Penrose

Abstract

A stochastic differential equation is conjectured for approximately modelling the fluctuating size changes of an individual droplet in a fluid that is metastable with respect to nucleation of a new phase, in the limit when the critical droplet size is very large. The Freidlin–Wentzell formula for this SDE is used to make estimates of large-deviation type for probabilities of such events as the formation of a critical droplet at a specified time. A relation is obtained connecting these estimates to the nucleation rate predicted by the well-established theory of Becker and Döring.

12.1 Introduction

Nucleation is the initiation of a phase transition (such as the transition from gas to liquid or liquid to gas) when a significant droplet of the new phase forms. For example, when the atmospheric temperature drops in the evening, or when a stream of air cools on going up a mountainside, a phase transition becomes possible in which the water vapour mixed with the air will change from gas to liquid. At first only very small droplets of liquid water are formed. The droplets may never grow large enough to be seen; but if the atmospheric conditions are right the droplets can eventually become large enough to be seen as a mist, fog, or cloud.

The growth of any individual droplet is a stochastic process: its size can either increase or decrease as molecules attach themselves to the droplet or detach themselves from it. There is a contest between, on the one hand, the general preference of the water molecules (at a sufficiently low temperature) to be in the liquid rather than the vapour phase, which tends to increase the size of the droplet and, on the other hand, the surface tension, which tries to reduce the surface area of the droplet and, in consequence, its size. The surface tension effect is stronger for small droplets than for large ones because of the greater curvature of the surface of a small droplet. There is a critical droplet size at which the surface tension exactly balances the water's preference for being in the liquid rather than the vapour phase. Droplets of supercritical size tend to grow, while those of subcritical size tend to shrink. At first, no supercritical droplets at all are present (except, perhaps, at places on the edge of the vapour, such as leaves on which dew may form); there is no mist, and the vapour is said to be *metastable*. If all the droplets followed the average behaviour, supercritical droplets would never form and the metastable state would last for ever. But this is a stochastic process, and the sizes of the droplets fluctuate. Eventually, as a result of these fluctutations, some droplets will reach and then surpass the critical size, and when enough of them have done so the mist will be visible.

This paper is concerned with the mathematical description of nucleation as a stochastic process. For mathematical simplicity we shall consider the limiting case of very large critical droplet size—in the physical example mentioned, this means that the temperature is only just below the dew point. The idea is to approximate the behaviour of the droplets by a stochastic differential equation, and to obtain quantitative information about the nucleation process by applying the Freidlin–Wentzell formula to this equation.

12.2 A mathematical description of nucleation

In 1935, R. Becker and W. Döring proposed a mathematical model of nucleation in which each droplet is considered to be fully described by its size, that is, by the number of molecules comprising it. The shapes and positions of the droplets are ignored. The size can be any positive integer, and by convention molecules of the vapour are treated as droplets of size 1 (usually called *monomers*). The Becker–Döring model includes two characteristic assumptions: (i) the only way the size of a droplet (other than a monomer) can change is by emitting or absorbing a monomer (ii) the probability per unit time that a given droplet will emit a monomer depends only on the size of that droplet, while the probability per unit time of absorbing a monomer depends on the size of the droplet and on the overall concentration of monomers.

Focusing attention on a particular droplet, let us denote its size (i.e., the number of molecules in it) at time t by $N(t)$. Then, following the assumptions of Becker and Döring, we can treat $N(t)$ is a stochastic process, in which the probability per unit time for N to increase by 1 is $a_n z$ and for it to decrease by 1 is b_n, where z is a parameter representing the overall concentration of monomers (i.e., the number of monomers per unit volume). In symbols, the transition probabilities are:

$$\mathbf{Pr}(N(t+\delta t) = n+1 \,|\, N(t) = n) = a_n z \delta t + O(\delta t)^2 \quad (n = 1, 2, \ldots)$$

$$\mathbf{Pr}(N(t+\delta t) = n-1 \,|\, N(t) = n) = b_n \delta t + O(\delta t)^2 \quad (n = 2, 3, \ldots)$$

$$\mathbf{Pr}(|N(t+\delta t) - N(t)| \geq 2 \,|\, N(t) = n) = O(\delta t)^2 \quad (12.1)$$

where the constants $a_1, a_2, \ldots, b_2, b_3, \ldots$ depend on the physical situation. Here we shall assume them to be given by the following approximate formulas, for which there is some physical justification (Lifshits and Slyozov 1961, Wagner 1961, Penrose and Buhagiar 1983, Penrose 1997)[1]:

$$\begin{aligned} a_n &= n^\gamma & (n = 1, 2, \ldots) \\ b_n &= n^\gamma(1 + \mu n^{-1/3}) & (n = 2, 3, \ldots) \end{aligned} \quad (12.2)$$

where μ is a positive constant proportional to the surface tension at the surface of a droplet and γ is a constant satisfying $\gamma \geq \frac{1}{3}$. Later in this paper we shall specialize to the case $\gamma = \frac{1}{3}$. The significance of the exponents $\frac{1}{3}$ and $-\frac{1}{3}$ is that the radius of a droplet is proportional to the cube root of its size. From the formulas (12.2) it can be seen that $a_n z < b_n$ if $n^{1/3}(z-1) < \mu$, but $a_n z > b_n$ if $n^{1/3}(z-1) > \mu$; so provided that $z > 1$ there is a critical droplet size:

$$n_c := \left(\frac{\mu}{z-1}\right)^3 \quad (12.3)$$

such that any droplet whose size exceeds this is more likely to grow than to shrink.

To finish specifying the model we need to say how z depends on time. In this paper, following the original paper of Becker and Döring (1935), we shall assume that z is a constant. Physically, z is the concentration of monomers, and is related to $\mathbf{Pr}\{N(t) = 1\}$, the probability that a randomly chosen droplet will be a monomer, by the formula:

$$z = c\mathbf{Pr}\{N(t) = 1\} \quad (12.4)$$

where c is the total number of molecules per unit volume. Thus the assumption of constant z can be arrived at physically by assuming that c is a constant and that nearly all the droplets are monomers, so that $\mathbf{Pr}\{N(t) = 1\} \approx 1$ and $z \approx c = \text{const}$. Alternatively one may assume that $\mathbf{Pr}\{N(t) = 1\}$ does change with time but that c changes in such a way that z remains constant: this will happen, for example if the process takes place at constant pressure. The case where c is constant but both z and $\mathbf{Pr}\{N(t) = 1\}$ change with time has many interesting features (see for example Lifshits and Slyozov 1961, Ball et al. 1986) but is not our concern here.

There are two ways to obtain information about metastability and nucleation from this stochastic model. The one devised by Becker and Döring was to

---

[1] The formulas in eqn (12.2) are often presented in a more general form such as $a_n = an^\gamma$, $b_n = an^\gamma z_s(1 + \mu n^{-1/d})$ where a and z_s are positive constants, d is the number of space dimensions (at least 3), and γ satisfies $\gamma \geq 1/d$. The version used in (12.2) can, however, be obtained from the more general version by setting $d = 3$ and making a suitable choice of time and length units.

study the average behaviour of the entire collection of droplets, described by the probability distribution:
$$p_n(t) := \mathbf{P}\{N(t) = n\} \tag{12.5}$$
They found a steady-state distribution in which, for each droplet size n, the rate of occurrence of events which increase the droplet size from n to $n+1$ slightly exceeds the rate for events which decrease the size from $n+1$ to n. The excess, which is independent of n, can be interpreted as at the rate at which droplets surpass the critical size; it is called the nucleation rate. This approach, now known as 'classical nucleation theory', is summarized in Section 12.5.

In 1984 Cassandro et al. introduced an alternative way of obtaining information about metastability and nucleation in stochastic models. In this method, the 'pathwise approach', we focus not on averages but on the stochastic behaviour of a single droplet. The pathwise approach has been used to study metastability in a variety of statistical mechanics models. A very thorough account of this body of work and the related theory is given in the book by Olivieri and Vares (2006); for a shorter account see den Hollander (2004).

It is the purpose of this article to apply a variant of the pathwise approach to the Becker–Döring model. The idea is to concentrate on a particular droplet and represent its size as a stochastic process. We can treat such questions as how likely it is that the size of a given droplet will reach or pass the critical size, and if it does so how long that is likely to take.

12.3 The proposed SDE

One can think of the process (12.1) as a biased random walk along the positive integer axis—or as a birth-and-death process. The expected rate of increase in the droplet size N is given by:
$$\mathbf{E}(N(t+\delta t) - N(t)|N(t) = n) = (a_n z - b_n)\delta t + O(\delta t)^2 \tag{12.6}$$
Thus there is a drift in the expected size of $N(t)$; the rate of drift is:
$$a_n z - b_n = n^\gamma((z-1) - \mu n^{-1/3}) \tag{12.7}$$
The rate of increase in the variance of n per unit time may be estimated as:
$$\lim_{\delta t \to 0} \frac{1}{\delta t}\mathbf{E}((N(t+\delta t) - N(t))^2|N(t) = n) = a_n z + b_n \tag{12.8}$$

The idea of the present work is to approximate this stochastic process by one in which the unit-size jumps are replaced by jumps with a Gaussian distribution having the same mean and variance. This approximating process corresponds to the stochastic differential equation:
$$dN = (a_n z - b_n)dt + \sqrt{a_n z + b_n}\,dW(t) \tag{12.9}$$
where $W(t)$ is a Wiener process.

Although such an SDE would not be a good approximation for the individual jumps, it may be a good one if we look at the process on a different scale, where the jumps look very small and there are a large number of them. There is an analogy with Khinchin's method of deriving the Central Limit Theorem [2], which involves rescaling the time and space variables in a similar way, although in his case the rescaled equation is deterministic (the heat equation) rather than stochastic. Our rescaling will use, in place of N, a new random variable X proportional to N/n_c. The jumps in X will then be proportional to $1/n_c$, so they will be small if we consider a limit in which n_c is large. As we have seen, n_c is proportional to $(z-1)^{-3}$ and so z is close to 1 in this limit; we shall write:

$$z := 1 + \epsilon$$
$$X := \epsilon^3 N = \mu^3 (N/n_c) \tag{12.10}$$

and consider the limit $\epsilon \to 0$.

In addition to rescaling the size of the jumps, it is useful to re-scale the time, so that X is regarded as a function of a re-scaled time variable τ rather than of the original time variable t. The advantage of rescaling the time is that the drift velocities of the two processes, each with respect to its own time scale, can be made comparable. According to eqn (12.6), the drift velocity of the random walk variable N is:

$$D_t N(t) := \lim_{\delta t \searrow 0} \mathbf{E} \left\{ \frac{N(t + \delta t) - N(t)}{\delta t} \bigg| N(t) \right\}$$
$$= a_{N(t)} z - b_{N(t)} = N(t)^\gamma (z - 1 - \mu N(t)^{-1/3}) \tag{12.11}$$

Consequently the drift velocity of X, with respect to the re-scaled time variable τ, is:

$$D_\tau X(\tau) := \lim_{\delta \tau \searrow 0} \mathbf{E} \left\{ \frac{X(\tau + \delta \tau) - X(\tau)}{\delta \tau} \bigg| X(\tau) \right\}$$
$$= \frac{dt}{d\tau} D_t(\epsilon^3 N(t)) = \epsilon^3 \frac{dt}{d\tau} N(t)^\gamma (z - 1 - \mu N(t)^{-1/3})$$
$$= \epsilon^3 \frac{dt}{d\tau} (\epsilon^{-3} X(\tau))^\gamma (\epsilon - \mu (\epsilon^{-3} X(\tau))^{-1/3})$$
$$= \epsilon^{4-3\gamma} \frac{dt}{d\tau} X(\tau)^\gamma (1 - \mu/X(\tau)^{1/3}) \tag{12.12}$$

On making the choice:

$$\tau := t \epsilon^{4-3\gamma} \tag{12.13}$$

this simplifies to:

$$D_\tau X(\tau) = X(\tau)^\gamma (1 - \mu/X(\tau)^{1/3}) \tag{12.14}$$

---

[2]There is a description of this method on page 10 of Ito and McKean's book (1996).

Comparison with (12.11) shows that the drift velocities of the two processes, each relative to its own time scale, are given by similar differential equations.

We can do a similar calculation for the variance of $X(\tau)$, using the formula (12.8):

$$\lim_{\delta\tau\searrow 0}\frac{\mathbf{E}\left\{[X(\tau+\delta\tau)-X(\tau)]^2\,\big|\,X(\tau)\right\}}{\delta\tau}=$$

$$=\lim_{\delta t\searrow 0}\frac{\epsilon^6\mathbf{E}\left\{[N(t+\delta t)-N(t)]^2\,\big|\,N(t)=\epsilon^{-3}X(\tau)\right\}}{\delta t}\frac{dt}{d\tau}=$$

$$=\epsilon^6[\epsilon^{-3}X(\tau)]^\gamma(2+\epsilon+\mu(\epsilon^{-3}X(\tau))^{-1/3})\epsilon^{3\gamma-4}=2\epsilon^2 X(\tau)^\gamma+O(\epsilon^3)$$
(12.15)

In the limit of small ϵ this is the same rate of change of variance as for a Brownian motion multiplied by $\sqrt{2}\epsilon X(\tau)^{\gamma/2}$. Adding together this Brownian motion and the drift given by (12.14), we may conjecture that in the limit of small ϵ the random variable X will obey the stochastic differential equation:

$$dX(\tau)=X(\tau)^\gamma(1-\mu/X(\tau)^{1/3})d\tau+\sqrt{2}\epsilon X(\tau)^{\gamma/2}dW(\tau) \quad (12.16)$$

where $W(\tau)$ is a Brownian motion.

12.4 Applying the Freidlin–Wentzell formula

12.4.1 *A formula for the action*

In this section we shall use ideas from the Freidlin–Wentzell theory (Freidlin and Wentzell 1998, Olivieri and Vares 2006) to estimate the probabilities for different ways in which the size of a droplet can change over time. The Freidlin–Wentzell theory applies to SDEs of the form:

$$dX(\tau)=v(X(\tau))d\tau+\epsilon\sigma(X(\tau))dW(\tau) \quad (12.17)$$

The fundamental object in this theory is the 'action' or rate function. For an arbitrary trajectory $x(\tau)$ the action is defined to be:

$$S(\tau_1,\tau_2):=\frac{1}{2}\int_{\tau_1}^{\tau_2}\left(\frac{\dot{x}(\tau)-v(x(\tau))}{\sigma(x(\tau))}\right)^2 d\tau \quad (12.18)$$

where $\dot{x}(\tau):=dx/d\tau$. The main property of the action is that, in the limit of small ϵ, the probability of executing the path $x(\tau)$, or one very similar to it, between times τ_1 and τ_2, conditional on starting at the given point $x(\tau_1)$ at time τ_1, is $\exp\{-\epsilon^{-2}S(\tau_1,\tau_2)+o(\epsilon^{-2})\}$.

The conjectural SDE (12.16) is of the Freidlin–Wentzell form, with:

$$v(x)=x^\gamma(1-\mu x^{-1/3}),\quad \sigma(x)=\sqrt{2}\,x^{\gamma/2} \quad (12.19)$$

and so the action for eqn (12.16) is:

$$S(\tau_1, \tau_2) := \frac{1}{2}\int_{\tau_1}^{\tau_2} \frac{[\dot{x}(\tau) - x(\tau)^\gamma(1 - \mu x(\tau)^{-1/3})]^2}{2x(\tau)^\gamma} d\tau$$

$$= \frac{1}{2}\int_{\tau_1}^{\tau_2} \frac{[\dot{x}(\tau) + x(\tau)^\gamma U'(x(\tau))]^2}{2x(\tau)^\gamma} d\tau \quad (12.20)$$

where:

$$U(x) := \frac{3}{2}\mu x^{2/3} - x \quad (12.21)$$

and the prime denotes a derivative, so that $U'(x) = \mu x^{-1/3} - 1$.

12.4.2 The minimization problem

Let (x_1, τ_1) and (x_2, τ_2) be given initial and final states of the rescaled process. The most probable path connecting them can be found by minimizing the action $S(\tau_1, \tau_2)$ subject to the constraints $x(\tau_1) = x_1, x(\tau_2) = x_2$, using the calculus of variations. To carry out the minimization we first multiply out the integrand in (12.20), obtaining:

$$2S(\tau_1, \tau_2) = \frac{1}{2}\int_{\tau_1}^{\tau_2} x(\tau)^{-\gamma}\dot{x}(\tau)^2 d\tau + \int_{\tau_1}^{\tau_2} U'(x(\tau))\dot{x}(\tau)d\tau +$$

$$+ \frac{1}{2}\int_{\tau_1}^{\tau_2} x(\tau)^\gamma U'(x(\tau))^2 d\tau$$

$$= \int_{\tau_1}^{\tau_2} L(\dot{x}(\tau), x(\tau))) d\tau + U(x_2) - U(x_1) \quad (12.22)$$

where L is a 'Lagrangian' defined by:

$$L(\dot{x}, x) := \frac{1}{2}x^{-\gamma}\dot{x}^2 + \frac{1}{2}x^\gamma U'(x)^2 \quad (12.23)$$

The minimizer satisfies the Euler–Lagrange equation:

$$\frac{d}{d\tau}\frac{\partial L}{\partial \dot{x}} = \frac{\partial L}{\partial x} \quad (12.24)$$

The following procedure gives a first integral of this equation, enabling its solution to be reduced to a quadrature. We define a 'momentum' p by [3]:

$$p := \frac{\partial L}{\partial \dot{x}} = x^{-\gamma}\dot{x} \quad (12.25)$$

---

[3] This use of the symbol p is traditional in Hamiltonian dynamics. It is hoped that there will be no confusion with the use of the same symbol for probability in other parts of this paper.

and a 'Hamiltonian' H by:

$$H := p\dot{x} - L(\dot{x}, x) = \frac{1}{2}x^\gamma p^2 - \frac{1}{2}x^\gamma U'(x)^2 \qquad (12.26)$$

These definitions imply $dH = \dot{x}dp + pd\dot{x} - (\partial L/\partial \dot{x})d\dot{x} - (\partial L/\partial x)dx = \dot{x}dp - (\partial L/\partial x)dx$. Consequently, looking on H as a function of p and x and using (12.24), we find that Hamilton's equations:

$$\frac{\partial H(p,x)}{\partial p} = \frac{dx}{d\tau}, \quad \frac{\partial H(p,x)}{\partial x} = -\frac{dp}{d\tau} \qquad (12.27)$$

are satisfied on the minimizer. It follows that $dH/d\tau = 0$, so that H is a constant along the minimizer. The value of this constant, which is analogous to the energy in mechanics, will be denoted by E.

12.4.3 Paths for which $E = 0$

The calculation of the action is particularly simple in the case $E = 0$. Setting $H = 0$ in eqn (12.26) yields $p = \pm U'(x)$ and then, from (12.25), $\dot{x} = \pm x^\gamma U'(x)$, so that the equation of the minimizer is:

$$\frac{dx}{d\tau} = \pm x^\gamma(\mu x^{-1/3} - 1) \qquad (12.28)$$

With the minus sign, this is the equation of the 'average' path which, according to equation (12.16), the droplet size would follow if there were no noise at all. For that path the value of S is zero (by eqn (12.20)) and so the the probability of the 'average' path, or one very similar to it, is $\exp\{-o(\epsilon^{-2})\}$.

With the plus sign, the path is an 'average' path traversed backwards. Since the right side is zero at the critical cluster size $x = \mu^3$ and varies approximately linearly with x nearby, with a negative derivative, the value of $x(\tau)$ for this path approaches the critical size asymptotically as $\tau \to \infty$. For a path of this type the action is, using first (12.20) and then the fact that $\dot{x} = x^\gamma U'(x)$ on this path:

$$\begin{aligned} S(\tau_1, \tau_2) &:= \frac{1}{2}\int_{\tau_1}^{\tau_2} \frac{[\dot{x}(\tau) + x(\tau)^\gamma U'(x(\tau))]^2}{2x(\tau)^\gamma} d\tau \\ &= \frac{1}{2}\int_{\tau_1}^{\tau_2} \frac{[2x(\tau)^\gamma U'(x(\tau))][2\dot{x}(\tau)]}{2x(\tau)^\gamma} d\tau \\ &= \int_{\tau_1}^{\tau_2} U'(x(\tau))\dot{x}(\tau)d\tau = U(x_2) - U(x_1) \end{aligned} \qquad (12.29)$$

In particular, if the initial scaled droplet size is small and the final scaled size is close to the scaled critical size, which is μ^3, the action is $U(\mu^3) - U(0) = \frac{1}{2}\mu^3$, so that the probability of the droplet's reaching the critical size by a path close to the $E = 0$ path is $\exp\{-\frac{1}{2}\epsilon^{-2}\mu^3 + o(\epsilon^{-2})\}$.

12.4.4 General values of E

Putting $H = E$ in eqn (12.26), solving for p and then using (12.25) we get:

$$x^{-\gamma}\dot{x} = p = \pm\sqrt{2Ex^{-\gamma} + (\mu x^{-1/3} - 1)^2}$$
i.e. $\quad\dot{x} = x^{\gamma}p = \pm\sqrt{2Ex^{\gamma} + (\mu x^{\gamma-1/3} - x^{\gamma})^2}\quad$ (12.30)

The qualitative features of the solution depend on the value of E. They can be worked out by studying how the radicand (the expression under the radical sign) depends on x. We give below the analysis for the important case $\gamma = \frac{1}{3}$, in which eqn (12.30) simplifies to:

$$\dot{x} = \pm\sqrt{\mu^2 + 2(E - \mu)x^{1/3} + x^{2/3}} \quad (12.31)$$

1. If $E < 0$ the radicand (now a quadratic polynomial in $x^{1/3}$) is zero for two positive values of x, whose geometric mean is μ^3. Between these two roots, it is negative. The solution can have a maximum at the root lying between $x = 0$ and $x = \mu^3$, or a minimum at the root above μ^3. Values of x between these two roots are impossible.

2. If $E = 0$ the radicand has a double zero at $x = \mu^3$. The formula (12.31) simplifies to:

$$\dot{x} = \pm(\mu - x^{1/3}) \quad \text{whence} \quad \tau = \text{const} \pm \int (\mu - x^{1/3})^{-1} dx \quad (12.32)$$

The solutions are monotonic and have an asymptote $x = \mu^3$. The most probable path through any given point (x_0, τ_0) is of this type (with the negative sign chosen if $0 < x_0 < \mu^3$). For the calculation of the action in this case, see eqn (12.29).

3. If $0 < E < 2\mu$ the radicand is positive for all x. The solutions go monotonically from $-\infty$ to $+\infty$ or vice versa.

4. If $E = 2\mu$ the radicand has a double zero at $x = -\mu^3$. The formula (12.31) simplifies to:

$$\dot{x} = \pm(\mu + x^{1/3}) \quad \text{whence} \quad \tau = \text{const} \pm \int (\mu + x^{1/3})^{-1} dx \quad (12.33)$$

The solutions are monotonic and have an (unphysical) asymptote $x = -\mu^3$.

5. If $E > 2\mu$ the radicand is zero for two different values of x, both negative, whose geometric mean is $-\mu^3$. The physically meaningful solutions (i.e. those for which x is positive) are monotonic.

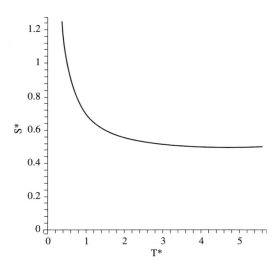

FIG. 12.1: The action (approximately ϵ^2 times the negative logarithm of the probability) of the most probable path taking a small cluster to one of critical size in a given time, plotted as a function of that given time. The abscissa, labelled T^*, is μ^{-2} times the integral in eqn (12.36), which in turn is ϵ^3 times the physical time for the droplet to grow. The ordinate, labelled S^*, is μ^{-3} times the action $S(\tau_1, \tau_2)$ as given by eqn (12.37). At large times, S^* approaches the limit $\frac{1}{2}$, in agreement with eqn (12.42).

12.4.5 The probability of a given droplet becoming critical in a given time

To get a path which increases from a subcritical value $(x_1 < \mu^3)$ to a supercritical one $(x_2 > \mu^3)$ we need the positive sign in the formula (12.30) and we also need $E > 0$. The time to get from x_1 to x_2 along such a path is, by (12.30):

$$\tau_2 - \tau_1 = \int_{x_1}^{x_2} \frac{dx}{\dot{x}} = \int_{x_1}^{x_2} \frac{dx}{x^\gamma \sqrt{2Ex^{-\gamma} + (\mu x^{-1/3} - 1)^2}} \quad (12.34)$$

If we assume $x_1 < \mu^3 \le x_2$ and $\tau_1 < \tau_2$, then, as E increases from 0 to $+\infty$, the value of the integral decreases (at fixed x_1, x_2) from $+\infty$ to zero, and so the given values of x_1, x_2, τ_1, τ_2 determine a unique positive value of E.

The action along this path is, by (12.20) and (12.30)

$$S = \frac{1}{4} \int_{x_1}^{x_2} x^{-\gamma} [\dot{x} + x^\gamma U'(x)]^2 \frac{dx}{\dot{x}} = \frac{1}{4} \int_{x_1}^{x_2} x^\gamma [x^{-\gamma} \dot{x} + U'(x)]^2 \frac{dx}{\dot{x}}$$

$$= \frac{1}{4} \int_{x_1}^{x_2} \frac{\{\sqrt{2Ex^{-\gamma} + (\mu x^{-1/3} - 1)^2} + \mu x^{-1/3} - 1\}^2 dx}{\sqrt{2Ex^{-\gamma} + (\mu x^{-1/3} - 1)^2}} \quad (12.35)$$

12.4.6 Exact solutions for the case $\gamma = 1/3$

In the physically important case $\gamma = 1/3$, the integrals in (12.34) and (12.35) simplify to:

$$T := \tau_2 - \tau_1 = \int_{x_1}^{x_2} \frac{dx}{\sqrt{2Ex^{1/3} + (\mu - x^{1/3})^2}} \qquad (12.36)$$

$$S(\tau_1, \tau_2) = \frac{1}{4} \int_{x_1}^{x_2} \frac{[\sqrt{2Ex^{1/3} + (\mu - x^{1/3})^2} + \mu - x^{1/3}]^2 \, dx}{x^{1/3}\sqrt{2Ex^{1/3} + (\mu - x^{1/3})^2}} \qquad (12.37)$$

Both integrals can be done analytically.

Of particular interest is the case where $x_1 = 0$, $x_2 = \mu^3$, in which the droplet starts out very small and ends up at the critical size. Fig. 12.1 shows a graph of S as a function T for this case, obtained by eliminating E between eqns (12.36) and (12.37). The interpretation of the graph is that the probability for a very small droplet to reach the critical size after a time T is $\exp\{-\epsilon^{-2}S + o(\epsilon^{-2})\}$.

12.5 Relation of the results of section 12.4 to the 'classical' Becker–Döring nucleation theory

The method used by Becker and Döring to estimate nucleation rates was based on the time evolution equations for the probability $p_n(t)$ that the size of the droplet at time t is n, i.e., $p_n(t) = \mathbf{Pr}(N(t) = n)$, as in eqn (12.5). The evolution equations for $p_2(t), p_3(t), \ldots$ can be written:

$$dp_n/dt = J_{n-1} - J_n \qquad (n = 2, 3, \ldots)$$
$$J_n = a_n z p_n - b_{n+1} p_{n+1} \qquad (n = 1, 2, 3, \ldots) \qquad (12.38)$$

Becker and Döring looked for a solution which was stationary, in the sense that $p_2, p_3, \ldots$ are independent of time, and p_1 (whose evolution equation they did not discuss[4]) varies only very slowly. For a solution that is stationary in this sense, J_n must be independent of n. The common value of J_n is interpreted as the rate of nucleation. Denoting this common value by J, we can solve the equations (12.38) successively to get:

$$p_1 = \frac{J + b_2 p_2}{a_1 z} = \frac{J}{a_1 z} + \frac{b_2}{a_1 z}\left(\frac{J + b_3 p_3}{a_2 z}\right) = \frac{J}{a_1 z} + \frac{b_2 J}{a_1 a_2 z^2} + \frac{b_2 b_3}{a_1 a_2 z^2}\left(\frac{J + b_4 p_4}{a_3 z}\right)$$

$$= J\left(\frac{1}{a_1 z} + \frac{b_2}{a_1 a_2 z^2} + \frac{b_2 b_3}{a_1 a_2 a_3 z^3} + \cdots\right)$$

$$= J \sum_{n=1}^{\infty} \frac{Q_1}{Q_n a_n z^n} \qquad (12.39)$$

---
[4] A treatment of nucleation rates which does allow for the time evolution of p_1 is given by Penrose (1989).

where:
$$Q_1 := 1, \quad Q_n := \frac{a_1 a_2 \ldots a_{n-1}}{b_2 \ldots b_n} \sim \text{const.} \exp(-\frac{3}{2}\mu n^{2/3}) \quad (n = 1, 2, \ldots) \quad (12.40)$$

The series in (12.39) converges for $z > 1$ and its largest term occurs at the value of n satisfying $a_n z = b_n$, i.e., the critical droplet size. Under the approximations used earlier, this size is $(\mu/\epsilon)^3$ and the size of the corresponding term in the series is:

$$\frac{Q_1}{Q_n a_n z^n} \approx \text{const.} \exp\left(\frac{3}{2}\mu \left(\frac{\mu}{\epsilon}\right)^2 - \left(\frac{\mu}{\epsilon}\right)^3 \log z\right) \approx \text{const.} \exp\left(\frac{\mu^3}{2\epsilon^2}\right) \quad (12.41)$$

since $\log z \approx \epsilon$. The rate of nucleation J, calculated from (12.39), is therefore equal to the reciprocal of this expression, multiplied by a factor whose logarithm is $o(1)$, so that $J = \exp\left(-\mu^3/2\epsilon^2 + o(1)\right)$. This formula agrees with the large-deviation estimate obtained in Section 12.4 for the probability that a given droplet will escape, which (after division by the time needed for the escape to take place) can also be thought of as a rate of nucleation. For large times, this probability, according to (12.29) with $U(x)$ given by (12.21), is:

$$\exp(-\epsilon^{-2}\{U(\mu^3) - U(0)\}) = \exp(-\mu^3/2\epsilon^2) \quad (12.42)$$

agreeing with the above estimate of J.

12.6 Conclusions

The stochastic differential equation (12.16) proposed in Section 12.3 provides a possible mathematical model for the way that the sizes of droplets in a metastable thermodynamic phase, such as a supersaturated vapour, can change with time, including the way that nuclei of the new phase can form. The action formula (12.20), obtained in Section 12.4 by applying the ideas of Freidlin and Wentzell to this SDE, provides a convenient estimate of large-deviation type for the probabilities of the various different ways that the rescaled droplet size can vary as a function of rescaled time. Applied to the problem of estimating nucleation rates, this method gives a result that is consistent with the estimate provided by Becker and Döring's classical theory of nucleation.

This work raises the problem of whether the use of the SDE (12.16), and/or the action formula (12.20) for the logarithms of path probabilities, can be justified by a rigorous argument. One way of trying to justify the method is to treat the jump rates as approximately independent of particle size over a short time interval and using a Gaussian approximation (as in the Central Limit Theorem) for the probability distribution of the resulting biased random walk. Unfortunately, however, it seems to be impossible to choose the time interval so that it is short enough to make the accumulating error due to the approximation of constant jump rates small and yet also long enough to make the errors in the Gaussian approximation small. Thus the problem of justifying (12.16) and (12.20) or, if these formulas are incorrect, of finding the correct formulas, remains unsolved.

Acknowledgements

I am grateful to Antonio Galves for the initial impetus for this work, and to the IHES in Bures-sur-Yvette, France for its hospitality at that time. I also thank Enzo Olivieri, Mathew Penrose, Anatolii Puhalskii, and Bálint Tóth for helpful discussions and information.

References

Becker, R. and Döring, W. (1935), Kinetische Behandlung der Keimbildung in übersättigten Dämpfern, *Ann. Phys* (Leipzig) **24**, 719–52.

Ball, J. M., Carr, J., and Penrose, O. (1986) The Becker–Döring cluster equations: Basic properties and asymptotic behaviour of solutions. *Commun. Math. Phys.* **104**, 657–92.

Cassandro, M., Galves, A., Olivieri, E., and Vares, M. E. (1984), Metastable behaviour of stochastic dynamics: a pathwise approach, *J. Stat. Phys.* **35**, 603–34.

Freidlin, M. I. and Wentzell, A. D. (1998), *Random Perturbations of Dynamical Systems*, Springer: Berlin.

den Hollander, F. (2004), Metastability under stochastic dynamics, *Stochastic Processes and their Applications* **114**, 1–26.

Ito, K. and McKean, H P. (1996), *Diffusion Processes and Their Sample Paths*, Springer: Berlin.

Lifshits, I. M. and Slyozov, V. (1961), Kinetics of precipitation from supersaturated solid solutions, *J. Phys. Chem. Sol.* **19**, 35–50.

Olivieri, E. and Vares, M. E. (2006), *Large Deviations and Metastability*, Cambridge University Press: Cambridge.

Penrose, O. and Buhagiar, A. (1983), Kinetics of nucleation in a lattice gas model: microscopic theory and simulation compared, *J. Stat. Phys.* **30**, 219–41 (1983).

Penrose, O. (1989), Metastable states for the Becker-Doering cluster equations, *Commun. Math. Phys.* **124**, 515–41.

Penrose, O. (1997), The Becker–Döring equations at large times and their connection with the LSW theory of coarsening, *J. Stat. Phys.* **89**, 305–20.

Wagner, C. (1961), Theorie der Alterung von Niederschlagen durch Umlösen (Ostwald-Reifung) *Z. Electrochem.* **65**, 581–91.

B
APPLICATIONS IN PHYSICS

13

ON THE STOCHASTIC BURGERS EQUATION AND SOME APPLICATIONS TO TURBULENCE AND ASTROPHYSICS

A. D. Neate and A. Truman

Abstract

We summarize a selection of results on the inviscid limit of the stochastic Burgers equation emphasizing geometric properties of the caustic, Maxwell set and Hamilton–Jacobi level surfaces and relating these results to a discussion of stochastic turbulence. We show that for small viscosities there exists a vortex filament structure near to the Maxwell set. We discuss how this vorticity is directly related to the adhesion model for the evolution of the early universe and include new explicit formulas for the distribution of mass within the shock.

13.1 Introduction

The Burgers equation was first introduced by J. M. Burgers as a model for pressureless gas dynamics. It has since provided a tool for studying turbulence in fluids (Frisch and Bec 2001), for obtaining detailed asymptotics for stochastic Schrödinger and heat equations (Truman and Zhao 1996a,b; Elworthy et al. 2001) and has played a part in Arnol'd's work on caustics (Arnol'd 1989, 1990, 1992) and Maslov's works in semiclassical quantum mechanics (Maslov and Fedoriuk 1981). It has also been used for studying the formation of galaxies in the early universe in the Zeldovich approximation and also the adhesion model (Arnol'd et al. 1982; Shandarin and Zel'dovich 1989). A detailed explanation of these applications as well as a complete history of the Burgers equation can be found in (Bec and Khanin 2007).

In this article we will summarize a selection of results on the inviscid limit of the stochastic Burgers equation and outline some applications of these results to turbulence and the adhesion model. We begin in Section 2 with a summary of results on deterministic Hamilton–Jacobi theory for the heat and Burgers equation.

In Sections 3 to 5 we present some geometric and analytic results first developed by Davies, Truman, and Zhao (Davies et al. 2002, 2005) and later extended by Truman and Neate (Neate and Truman 2005, 2007a). These results relate the geometry of the caustic, Hamilton–Jacobi level surfaces and Maxwell set to that of their algebraic pre-images under the inviscid classical mechanical flow map Φ_t which will be defined in Section 3. In two dimensions these results show that a Hamilton–Jacobi level surface, or Maxwell set, can only have a cusp where their pre-images intersect the pre-caustic and so can only have cusps on the caustic. They also allow us to give conditions for the formation of swallowtails on both caustics and level surfaces which in turn have implications for the geometry of the Maxwell set.

We also introduce a reduced (one dimensional) action function which was developed by Reynolds, Truman, and Williams (Truman et al. 2003) under the assumption that only singularities of A_k type occur (Arnol'd 1992). Using this, we can find explicit equations for the caustic, level surfaces and Maxwell set and their pre-images. In Section 6 we use this to write down an explicit stochastic process whose zeros give 'turbulent times' at which cusps on the Hamilton–Jacobi level surfaces appear and disappear infinitely rapidly.

Finally, in Sections 7 and 8, we summarize results showing that the fluid has nonzero vorticity in some neighbourhood of the Maxwell set (Neate and Truman 2007a). We show that this vorticity disappears under the assumptions required for the adhesion model for the evolution of the early universe and outline a new formula for the mass which adheres to the shock (the Maxwell set).

Notation: throughout this paper x, x_0, x_t etc will denote vectors (usually in $\mathbb{R}^d$). Cartesian coordinates of these will be indicated using a sub/superscript where relevant; thus $x = (x_1, x_2, \ldots, x_d)$, $x_0 = (x_0^1, x_0^2, \ldots, x_0^d)$ etc. The only exception will be in discussions of explicit examples in two and three dimensions when we will use (x, y) and (x_0, y_0) etc to denote the vectors.

13.2 Elements of Hamilton–Jacobi theory

We begin by considering a deterministic classical mechanical system consisting of a unit mass moving under the influence of a conservative force, $-\nabla V$. This system has Hamiltonian:

$$H(q,p) = \frac{1}{2}p^2 + V(q),$$

where $p, q \in \mathbb{R}^d$. Let us assume that the system has a given initial velocity field ∇S_0 for some function $S_0 : \mathbb{R}^d \to \mathbb{R}$.

The evolution of this system will be given by the classical mechanical flow map, $\Phi_s : \mathbb{R}^d \to \mathbb{R}^d$ defined by:

$$\frac{d^2 \Phi_s}{ds^2} = -\nabla V(\Phi_s),$$

with initial condition:
$$\Phi_0 = I_d, \qquad \dot{\Phi}_0 = \nabla S_0,$$

where I_d denotes the d-dimensional identity map. Thus, if $X(s)$ is a classical mechanical path with $X(0) = x_0$, then:

$$X(s) = \Phi_s(x_0), \quad \dot{X}(0) = \nabla S_0(x_0).$$

Usually we also demand that $X(t) = x$ for fixed x and t. If S_0 and V are twice continuously differentiable with bounded second order derivatives, then there exists a caustic time $t_c > 0$, such that for all $t \in (0, t_c)$ the classical mechanical flow map is a diffeomorphism. This is a simple consequence of the global inverse function theorem (Abraham and Marsden 1978). Therefore we can define:

$$x_0(x, t) := \Phi_t^{-1}(x),$$

to be the unique pre-image of the point x reached by the path $X(s)$ at time t. If we now define:

$$S(x, t) := S_0(x_0(x, t)) + \int_0^t \left(\frac{1}{2} \dot{X}^2(s) - V(X(s)) \right) ds,$$

then it can be easily shown that $S_t(x) := S(x, t)$ satisfies the Hamilton–Jacobi equation:

$$\frac{\partial S_t}{\partial t} + H(x, \nabla S_t) = 0, \quad S_{t=0}(x) = S_0(x). \tag{13.1}$$

We now show how the function S_t can be used to construct a semi-classical solution to a corresponding heat equation (Truman 1977; Truman and Zhao 1996a, 1998).

Consider the heat equation for $u^\mu(x, t) \in \mathbb{R}$ where $x \in \mathbb{R}^d$ and $t > 0$:

$$\frac{\partial u^\mu}{\partial t} = \frac{\mu^2}{2} \Delta u^\mu + \mu^{-2} V(x) u^\mu, \tag{13.2}$$

with initial condition:

$$u^\mu(x, 0) = \exp\left(-\frac{S_0(x)}{\mu^2}\right) T_0(x). \tag{13.3}$$

Let $B_s \in \mathbb{R}^d$ be a d-dimensional Wiener process on the space $(\Omega, \mathcal{F}, \mathbb{P})$ with $\mathbb{E}\{B(s)B(t)\} = \min(s, t)$. Define an Ito diffusion $X_s^\mu \in \mathbb{R}^d$ and an Ito process $Y_s^\mu \in \mathbb{R}^d$ by:

$$dX_s^\mu = -\nabla S_{t-s}(X_s^\mu) \, ds + \mu \, dB_s, \quad X_0^\mu = x, \tag{13.4}$$
$$dY_s^\mu = \mu \, dB_s, \quad Y_0^\mu = x, \tag{13.5}$$

where $0 < s \leq t < t_c$. The time reversal in S_{t-s} allows us to effectively consider a diffusion process which will reach the point x at time t. Define $h(s, \omega) := h_0(Y_s^\mu(\omega), s)$ where:

$$h_0(Y_s^\mu, s) := -\mu^{-1} \nabla S_{t-s}(Y_s^\mu).$$

Since h satisfies the Novikov condition:

$$\mathbb{E}_\mathbb{P} \left\{ \exp\left(\frac{1}{2} \int_0^{t_c} h^2(s, \omega) \, ds\right) \right\} < \infty,$$

where $\mathbb{E}_{\mathbb{P}}$ denotes expectation with respect to the measure $\mathbb{P}$, it follows that:

$$M_s := \exp\left(-\int_0^s h(u,\omega)\,dB_u - \frac{1}{2}\int_0^s h^2(u,\omega)\,du\right),$$

is a martingale with respect to $\mathcal{F}_s = \sigma(B_s)$ and $\mathbb{P}$.

Using the Girsanov theorem, we can now define a new measure $\tilde{\mathbb{P}}$ on $(\Omega, \mathcal{F})$:

$$d\tilde{\mathbb{P}}(\omega) = M_{t_c}(\omega)\,d\mathbb{P}(\omega),$$

and then:

$$\tilde{B}_s := \int_0^s h(u,\omega)\,du + B_s,$$

is a Brownian motion with respect to $\tilde{\mathbb{P}}$. Therefore, $(Y_s^\mu, \tilde{B}_s)$, where Y_s^μ is defined in (13.5), forms a weak solution to equation (13.4). That is:

$$dY_s^\mu = -\nabla \mathcal{S}_{t-s}(Y_s^\mu)\,ds + \mu\,d\tilde{B}_s,$$

and conseqeuntly:

$$\mathbb{E}_{\mathbb{P}}\{f(X_s^\mu)\} = \mathbb{E}_{\tilde{\mathbb{P}}}\{f(Y_s^\mu)\} = \mathbb{E}_{\mathbb{P}}\{M_s f(B_s)\}. \tag{13.6}$$

It follows from the Feynmann–Kac formula that the heat equation (13.2) has a solution given by:

$$u^\mu(x,t) = \mathbb{E}_{\tilde{\mathbb{P}}}\left\{T_0(Y_t^\mu)\exp\left(-\mu^{-2}S_0(Y_t^\mu) + \mu^{-2}\int_0^t V(Y_s^\mu)\,ds\right)\right\},$$

and so by equation (13.6):

$$u^\mu(x,t) = \mathbb{E}_{\mathbb{P}}\left\{T_0(X_t^\mu)\exp\left(-\mu^{-2}S_0(X_t^\mu) + \mu^{-2}\int_0^t V(X_s^\mu)\,ds\right)\frac{d\tilde{\mathbb{P}}}{d\mathbb{P}}\right\}$$

$$= \mathbb{E}_{\mathbb{P}}\left\{T_0(X_t^\mu)\exp\left(-\mu^{-2}S_0(X_t^\mu) + \mu^{-2}\int_0^t V(X_s^\mu)\,ds\right.\right.$$

$$\left.\left. + \mu^{-1}\int_0^t \nabla \mathcal{S}_{t-s}(X_s^\mu)\,dB_s - \frac{1}{2\mu^2}\int_0^t |\nabla \mathcal{S}_{t-s}(X_s^\mu)|^2\,ds\right)\right\}. \tag{13.7}$$

Now, using Ito's formula:

$$\mathcal{S}(X_t^\mu, 0) = \mathcal{S}(x,t) + \int_0^t \left(\frac{\partial \mathcal{S}_{t-s}}{\partial s}(X_s^\mu) - |\nabla \mathcal{S}_{t-s}(X_s^\mu)|^2 + \frac{\mu^2}{2}\Delta \mathcal{S}_{t-s}(X_s^\mu)\right)ds$$

$$+ \mu\int_0^t \nabla \mathcal{S}_{t-s}(X_s^\mu)\,dB_s,$$

and so substituting into equation (13.7) for $\int_0^t \nabla \mathcal{S}_{t-s}(X_s^\mu)\, dB_s$ gives:

$$u^\mu(x,t) = e^{-\frac{\mathcal{S}_t(x)}{\mu^2}} \mathbb{E}_\mathbb{P}\left\{T_0(X_t^\mu)\exp\left(-\frac{1}{2}\int_0^t \Delta\mathcal{S}_{t-s}(X_s^\mu)\,ds\right.\right.$$
$$\left.\left. - \mu^{-2}\int_0^t \left(\frac{\partial \mathcal{S}_{t-s}}{\partial s}(X_s^\mu) - \frac{1}{2}|\nabla \mathcal{S}_{t-s}(X_s^\mu)|^2 - V(X_s^\mu)\right)ds\right)\right\}.$$

But $\mathcal{S}_t$ satisfies the Hamilton–Jacobi equation (13.1), and so, by reversing time in the diffusion X^μ, we have:

$$u^\mu(x,t) = \exp\left(-\frac{\mathcal{S}_t(x)}{\mu^2}\right)\mathbb{E}_x\left\{T_0(X_0^\mu)\exp\left(-\frac{1}{2}\int_0^t \Delta\mathcal{S}_{t-s}(X_s^\mu)\,ds\right)\right\}. \quad (13.8)$$

Using the logarithmic Hopf–Cole transformation (Hopf 1950):

$$v^\mu(x,t) = -\mu^2 \nabla \ln u^\mu(x,t), \quad (13.9)$$

the heat equation (13.2) becomes the Burgers equation for velocity field $v^\mu(x,t) \in \mathbb{R}^d$ where μ^2 is now the coefficient of viscosity:

$$\frac{Dv^\mu}{Dt} = \frac{\partial v^\mu}{\partial t} + (v^\mu \cdot \nabla)v^\mu = \frac{\mu^2}{2}\Delta v^\mu - \nabla V, \quad (13.10)$$

with initial condition:

$$v^\mu(x,0) = \nabla S_0(x) + \mathrm{O}(\mu^2).$$

We will be particularly interested in the behaviour of v^μ for small values of μ. In the remainder of this paper we will focus on the discontinuities that develop in v^μ as $\mu \to 0$.

The convergence factor T_0 in the initial condition (13.3) is related to the square root of the Burgers fluid mass density $\rho_t^{\frac{1}{2}}$:

$$T_0(x_0(x,t))\left|\left(\frac{\partial x_0}{\partial x}(x,t)\right)\right|^{\frac{1}{2}} = \rho_t^{\frac{1}{2}}(x). \quad (13.11)$$

For $t \in (0,t_c)$ it can be seen that mass is conserved:

$$\text{total mass} = \int \rho_t(x)\,dx = \int T_0^2(x_0)\,dx_0 = \int \rho_0(x)\,dx.$$

The next lemma will be key to our treatment of the solution for the Burgers equation.

Lemma 13.1 *Consider the above C^2 Hamiltonian dynamical system with Hamiltonian $H(q,p)$ and Hamilton–Jacobi function S_t satisfying:*

$$\frac{\partial S_t}{\partial t} + H(x, \nabla S_t) = 0, \qquad S_{t=0}(x) = S_0(x),$$

so that:

$$\dot{X}(t) = \nabla S_t(X(t)), \qquad \dot{X}(0) = \nabla S_0(X(0)).$$

Then:

$$\exp\left\{-\frac{1}{2}\int_0^t \Delta S_s(X(s))\,\mathrm{d}s\right\} = \left|\frac{\partial X(0)}{\partial X(t)}\right|^{\frac{1}{2}},$$

where the right hand side is a Jacobian determinant.

In particular it follows from Lemma 13.1, for $t \in (0, t_c)$, that by considering an asymptotic expansion of the diffusion X_s^μ in the solution to the heat equation (13.8):

$$u^\mu(x,t) = \exp\left(-\frac{S_t(x)}{\mu^2}\right) T_0(x_0(x,t)) \times \left|\left(\frac{\partial x_0}{\partial x}(x,t)\right)\right|^{\frac{1}{2}} (1 + \mathrm{O}(\mu^2)), \quad (13.12)$$

where $x_0(x,t)$ is the unique start point of X_s^0 with:

$$\dot{X}_s^0 = \nabla S_s(X_s^0), \qquad X_t^0 = x.$$

Consequently, the Burgers velocity field is given by:

$$v^\mu = v^\mu(x,t) \sim \nabla S_t(x) + \mathrm{O}(\mu^2).$$

13.3 The stochastic case

We now consider the behaviour of a Burgers equation with stochastic forcing. That is for $v^\mu(x,t) \in \mathbb{R}^d$:

$$\frac{\partial v^\mu}{\partial t} + (v^\mu \cdot \nabla) v^\mu = \frac{\mu^2}{2} \Delta v^\mu - \nabla V(x) - \epsilon \nabla k_t(x) \dot{W}_t, \quad (13.13)$$

with initial condition $v^\mu(x,0) = \nabla S_0(x) + \mathrm{O}(\mu^2)$, where $\dot{W}_t$ denotes white noise.

Using the logarithmic Hopf–Cole transformation (13.9), the Burgers equation (13.13) becomes the Stratonovich heat equation:

$$\frac{\partial u^\mu}{\partial t} = \frac{\mu^2}{2}\Delta u^\mu + \mu^{-2} V(x) u^\mu + \frac{\epsilon}{\mu^2} k_t(x) u^\mu \circ \dot{W}_t, \quad (13.14)$$

with initial condition $u^\mu(x,0) = \exp\left(-\frac{S_0(x)}{\mu^2}\right) T_0(x)$.

Now let:

$$A[X] := \frac{1}{2}\int_0^t \dot{X}^2(s)\,\mathrm{d}s - \int_0^t V(X(s))\,\mathrm{d}s - \epsilon \int_0^t k_s(X(s))\,\mathrm{d}W_s,$$

and select a path X with $X(t) = x$ which minimizes $A[X]$. This requires:

$$\mathrm{d}\dot{X}(s) + \nabla V(X(s))\,\mathrm{d}s + \epsilon \nabla k_s(X(s))\,\mathrm{d}W_s = 0.$$

We then define the stochastic action, $A(X(0), x, t) := \inf_X \{A[X] : X(t) = x\}$. Setting:

$$\mathcal{A}(X(0), x, t) := S_0(X(0)) + A(X(0), x, t),$$

and then minimizing $\mathcal{A}$ over $X(0)$, gives $\dot{X}(0) = \nabla S_0(X(0))$. Moreover, it follows that:

$$\mathcal{S}_t(x) := \inf_{X(0)} \{\mathcal{A}(X(0), x, t)\},$$

is the minimal solution of the Hamilton–Jacobi equation:

$$\mathrm{d}\mathcal{S}_t + \left(\frac{1}{2}|\nabla \mathcal{S}_t|^2 + V(x)\right)\mathrm{d}t + \epsilon k_t(x)\,\mathrm{d}W_t = 0, \qquad \mathcal{S}_{t=0}(x) = S_0(x).$$

Following the work of Freidlin and Wentzell (Freidlin and Wentzell 1998):

$$-\mu^2 \ln u^\mu(x,t) \to \mathcal{S}_t(x),$$

as $\mu \to 0$. This gives the inviscid limit of the minimal entropy solution of the Burgers equation as $v^0(x,t) = \nabla \mathcal{S}_t(x)$ (Dafermos 2005).

Define the classical flow map $\Phi_s : \mathbb{R}^d \to \mathbb{R}^d$ by:

$$\mathrm{d}\dot{\Phi}_s + \nabla V(\Phi_s)\,\mathrm{d}s + \epsilon \nabla k_s(\Phi_s)\,\mathrm{d}W_s = 0, \qquad \Phi_0 = \mathrm{id}, \qquad \dot{\Phi}_0 = \nabla S_0.$$

Since $X(t) = x$ it follows that $X(s) = \Phi_s(\Phi_t^{-1}(x))$, where the pre-image $x_0(x,t) = \Phi_t^{-1}(x)$ is not necessarily unique.

Given some regularity and boundedness, the global inverse function theorem gives a random caustic time $t_c(\omega)$ such that for $0 < t < t_c(\omega)$, the pre-image, $x_0(x,t)$, if it exists, is unique and Φ_t is a random diffeomorphism. Thus, before the caustic time $v^0(x,t) = \dot{\Phi}_t(\Phi_t^{-1}(x))$ is the inviscid limit of a solution of the Burgers equation with probability one (Truman and Zhao 1996a, 1998).

The method of characteristics suggests that discontinuities in $v^0(x,t)$ are associated with the nonuniqueness of the real pre-image $x_0(x,t)$. In the situation we consider, when this occurs the classical flow map Φ_t focuses an infinitesimal volume of points $\mathrm{d}x_0$ into a zero volume $\mathrm{d}X(t)$.

Definition 13.1 *The caustic at time t is defined to be the set:*
$$C_t = \left\{ x : \det\left(\frac{\partial X(t)}{\partial x_0}\right) = 0 \right\}.$$

Assume that after the caustic time $t_c(\omega) > 0$, x has n real pre-images:
$$\Phi_t^{-1}\{x\} = \{x_0(1)(x,t), x_0(2)(x,t), \ldots, x_0(n)(x,t)\},$$
where each $x_0(i)(x,t) \in \mathbb{R}^d$. Then the Feynman–Kac formula and Laplace's method in infinite dimensions give for a nondegenerate critical point (Davies and Truman 1983, 1984):
$$u^\mu(x,t) = \sum_{i=1}^n \theta_i \exp\left(-\frac{S_0^i(x,t)}{\mu^2}\right), \qquad (13.15)$$
where $S_0^i(x,t) := S_0(x_0(i)(x,t)) + A(x_0(i)(x,t), x, t)$, and θ_i is an asymptotic series in μ^2. An asymptotic series in μ^2 can also be found for $v^\mu(x,t)$ (Truman and Zhao 1998). Note that $\mathcal{S}_t(x) = \min\{S_0^i(x,t) : i = 1, 2, \ldots, n\}$.

Definition 13.2 *The Hamilton–Jacobi level surface is the set:*
$$H_t^c = \{x : S_0^i(x,t) = c \text{ for some } i\}.$$

As $\mu \to 0$, the dominant term in the expansion (13.15) comes from the minimizing $x_0(i)(x,t)$ which we denote $\tilde{x}_0(x,t)$. Assuming $\tilde{x}_0(x,t)$ is unique, we obtain the inviscid limit of the Burgers fluid velocity as the minimal entropy $v^0(x,t) = \dot{\Phi}_t(\tilde{x}_0(x,t))$.

If the minimizing pre-image $\tilde{x}_0(x,t)$ suddenly changes value between two pre-images $x_0(i)(x,t)$ and $x_0(j)(x,t)$, a jump discontinuity will occur in $v^0(x,t)$. There are two distinct ways in which the minimizer can change; either two pre-images coalesce and disappear (become complex), or the minimizer switches between two pre-images at the same action value. The first of these occurs as x crosses the caustic. When this results in the minimizer disappearing the caustic is said to be cool. The second occurs as x crosses the Maxwell set and again, when the minimizer is involved, the Maxwell set is said to be cool.

Definition 13.3 *The Maxwell set is*:
$$M_t = \{x : \exists\, x_0, \check{x}_0 \in \mathbb{R}^d \text{ s.t.}$$
$$x = \Phi_t(x_0) = \Phi_t(\check{x}_0),\ x_0 \neq \check{x}_0 \text{ and } \mathcal{A}(x_0, x, t) = \mathcal{A}(\check{x}_0, x, t)\}.$$

We illustrate this in one dimension by considering the integral:
$$I(x,t) = \int_{\mathbb{R}} G(x_0) \exp\left(i\frac{F(x_0, x, t)}{\mu^2}\right) dx_0, \qquad (13.16)$$
where $G \in C_0^\infty(\mathbb{R})$, $x \in \mathbb{R}^d$ and $i = \sqrt{-1}$. Consider the graph of the phase function, $F_{(x,t)}(x_0) = F(x_0, x, t)$, as x crosses the caustic and Maxwell set (see Fig. 13.1).

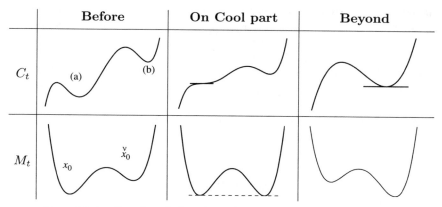

FIG. 13.1: Graphs of the phase function as x crosses C_t and M_t.

As we cross the caustic, the critical point at (a) becomes an inflexion which disappears causing $\tilde{x}_0(x,t)$ to jump from (a) to (b). This only causes a jump in $v^\mu(x,t)$ when the point of inflexion is the global minimizer of F. As we cross the Maxwell set, the critical points at x_0 and $\tilde{x}_0$ move so that $F_{(x,t)}(x_0) = F_{(x,t)}(\tilde{x}_0)$. If this pair of critical points also minimize the phase function, then the inviscid limit of the solution to the Burgers equation will jump.

13.4 The reduced action function

In this section we will find the phase function F in equation (13.16). We briefly summarize some results of Davies, Truman, and Zhao (Davies et al. 2002, 2005). As before, let the stochastic action be defined as:

$$A(x_0, p_0, t) = \frac{1}{2}\int_0^t \dot{X}(s)^2 \, ds - \int_0^t \left[V(X(s)) \, ds + \epsilon k_s(X(s)) \, dW_s \right],$$

where $X(s) = X(s, x_0, p_0) \in \mathbb{R}^d$ and for $s \in [0,t]$ with $x_0, p_0 \in \mathbb{R}^d$:

$$d\dot{X}(s) = -\nabla V(X(s)) \, ds - \epsilon \nabla k_s(X(s)) \, dW_s, \quad X(0) = x_0, \quad \dot{X}(0) = p_0.$$

We assume $X(s)$ is unique and let $\mathcal{F}_s$ denote the sigma algebra generated by $X(u)$ up to time s. It follows from Kunita (Kunita 1984):

Lemma 13.2 *Assume $S_0, V \in C^2$ and $k_t \in C^{2,0}$, $\nabla V, \nabla k_t$ Lipschitz with Hessians $\nabla^2 V, \nabla^2 k_t$ and all second derivatives with respect to space variables of V and k_t bounded. Then for p_0, possibly x_0 dependent:*

$$\frac{\partial A}{\partial x_0^\alpha}(x_0, p_0, t) = \dot{X}(t) \cdot \frac{\partial X(t)}{\partial x_0^\alpha} - \dot{X}_\alpha(0), \quad \alpha = 1, 2, \ldots, d.$$

The methods of (Kolokol'tsov et al. 2004) guarantee that for small t the map $p_0 \mapsto X(t, x_0, p_0)$ is onto for all x_0. Therefore, we can define $\mathcal{A}(x_0, x, t) := A(x_0, p_0(x_0, x, t), t)$ where $p_0 = p_0(x_0, x, t)$ is the random minimizer (which we assume to be unique) of $A(x_0, p_0, t)$ when $X(t, x_0, p_0) = x$.

Thus, the stochastic action corresponding to the initial momentum $\nabla S_0(x_0)$ is $\mathcal{A}(x_0, x, t) := A(x_0, x, t) + S_0(x_0)$.

Theorem 13.1 *If Φ_t is the stochastic flow map then:*

$$\Phi_t(x_0) = x \quad \Leftrightarrow \quad \frac{\partial}{\partial x_0^\alpha}[\mathcal{A}(x_0, x, t)] = 0, \qquad \alpha = 1, 2, \ldots, d.$$

Using this we can create a one dimensional *reduced action function*. This is done by finding a series of functions $x_0^\alpha(x_0^1, \ldots, x_0^{\alpha-1}, x, t)$ for decreasing $\alpha = d, d-1, \ldots, 2$ by systematically locally solving the equations:

$$\frac{\partial \mathcal{A}}{\partial x_0^\alpha}(x_0^1, \ldots, x_0^\alpha, x_0^{\alpha+1}(\ldots), \ldots, x_0^d(\ldots), x, t) = 0.$$

At each stage this eliminates one more coordinate from x_0 until only x_0^1 remains. This gives *local reducibility* on the assumption that $\partial^2 \mathcal{A}/(\partial x_0^\alpha)^2 \neq 0$ for $\alpha = 2, 3, \ldots, d$ and also some mild regularity conditions (Truman et al. 2003).

Definition 13.4 *The reduced action function is the univariate function:*

$$f_{(x,t)}(x_0^1) := \mathcal{A}(x_0^1, x_0^2(x_0^1, x, t), \ldots, x_0^d(x_0^1, x_0^2(\cdot), \ldots, x_0^{d-1}(\cdot), x, t), x, t).$$

The Hamilton–Jacobi level surface H_t^c is found by eliminating x_0 between:

$$\mathcal{A}(x_0, x, t) = c, \qquad \nabla_{x_0} \mathcal{A}(x_0, x, t) = 0.$$

Alternatively, if we eliminate x to give an expression in x_0, we have the pre-level surface $\Phi_t^{-1} H_t^c$. Similarly the caustic C_t (and pre-caustic $\Phi_t^{-1} C_t$) are obtained by eliminating x_0 (or x) between:

$$\det\left(\frac{\partial^2 \mathcal{A}}{\partial x_0^\alpha \partial x_0^\beta}(x_0, x, t)\right)_{\alpha, \beta = 1, 2, \ldots, d} = 0, \qquad \nabla_{x_0} \mathcal{A}(x_0, x, t) = 0.$$

The Maxwell set M_t (and pre-Maxwell set $\Phi_t^{-1} M_t$) are obtained by eliminating x_0 and $\check{x}_0$ (or x and $\check{x}_0$) between the four equations,

$$\nabla_{x_0} \mathcal{A}(x_0, x, t) = 0, \quad \nabla_{x_0} \mathcal{A}(\check{x}_0, x, t) = 0, \quad \mathcal{A}(x_0, x, t) = \mathcal{A}(\check{x}_0, x, t) = c.$$

The pre-images are calculated algebraically and in the case of the pre-level surfaces are not necessarily the topological inverse images. This can be done in the free case or when the relevant functions are polynomials in all variables which is an implicit assumption in what follows.

For polynomial $\mathcal{A}$, the eliminations involved with the Hamilton–Jacobi level surfaces and caustics are fairly simple to complete using the reduced action function with resultants and discriminants which can be calculated via Sylvester determinants (van der Waerden 1949). The Maxwell set is more complicated to find as eliminating pre-images leads to a surface involving both real and complex pre-images termed the 'Maxwell–Klein set' (Neate and Truman 2005). It is easier to find the pre-Maxwell set and then use the flow map to parameterize the Maxwell set. Parameterizing in this manner allows one to restrict the pre-image of the Maxwell set to have only real values. In the polynomial case we have the following lemma.

Lemma 13.3 Let D^x denote the polynomial discriminant taken with respect to x. The set of all singularities is:

$$D^c(D^{\lambda_1}(f_{(x,t)}(\lambda_1) - c)) = 0,$$

which factorizes as:

$$k \times B_t(x)^2 \times C_t(x)^3 = 0,$$

where $B_t = 0$ is the equation of the Maxwell–Klein set, $C_t = 0$ is the equation of the caustic and k is some nonzero constant.

The pre-Maxwell set is given by:

$$D^{\lambda_1}\left(\frac{f_{(\Phi_t(x_0),t)}(x_0^1) - f_{(\Phi_t(x_0),t)}(\lambda_1)}{(x_0^1 - \lambda_1)^2}\right) = 0.$$

The reduced action function can also be used to identify the cool (singular) parts of the Maxwell set and caustic (Neate and Truman 2008b).

13.5 Geometric Results

The results in this section are taken from (Davies et al. 2002; Neate and Truman 2005, 2007a). Assume that $\mathcal{A}(x_0, x, t)$ is C^4 in space variables with $\det\left(\frac{\partial^2 \mathcal{A}}{\partial x_0^\alpha \partial x^\beta}\right) \neq 0$.

Lemma 13.4 Let Φ_t denote the stochastic flow map and $\Phi_t^{-1}\Gamma_t$ and Γ_t be some surfaces where if $x_0 \in \Phi_t^{-1}\Gamma_t$ then $x = \Phi_t(x_0) \in \Gamma_t$. Then, Φ_t is a differentiable map from $\Phi_t^{-1}\Gamma_t$ to Γ_t with Frechet derivative:

$$(D\Phi_t)(x_0) = \left(-\frac{\partial^2 \mathcal{A}}{\partial x \partial x_0}(x_0, x, t)\right)^{-1}\left(\frac{\partial^2 \mathcal{A}}{(\partial x_0)^2}(x_0, x, t)\right).$$

Let $n_H(x_0)$, $n_C(x_0)$ and $n_M(x_0)$ denote the normal at x_0 to the pre-level surface, pre-caustic and pre-Maxwell set respectively. Using Lemma 13.4 we can show the following.

Theorem 13.2 *The normal to the pre-level surface is, to within a scalar multiplier, given by:*

$$n_H(x_0) = -\left(\frac{\partial^2 \mathcal{A}}{(\partial x_0)^2}\right)\left(\frac{\partial^2 \mathcal{A}}{\partial x_0 \partial x}\right)^{-1} \dot{X}(t, x_0, \nabla S_0(x_0)).$$

Theorem 13.3 *Assume that a point x on the Maxwell set corresponds to exactly two pre-images on the pre-Maxwell set, x_0 and $\breve{x}_0$. Then the normal to the pre-Maxwell set at x_0 is, to within a scalar multiplier, given by:*

$$n_M(x_0) = -\left(\frac{\partial^2 \mathcal{A}}{(\partial x_0)^2}(x_0, x, t)\right)\left(\frac{\partial^2 \mathcal{A}}{\partial x_0 \partial x}(x_0, x, t)\right)^{-1} \cdot$$
$$\left(\dot{X}(t, x_0, \nabla S_0(x_0)) - \dot{X}(t, \breve{x}_0, \nabla S_0(\breve{x}_0))\right).$$

We now consider the two dimensional case.

Definition 13.5 *Let $x = x(\gamma) = (x_1, x_2)(\gamma)$ denote a curve where γ is some intrinsic parameter (e.g., arc length) with $\gamma \in (\gamma_0 - \delta, \gamma_0 + \delta)$ for $\gamma_0 \in \mathbb{R}$ and $\delta > 0$. Then the curve is said to have a generalized cusp when $\gamma = \gamma_0$ if,*

$$\frac{dx}{d\gamma}(\gamma_0) = \left(\frac{dx_1}{d\gamma}(\gamma_0), \frac{dx_2}{d\gamma}(\gamma_0)\right) = 0.$$

It then follows from Theorems 13.2 and 13.3 that:

Theorem 13.4 *Assume that in two dimensions at $x_0 \in \Phi_t^{-1} H_t^c$ the normal $n_H(x_0) \neq 0$ so that the pre-level surface does not have a generalized cusp at x_0. Then, the level surface can only have a cusp at $\Phi_t(x_0)$ if $\Phi_t(x_0) \in C_t$. Moreover, if:*

$$x = \Phi_t(x_0) \in \Phi_t\left\{\Phi_t^{-1} C_t \cap \Phi_t^{-1} H_t^c\right\},$$

the level surface will have a generalized cusp at x.

Theorem 13.5 *Assume that in two dimensions at $x_0 \in \Phi_t^{-1} M_t$ the normal $n_M(x_0) \neq 0$ so that the pre-Maxwell set does not have a generalized cusp at x_0. Then, the Maxwell set can only have a cusp at $\Phi_t(x_0)$ if $\Phi_t(x_0) \in C_t$. Moreover, if:*

$$x = \Phi_t(x_0) \in \Phi_t\left\{\Phi_t^{-1} C_t \cap \Phi_t^{-1} M_t\right\},$$

the Maxwell set will have a generalized cusp at x.

These results lead to a range of conclusions relating to the geometry of these curves. In particular, they allow us to characterize when swallowtails will form (a swallowtail perestroika). The appearance of a swallowtail is related to the existence of points with complex pre-images which are discussed in detail in (Neate and Truman 2005).

Corollary 13.1 *Assume that at $x_0 \in \Phi_t^{-1} H_t^c \cap \Phi_t^{-1} C_t$, $n_H(x_0) \neq 0$ and $n_C(x_0) \neq 0$. Then at $\Phi_t(x_0)$ there is a cusp on the caustic if and only if $\Phi_t^{-1} H_t^c$ touches $\Phi_t^{-1} C_t$ at x_0. Moreover, it follows that, $x_0 \in \Phi_t^{-1} M_t$ and that $\Phi_t^{-1} H_t^c$ touches $\Phi_t^{-1} M_t$ at x_0. Also, at $\Phi_t(x_0)$, M_t will have a generalized cusp parallel to the cusp on C_t.*

Corollary 13.2 *Assume that at $x_0 \in \Phi_t^{-1} M_t \cap \Phi_t^{-1} C_t$, $n_M(x_0) \neq 0$ and $n_C(x_0) \neq 0$. Then, there is a cusp on the Maxwell set where it intersects the caustic at $x = \Phi_t(x_0)$ and the pre-Maxwell set touches a pre-level surface $\Phi_t^{-1} H_t^c$ at x_0. Moreover, if the cusp on the Maxwell set intersects the caustic at a regular point of the caustic, then there will be a cusp on the pre-Maxwell set which also meets the same pre-level surface $\Phi_t^{-1} H_t^c$ at another point $\check{x}_0$.*

Corollary 13.3 *Assume that at $x_0 \in \Phi_t^{-1} H_t^c \cap \Phi_t^{-1} C_t$, $n_H(x_0) \neq 0$ and $n_C(x_0) \neq 0$. Then at $\Phi_t(x_0)$ there is a point of swallowtail perestroika on the level surface H_t^c if and only if there is a generalized cusp on the caustic C_t at $\Phi_t(x_0)$.*

The results in this section have natural extensions to three dimensions where the cusps are replaced by curves of cusps. We now give some two dimensional examples.

Example 13.1 (The generic cusp) *We consider a two dimensional deterministic free example ($V \equiv 0$, $\epsilon = 0$). In general for such a system the flow map is given by:*

$$\Phi_t(x_0) = x_0 + t \nabla S_0(x_0),$$

with derivative map $D\Phi_t(x_0) = (I + t \nabla^2 S_0(x_0))$. The pre-level surface is then given by the eikonal equation:

$$\frac{t}{2} |\nabla S_0(x_0)|^2 + S_0(x_0) = c,$$

where the key identity is:

$$\nabla_{x_0} \left\{ \frac{t}{2} |\nabla S_0(x_0)|^2 + S_0(x_0) \right\} = (I + t \nabla^2 S_0(x_0)) \nabla S_0(x_0).$$

The generic cusp initial condition, $S_0(x_0, y_0) = x_0^2 y_0$, gives a simple cusped caustic (see Fig. 13.2).

Example 13.2 (The polynomial swallowtail) *Let $V(x,y) = 0$, $k_t(x,y) = x$ and $S_0(x_0, y_0) = x_0^5 + x_0^2 y_0$. The noisy potential does not affect either the pre-caustic or pre-Maxwell set. Consequently at time t the noise will have shifted the*

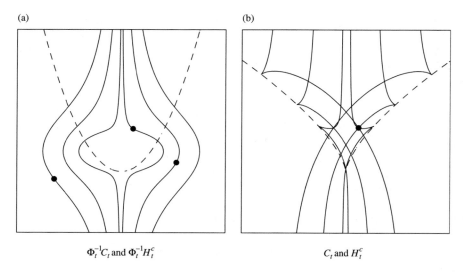

FIG. 13.2: The generic cusp caustic (dashed) with three level surfaces (solid line).

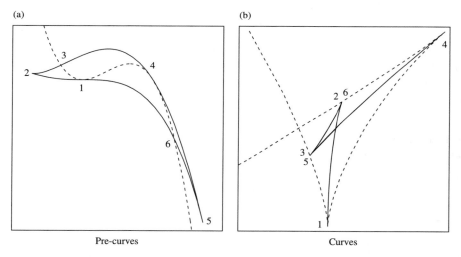

FIG. 13.3: The polynomial swallowtail caustic (dashed) and Maxwell set (solid line).

deterministic caustic and Maxwell set by $-\epsilon \int_0^t W(u)\,du$ in the x direction. This point will be returned to in Section 6.

From Theorem 13.5, the cusps on the Maxwell set correspond to the intersections of the pre-curves (points 3 and 6 on Fig. 13.3). But from Corollary 13.2, the cusps on the Maxwell set also correspond to the cusps on the

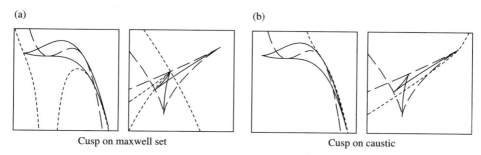

FIG. 13.4: The caustic (long dash) and Maxwell set (solid line) with the level surfaces (short dash) through special points.

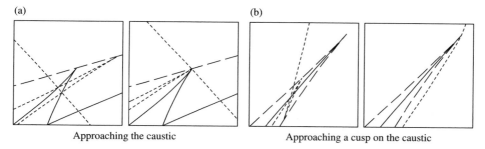

FIG. 13.5: The caustic (long dash) Maxwell set (solid line) and level surface (short dash).

pre-Maxwell set (points 2 and 5 on Fig. 13.3 and also Fig. 13.4). The Maxwell set terminates when it reaches the cusps on the caustic. These points satisfy the condition for a generalized cusp but, instead of appearing cusped, the curve stops and maps back exactly onto itself. At such points the pre-surfaces all touch (Fig. 13.4).

These two different forms of cusps correspond to very different geometric behaviours of the level surfaces. From the definition of a Maxwell set it is clear that any point on M_t is a point of self-intersection of some level surface. Where the Maxwell set stops or cusps corresponds to the disappearance of a point of self-intersection on a level surface. There are two distinct ways in which this can happen. Firstly, the level surface will have a point of swallowtail perestroika when it meets a cusp on the caustic. At such a point only one point of self-intersection will disappear, and so there will be only one path of the Maxwell set which will terminate at that point. However, when we approach the caustic at a regular point, the level surface must have a cusp but not a swallowtail perestoika. This corresponds to the collapse of two points of self-intersection and so two paths of the Maxwell set must approach the point and produce the cusp (see Fig. 13.5).

13.6 Recurrence of stochastic turbulence

Following (Truman et al. 2003; Neate and Truman 2007b), the geometric results of Section 5 can be used to characterize a sequence of turbulent times.

Definition 13.6 *Real turbulent times are defined to be times t at which there exist real points where the pre-level surface $\Phi_t^{-1} H_t^c$ and pre-caustic $\Phi_t^{-1} C_t$ touch.*

Real turbulent times correspond to times at which there is a change in the number of cusps or cusped curves on the level surface H_t^c. In d-dimensions, assuming Φ_t is globally reducible, let $f_{(x,t)}(x_0^1)$ denote the reduced action function and $x_t(\lambda)$ denote the caustic parameterized using the pre-caustic and flow map.

Theorem 13.6 *The real turbulent times t are given by the zeros of the zeta process ζ_t^c where:*

$$\zeta_t^c := f_{(x_t(\lambda),t)}(\lambda_1) - c,$$

λ *satisfies:*

$$\frac{\partial}{\partial \lambda_\alpha} f_{(x_t(\lambda),t)}(\lambda_1) = 0 \quad \text{for } \alpha = 1, 2, \ldots, d, \tag{13.17}$$

and $x_t(\lambda)$ is on the cool part of the caustic.

The term 'real' is used in (Neate and Truman 2005) to distinguish this form of turbulence from 'complex' turbulence where swallowtail perestroikas occur on the caustic. We shall not discuss the details of complex turbulence in this chapter.

We now consider the stochastic Burgers equation with white noise forcing in d-orthogonal directions:

$$\frac{\partial v^\mu}{\partial t} + (v^\mu \cdot \nabla) v^\mu = \frac{\mu^2}{2} \Delta v^\mu - \epsilon \dot{W}(t), \tag{13.18}$$

where $W(t) = (W_1(t), W_2(t), \ldots, W_d(t))$ is a d-dimensional Wiener process.

Proposition 13.1 *The stochastic action corresponding to the Burgers equation (13.18) is:*

$$\mathcal{A}(x_0, x, t) = \frac{|x - x_0|^2}{2t} + \frac{\epsilon}{t}(x - x_0) \cdot \int_0^t W(s)\, ds - \epsilon x \cdot W(t)$$

$$- \frac{\epsilon^2}{2} \int_0^t |W(s)|^2\, ds + \frac{\epsilon^2}{2t} \left| \int_0^t W(s)\, du \right|^2 + S_0(x_0).$$

Lemma 13.5 *If $x_t^\epsilon(\lambda)$ denotes the random caustic for the stochastic Burgers equation (13.18) and $x_t^0(\lambda)$ denotes the deterministic caustic (the $\epsilon = 0$ case) then:*

$$x_t^\epsilon(\lambda) = x_t^0(\lambda) - \epsilon \int_0^t W(u)\,du.$$

Using Proposition 13.1 and Lemma 13.5, we can find the zeta process explicitly.

Theorem 13.7 *In d-dimensions, the zeta process for the stochastic Burgers equation (13.18) is:*

$$\zeta_t^c = f^0_{(x_t^0(\lambda),t)}(\lambda_1) - \epsilon x_t^0(\lambda) \cdot W(t) + \epsilon^2 W(t) \cdot \int_0^t W(s)\,ds - \frac{\epsilon^2}{2}\int_0^t |W(s)|^2\,ds - c,$$

where $f^0_{(x,t)}(\lambda_1)$ is the deterministic reduced action function, $x_t^0(\lambda)$ is the deterministic caustic and λ must satisfy the stochastic equation:

$$\nabla_\lambda \left(f^0_{(x_t^0(\lambda),t)}(\lambda_1) - \epsilon x_t^0(\lambda) \cdot W(t)\right) = 0. \tag{13.19}$$

Equation (13.19) shows that the value of λ used in the zeta process may be either deterministic or random. In the two dimensional case this gives:

$$0 = \left(\nabla_x f^0_{(x_t^0(\lambda),t)}(\lambda_1) - \epsilon W(t)\right) \cdot \frac{dx_t^0}{d\lambda}(\lambda), \tag{13.20}$$

which has a deterministic solution for λ corresponding to a cusp on the deterministic caustic.

Using the law of the iterated logarithm, it is a simple matter to show formally that if there is a time τ such that $\zeta_\tau^c = 0$, then there will be infinitely many zeros of ζ_t^c in some neighbourhood of τ. This suggests that the set of zeros of ζ_t^c are a perfect set which can be rigorously proved in some generality (Reynolds 2002).

The intermittence of turbulence will be demonstrated if we can show that there is an unbounded increasing sequence of times at which the zeta process is zero. This can be done using an idea of David Williams and the Strassen form of the law of the iterated logarithm (Truman et al. 2003).

Theorem 13.8 *There exists an unbounded increasing sequence of times t_n for which $Y_{t_n} = 0$, almost surely, where:*

$$Y_t = W(t) \cdot \int_0^t W(s)\,ds - \frac{1}{2}\int_0^t |W(s)|^2\,ds,$$

and $W(t)$ is a d-dimensional Wiener process.

Corollary 13.4 *Let $h(t) = (2t\ln\ln(t))^{-\frac{1}{2}}$. If $h(t)^2 t^{-1} f^0_{(x_t^0(\lambda),t)}(\lambda_1) \to 0$ and $h(t)t^{-1}\sum_{i=0}^d x_t^{0_i}(\lambda) \to 0$, then the zeta process ζ_t^c is recurrent.*

13.7 A vortex line sheet on the Maxwell set

We now summarize results of (Neate and Truman 2007a) which show that for compressible flow and small viscosity, with appropriate initial conditions, there is a vortex filament structure in the neighbourhood of the cool part of the Maxwell set. This result is valid for both deterministic and stochastic cases.

We first compute the Burgers fluid velocity on the cool part of the Maxwell set where typically for $x \in M_t$ the pre-images $x_0(x,t)$ and $\check{x}_0(x,t)$ are well behaved functions. Recall from equations (13.11) and (13.12) that for $t < t_c(\omega)$ and small μ:

$$u^\mu(x,t) \sim \exp\left(-\frac{\mathcal{S}_t(x)}{\mu^2}\right) \rho_t^{\frac{1}{2}}(x)(1 + \mathrm{O}(\mu^2)).$$

For $t > t_c(\omega)$, the analogous result for v^μ is:

Lemma 13.6 *Let $x \in \mathrm{Cool}(M_t)$, so that $x = \Phi_t(x_0) = \Phi_t(\check{x}_0)$ where $x_0 \neq \check{x}_0$ and $\mathcal{A}(x_0, x, t) = \mathcal{A}(\check{x}_0, x, t)$. Then:*

$$v^\mu(x,t) \sim \frac{\rho_t^{\frac{1}{2}}(x)\nabla \mathcal{S}_t(x) + \check{\rho}_t^{\frac{1}{2}}(x)\nabla \check{\mathcal{S}}_t(x)}{\rho_t^{\frac{1}{2}}(x) + \check{\rho}_t^{\frac{1}{2}}(x)} + \mathrm{O}(\mu^2),$$

where:

$$\rho_t^{\frac{1}{2}}(x) = T_0(x_0(x,t))\left|\frac{\partial x_0}{\partial x}(x,t)\right|^{\frac{1}{2}}, \quad \check{\rho}_t^{\frac{1}{2}}(x) = T_0(\check{x}_0(x,t))\left|\frac{\partial \check{x}_0}{\partial x}(x,t)\right|^{\frac{1}{2}},$$

and $\mathcal{S}_t(x) = \mathcal{A}(x_0(x,t), x, t)$, $\check{\mathcal{S}}_t(x) = \mathcal{A}(\check{x}_0(x,t), x, t)$.

We denote by $v^0(x,t)$ the leading term for the behaviour of $v^\mu(x,t)$ on the Maxwell set and choose orthogonal curvilinear coordinates on M_t denoted by (ξ_1, ξ_2). Let the unit normal in a coordinate patch on M_t be denoted n.

Theorem 13.9 *If $v^0(x,t)$ is the leading behaviour of $v^\mu(x,t)$ for $x \in M_t$, then:*

$$v^0(x,t) = \frac{1}{2}\left\{\nabla(\mathcal{S}_t(x) + \check{\mathcal{S}}_t(x)) + \frac{\check{\rho}_t^{\frac{1}{2}}(x) - \rho_t^{\frac{1}{2}}(x)}{\check{\rho}_t^{\frac{1}{2}}(x) + \rho_t^{\frac{1}{2}}(x)}\left(\frac{\partial \check{\mathcal{S}}_t}{\partial n}(x) - \frac{\partial \mathcal{S}_t}{\partial n}(x)\right)n\right\},$$

where $\dfrac{\partial}{\partial n}$ denotes the normal derivative $(n \cdot \nabla)$ on M_t.

Definition 13.7 *The inviscid limit of the vorticity ω^0 is defined to be:*

$$\omega^0(x,t) := \nabla \wedge v^0(x,t),$$

where $v^0(x,t)$ denotes the leading behaviour of $v^\mu(x,t)$ as $\mu \to 0$.

Since the first term in v^0 is C^1 on M_t, we obtain:

Corollary 13.5 *For $x \in \text{Cool}(M_t)$:*

$$\omega^0 = \nabla \wedge \left\{ \frac{\check{\rho}_t^{\frac{1}{2}}(x) - \rho_t^{\frac{1}{2}}(x)}{\check{\rho}_t^{\frac{1}{2}}(x) + \rho_t^{\frac{1}{2}}(x)} \left(\frac{\partial \check{\mathcal{S}}_t}{\partial n}(x) - \frac{\partial \mathcal{S}_t}{\partial n}(x) \right) n \right\} \in T_x M_t.$$

From Corollary 13.5 it follows that $\omega^0 \neq 0$. Hence, we expect that for small viscosity even though initially there was zero vorticity, once a Maxwell set appears, the flow is no longer irrotational.

The above result shows that for small viscosity a vortex filament structure will appear in a neighbourhood of the cool part of the Maxwell set. The limit of this vortex filament structure is a sheet of vortex lines on the cool part of the Maxwell set. We now give the equation of these limiting vortex lines on the Maxwell set in terms of orthogonal coordinates (ξ_1, ξ_2).

Theorem 13.10 *The limiting vortex lines in a coordinate patch of M_t have equations:*

$$h_1(\xi) h_2(\xi) (\check{\rho}_t^{\frac{1}{2}}(x) - \rho_t^{\frac{1}{2}}(x)) \left(\frac{\partial}{\partial n} \left(f_{(x,t)}(\check{x}_0^1(x,t)) - f_{(x,t)}(x_0^1(x,t)) \right) \right)$$
$$= c(\check{\rho}_t^{\frac{1}{2}}(x) + \rho_t^{\frac{1}{2}}(x)),$$

where c is a real constant and $\frac{\partial}{\partial n}$ denotes the normal derivative $(n \cdot \nabla)$ on M_t.

This confirms that for our initial conditions, for small viscosity, and for compressible flow, vortex filaments will inevitably appear in a neighbourhood of the cool part of the Maxwell set. Given the rotational effects at work in the universe perhaps this suggests that we should consider the Burgers equation with vorticity from the outset. Kinematical considerations and Galilean invariance suggest that the appropriate equation is a Burgers equation with a vector potential. We hope to discuss this in a future paper (Neate and Truman 2008a).

13.8 The adhesion model

The adhesion model for the formation of the early universe is a refinement of the Zeldovich approximation (Arnol'd et al. 1982). In the original adhesion model there is no noise and a variational principle is assumed which forces the mass to move perpendicular to the cool part of the Maxwell set (often referred to as the shock) with the same velocity as the Maxwell set itself. This clearly results in mass adhering to the cool Maxwell set leading to an accumulation of mass at certain points (Bec and Khanin 2007; Bogaevsky 2004). A simple calculation gives the velocity of the Maxwell set as:

$$\frac{1}{2} \{ \nabla (\mathcal{S}_t(x) + \check{\mathcal{S}}_t(x)) \},$$

and a comparison with Theorem 13.9 and Corollary 13.5, reveals that this adhesion is sufficient to precisely destroy the vorticity near the Maxwell set. When adhesion occurs, an interesting question to consider is how rapidly mass will accrete on the Maxwell set as $\mu \to 0$ (Bec and Khanin 2007). This would relate directly to the mass involved in the formation of galaxies on the shock. Here we prove an inequality for the magnitude of the accumulated mass.

For this analysis we need to first mollify the Nelson diffusion process introduced in Section 2, equation (13.4),

$$dX_u^\mu = \nabla \mathcal{S}_u(X_u^\mu)\,du + \mu\,dB(u), \quad X_t^\mu = x.$$

This mollification will remove the discontinuities in the drift caused by the Maxwell set and caustic.

Let $0 < t < T$, for some fixed T, and let $u \in (0,t)$. Assume that we can mollify the minimizing Hamilton–Jacobi function $\mathcal{S}_u$ such that, $\mathcal{S}_u = \mathcal{S}_u^{\text{moll}}$ off some thin open set $\tau_u(\mu)$ surrounding the cool Maxwell set M_u^{Cool} and the cool caustic C_u^{Cool} where the Lebesgue measure $|\tau_u(\mu)| = O(\mu)$, and that $\nabla \mathcal{S}_u^{\text{moll}}$ is uniformly Lipschitz in space and bounded for $u \in (0,t)$.

We can then define a mollified potential, $V_u^{\text{moll}}(x)$, corresponding to this new system such that:

$$\frac{\partial \mathcal{S}_u^{\text{moll}}}{\partial u} + \frac{1}{2}|\nabla \mathcal{S}_u^{\text{moll}}|^2 + V_u^{\text{moll}} = 0,$$

so that as $\mu \to 0$, $(V_u^{\text{moll}} - V)$ is a surface potential.

This system then has a corresponding Burgers equation:

$$\frac{Dv^{\text{moll}}}{Dt} = \frac{\mu^2}{2}\Delta v^{\text{moll}} - \nabla V^{\text{moll}},$$

where this is the physically important velocity field in the limit $\mu \to 0$.

Let $\mu = \mu_n$ where μ_n is a sequence of real values such that $\mu_n \to 0$ as $n \to \infty$. Let the diffusion associated with $\mu = \mu_n$ after mollification be denoted by X^n and let:

$$A_t^n = \left\{\omega : X^n(\omega) \text{ avoids } C_u^{\text{Cool}} \cup M_u^{\text{Cool}} \text{ at all times } u \in (0,t)\right\}.$$

It is important to note that, formally at least:

$$\mathbb{P}(A_t^n) = \lim_{\lambda \to \infty} \mathbb{E}_x \left[\exp\left(-\lambda \int_0^t \chi_{C_u^{\text{Cool}}(\lambda^{-1})}(X_{\text{trev}}^n(u))\,du\right) \right.$$
$$\left. \times \exp\left(-\lambda \int_0^t \chi_{M_u^{\text{Cool}}(\lambda^{-1})}(X_{\text{trev}}^n(u))\,du\right)\right],$$

where:

$$C_s^{\text{Cool}}(\lambda^{-1}) := \left\{y : d(y, C_s^{\text{Cool}}) < \lambda^{-1}\right\},$$
$$M_s^{\text{Cool}}(\lambda^{-1}) := \left\{y : d(y, M_s^{\text{Cool}}) < \lambda^{-1}\right\}.$$

Here trev denotes time reversal so that:

$$X^n_{\text{trev}}(s) := X^n(t-s), \qquad X^n_{\text{trev}}(0) = x.$$

We need to assume that for sufficiently large n, $\mathbb{P}(A^n_t) > \delta > 0$.

Now assume that at any point x on the cool Maxwell set, M^{Cool}_t divides space into $m(x)$ parts. Therefore, x has $m(x)$ minimizing pre-images $x_0(i)(x,t)$, ($i = 1, 2, \ldots, m$) and for each pre-image there is a corresponding $S = S(i)$ given by:

$$\nabla S_u(i)(x) = \dot{X}(x_0, \nabla S_0(x_0), u)\Big|_{x_0 = x_0(i)(x,t)}.$$

We assume that all $\nabla S(i)$ are bounded and uniformly Lipschitz in space away from the caustic. Note that at a regular point of the Maxwell set $m = 2$ which is the case in which we are interested. In this case $x_0(i)(x,t)$ for $i = 1, 2$ correspond to matter arriving on different sides of the cool part of the Maxwell set. Arguing as in (Freidlin and Wentzell 1998), the Borel–Cantelli and Gronwall's lemmas give:

Lemma 13.7 *For $u \in (0, t)$ and $n \in \mathbb{N}$ ($i = 1, 2$) let:*

$$dX^n_i(u) = \nabla S^n_u(i)(X^n_i(u))\, du + \mu_n\, dB(u),$$

and:

$$dX^0_i(u) = \nabla S_u(i)(X^0_i(u))\, du,$$

with $X^n_i(t) = X^0_i(t) = x$. Then, if $\sum \mu_n^2 < \infty$, for $0 < t < T$:

$$\mathbb{P}\left[\sup_{0 < u < t} |X^n_i(u) - X^0_i(u)| \to 0 \text{ as } n \to \infty \Big| A^n_t\right] = 1.$$

We can now give our results on the distribution of mass. We parameterize the cool Maxwell set using the pre-Maxwell set and flow map:

$$x_0 = (x^1_0, x^2_0, \ldots, x^{d-1}_0, x^d_0(x^1_0, \ldots, x^{d-1}_0, t)) \in \Phi^{-1}_t M^{\text{Cool}}_t.$$

Define $x_0 \in \left(\Phi^{-1}_t M^{\text{Cool}}_t\right)^{\sim}$ if the classical path from x_0 to M^{Cool}_t avoids C^{Cool}_u and M^{Cool}_u for all $u < t$. It now follows from our results in Section 2 that:

Theorem 13.11 *Let the mass adhering to the Maxwell set in time interval $(0, T)$ be $m(0, T)$. Then if:*

$$m_0(0, T) = \int_0^T \frac{dt}{2} \int_{x_0 \in (\Phi^{-1}_t M^{\text{Cool}}_t)^{\sim}} T_0^2(x_0) \left|\frac{\partial x^d_0}{\partial t}\right| dx^1_0 \ldots dx^{d-1}_0,$$

we obtain:
$$m(0,T) \geq m_0(0,T).$$

The volume of M_t^{Cool} is zero and $m(0,T)$ is O(1). Therefore, the mass density on M_t^{Cool} will be infinite in parts of M_t^{Cool} in the inviscid limit. As we stated earlier, the process of adhesion forces the velocity away from M_t to be zero, and consequently the adhering particles carry no vorticity.

We now conjecture what happens when particles hit C_t^{Cool} first. Taking limits, $m_0(0,T)$ is the contribution to the mass in the shock from paths with no kinks which would be caused by hitting the caustic before time t.

It would therefore be reasonable to calculate $m_1(0,T)$, the contribution of mass from paths with one kink caused by a single intersection with the caustic before adhesion occurs. In this case we see that the shock causes a compression or decompression of mass. This comes from a generalized Ito formula for non C^2 functions.

Let $x_0 = x_0(t, u, x_0^1)$ where $x_0^1 \in k^1(u,t) \subset \mathbb{R}$ and $x_0 \in K^1(u,t)$ the curvilinear open set:

$$K^1(u,t) = \left(\Phi_u^{-1} C_u^{\text{Cool}}\right) \cap \left(\left(\lim_n \Phi_t^n\right)^{-1} M_t^{\text{Cool}}\right),$$

$$K^1(t) = \bigcup_{u \in (0,t)} K^1(u,t).$$

If we then integrate over $K^1(t)$, and use the generalized Ito formula for a discontinuous function due to Elworthy, Truman, and Zhao (Elworthy et al. 2007; Feng and Zhao 2006), we get:

$$m^1(0,T) = \int_0^T \frac{dt}{2} \int_0^t du \int_{k^1(u,t)} dx_0^1 \, T_0^2(x_0) \left|\frac{\partial x_0(t,u,x_0^1)}{\partial(t,u,x_0^1)}\right|$$
$$\times \exp\left\{-[\nabla_n \mathcal{S}_u(X(u^+)) - \nabla_n \mathcal{S}_u(X(u^-))]\right\},$$

where $\mathcal{S}_u(X(u^\pm))$ is the action evaluated at the appropriate minimizing preimage $x_0(i)(x,t)$ just above or below the caustic C_u and $\nabla_n \mathcal{S}$ is the normal derivative. The final factor in the integrand is the compression/decompression term coming from the adhesion process. In one dimension this result can be cast as a theorem. In higher dimensions it is only so far a conjecture.

Clearly, the mass involved in the formation of galaxies in the early universe would be given by a sum of terms involving $m^r(0,T)$ for $r = 0, 1, 2, \ldots$. As discussed at the start of this section, the complete adhesion of all matter to the cool part of the Maxwell set precisely destroys the rotation discussed in Section 7. However, if the adhesion were only partial, this would not totally remove the rotation, and could explain the formation of spiral galaxies in the early universe. We will discuss such properties and detailed examples in a forthcoming work on the Burgers equation with vorticity (Neate and Truman 2008a).

Acknowledgements

It is a pleasure for AN to thank the Welsh Institute for Mathematical and Computational Sciences (WIMCS) for their financial support in this research.

References

Abraham, R. and Marsden, J. E. (1978). *Foundations of Mechanics.* Benjamin/Cummings Publishing Co. Inc.: Reading, Mass. Advanced Book Program. Second edition.

Arnol'd, V. I. (1989). *Mathematical methods of Classical Mechanics,* Volume 60 of *Graduate Texts in Mathematics.* Springer-Verlag: New York. Translated from the 1974 Russian original by K. Vogtmann and A. Weinstein, corrected reprint of the second (1989) edition.

Arnol'd, V. I. (1990). *Singularities of Caustics and Wave Fronts,* Volume 62 of *Mathematics and its Applications (Soviet Series).* Kluwer Academic Publishers Group: Dordrecht.

Arnol'd, V. I. (1992). *Catastrophe Theory* (Third ed.). Springer-Verlag: Berlin. Translated from the Russian by G. S. Wassermann, based on a translation by R. K. Thomas.

Arnol'd, V. I., Zel'dovich, Y. B., and Shandarin, S. F. (1982). The large-scale structure of the universe. I. General properties. One-dimensional and two-dimensional models. *Geophys. Astrophys. Fluid Dynam.* **20**(1 & 2), 111–30.

Bec, J. and Khanin, K. (2007). Burgers turbulence. Pre-print arXiv:0704.1611v1.

Bogaevsky, I. A. (2004). Matter evolution in Burgulence. Pre-print arXiv:math-ph\0407073.

Dafermos, C. M. (2005). *Hyperbolic Conservation Laws in Continuum Physics* (Second ed.), Volume 325 of *Grundlehren der Mathematischen Wissenschaften [Fundamental Principles of Mathematical Sciences].* Springer-Verlag: Berlin.

Davies, I. and Truman, A. (1983). On the Laplace asymptotic expansion of conditional Wiener integrals and the Bender-Wu formula for x^{2N}-anharmonic oscillators. *J. Math. Phys.* **24**(2), 255–66.

Davies, I. M. and Truman, A. (1984). Laplace asymptotic expansions of conditional Wiener integrals and generalised Mehler kernel formulae for Hamiltonians on $L^2(\mathbf{R}^n)$. *J. Phys. A* **17**(14), 2773–89.

Davies, I. M., Truman, A., and Zhao, H. (2002). Stochastic heat and Burgers equations and their singularities. I. Geometrical properties. *J. Math. Phys.* **43**(6), 3293–328.

Davies, I. M., Truman, A., and H. Zhao (2005). Stochastic heat and Burgers equations and their singularities. II. Analytical properties and limiting distributions. *J. Math. Phys.* **46**(4), 043515, 31.

Elworthy, K., Truman, A., and Zhao, H. (2001). Stochastic elementary formulas on caustics. MRRS pre-print.

Elworthy, K. D., Truman, A., and Zhao, H (2007). Generalised Ito formulae and space-time Lebesgue-Stieltjes integrals of local times. In *Seminaire de Probabilities Strasbourg* XL Springer: Berlin.

Feng, C. and Zhao, H. (2006). Two-parameter p,q-variation paths and integrations of local times. *Potential Anal.* **25**(2), 165–204.

Freidlin, M. I. and Wentzell, A. D. (1998). *Random Perturbations of Dynamical Systems* (Second ed.), Volume 260 of *Grundlehren der Mathematischen Wissenschaften [Fundamental Principles of Mathematical Sciences]*. Springer-Verlag: New York. Translated from the 1979 Russian original by Joseph Szücs.

Frisch, U. and Bec, J. (2001). 'Burgulence'. In *Turbulence: Nouveaux Aspects/ New Trends in Turbulence (Les Houches, 2000)*, pp. 341–383. EDP Sci., Les Ulis.

Hopf, E. (1950). The partial differential equation $u_t + uu_x = \mu u_{xx}$. *Comm. Pure Appl. Math.* **3**, 201–30.

Kolokol'tsov, V. N., Schilling, R. L., and Tyukov, A. E. (2004). Estimates for multiple stochastic integrals and stochastic Hamilton-Jacobi equations. *Rev. Mat. Iberoamericana* **20**(2), 333–80.

Kunita, H. (1984). Stochastic differential equations and stochastic flows of homeomorphisms. In *Stochastic Analysis and Applications*, Volume 7 of *Adv. Probab. Related Topics*, pp. 269–291. Dekker: New York.

Maslov, V. P. and Fedoriuk, M. V. (1981). *Semiclassical Approximation in Quantum Mechanics*, Volume 7 of *Mathematical Physics and Applied Mathematics*. D. Reidel Publishing Co.: Dordrecht. Translated from the Russian by J. Niederle and J. Tolar, contemporary Mathematics, 5.

Neate, A. D. and Truman, A. (2005). A one-dimensional analysis of real and complex turbulence and the Maxwell set for the stochastic Burgers equation. *J. Phys. A* **38**(32), 7093–127.

Neate, A. D. and Truman, A. (2007a). Geometric properties of the Maxwell set and a vortex filament structure for Burgers equation. *Lett. Math. Phys.* **80**(1), 19–35.

Neate, A. D. and Truman, A. (2007b). A one-dimensional analysis of turbulence and its intermittence for the d-dimensional stochastic Burgers equation. *Markov Process. Related Fields* **13**(2), 213–38.

Neate, A. D. and Truman, A. (2008a). On the stochastic Burgers equation with vorticity. Forthcoming.

Neate, A. D. and Truman, A. (2008b). A one dimensional analysis of singularities and turbulence for the stochastic Burgers equation in d-dimensions. In *Seminar on Stochastic Analysis, Random Fields and Applications V*, Volume 59 of *Progress in Probability*. Birkhauser Verlag: Basel.

Reynolds, C. N. (2002). *On the polynomial swallowtail and cusp singularities of the stochastic Burgers equation*. Ph. D. thesis, University of Wales, Swansea.

Shandarin, S. F. and Zel'dovich, Y. B. (1989). The large-scale structure of the universe: turbulence, intermittency, structures in a self-gravitating medium. *Rev. Modern Phys.* **61**(2), 185–220.

Truman, A. (1977). Classical mechanics, the diffusion (heat) equation, and the Schrödinger equation. *J. Mathematical Phys.* **18**(12), 2308–315.

Truman, A., Reynolds, C. N., and Williams, D. (2003). Stochastic Burgers equation in D-dimensions—a one-dimensional analysis: hot and cool caustics and intermittence of stochastic turbulence. In *Probabilistic Methods in Fluids*, pp. 239–262. World Sci. Publ., River Edge, NJ.

Truman, A. and Zhao, H. Z. (1996a). On stochastic diffusion equations and stochastic Burgers' equations. *J. Math. Phys.* **37**(1), 283–307.

Truman, A. and Zhao, H. Z. (1996b). Quantum mechanics of charged particles in random electromagnetic fields. *J. Math. Phys.* **37**(7), 3180–97.

Truman, A. and Zhao, H. Z. (1998). Stochastic Burgers' equations and their semi-classical expansions. *Comm. Math. Phys.* **194**(1), 231–48.

van der Waerden, B. L. (1949). *Modern Algebra. Vol. I.* Frederick Ungar Publishing Co.: New York, N. Y. Translated from the second revised German edition by Fred Blum.

14

LIQUID CRYSTALS AND HARMONIC MAPS IN POLYHEDRAL DOMAINS

Apala Majumdar, Jonathan Robbins, and Maxim Zyskin

Abstract

Unit-vector fields **n** on a convex polyhedron P subject to tangent boundary conditions provide a simple model of nematic liquid crystals in prototype bistable displays. The equilibrium and metastable configurations correspond to minimizers and local minimizers of the Dirichlet energy, and may be regarded as S^2-valued harmonic maps on P. We consider unit-vector fields which are continuous away from the vertices of P. A lower bound for the infimum Dirichlet energy for a given homotopy class is obtained as a sum of minimal connections between fractional defects at the vertices of P. In certain cases, this lower bound can be improved by incorporating certain nonabelian homotopy invariants. For a rectangular prism, upper bounds for the infimum Dirichlet energy are obtained from locally conformal solutions of the Euler–Lagrange equations, with the ratio of the upper and lower bounds bounded independently of homotopy type. However, since the homotopy classes are not weakly closed, the infimum may not be realized; the existence and regularity properties of continuous local minimizers of given homotopy type are open questions. Numerical results suggest that some homotopy classes always contain smooth minimizers, while others may or may not depending on the geometry of P. Numerical results modelling a bistable device suggest that the observed nematic configurations may be distinguished topologically.

14.1 Introduction

Liquid crystals are intermediate phases of matter exhibiting partial ordering in the orientation and/or positions of their constituent particles. The constituents of nematic liquid crystals have a distinguished axis, and in the nematic phase these axes tend to align. The direction and degree of alignment can exhibit a rich variety of singularities. Standard references on liquid crystals include de Gennes and Prost (1995), Virga (1994), Kleman and Lavrentovich (2002), Stewart (2004).

The nematic phase is optically birefringent (light propagation is polarization-dependent). This, together with the fact that nematic ordering can be modified by external electric and magnetic fields, has led to a wide range of display applications. Most present-day liquid crystal displays (e.g., twisted nematic) are based on monostable cells, where, in the absence of external fields, the orientation

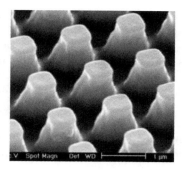

FIG. 14.1: The PABN cell (from Kitson and Geisow 2002).

assumes a single (spatially varying) equilibrium configuration which is effectively transparent to incident polarized light. To produce and maintain optical contrast, voltage pulses, which change the orientation, must be continually applied. There is considerable interest in developing bistable cells, which support two (and possibly more) stable configurations with contrasting optical properties. In bistable cells, power is needed only to switch between configurations. One mechanism for engendering bistability is to introduce microstructures into the geometry (Jones et al. 2000; Kitson and Geisow 2002; Tsakonas et al. 2007). Nematic liquid crystals in cells with polyhedral features (e.g., ridges, posts, wells) have been found to support multiple configurations. One such device, the PABN, or post-aligned bistable nematic cell, is shown in Fig. 14.1 (Kitson and Geisow 2002). It consists of a liquid crystal layer sandwiched between two planar substrates, with the lower substrate featured by an array of microscopic posts.

As a simple model for such systems, we consider nematic liquid crystals in a convex polyhedron $P \subset \mathbb{R}^3$ with orientation described by a director field, $\mathbf{n} : P \to RP^2$, taking values in the real projective plane. We consider the case of strong azimuthal anchoring, described by *tangent boundary conditions*. Tangent boundary conditions require that, on a face of P, $\mathbf{n}$ lies tangent to the face, but is otherwise unconstrained. It follows that on the edges of P, $\mathbf{n}$ is parallel to the edges, and therefore is necessarily discontinuous at the vertices. We are interested as to whether equilibria can be classified according to homotopy, and therefore restrict our attention to director fields which are continuous away from the vertices. For these, we can unambiguously assign an orientation to the director field (as P is simply connected), and regard $\mathbf{n}$ as a unit-vector field. We let $C_T^0(P, S^2)$ denote the space of continuous unit-vector fields on P satisfying tangent boundary conditions, or tangent unit-vector fields for short.

The elastic or Oseen-Frank energy of a configuration $\mathbf{n}$ is given by:

$$E = \int_P \left[K_1(\nabla \cdot \mathbf{n})^2 + K_2(\mathbf{n} \cdot (\nabla \times \mathbf{n}))^2 + K_3(\mathbf{n} \times (\nabla \times \mathbf{n}))^2 \right. \\ \left. + K_4 \nabla \cdot ((\mathbf{n} \cdot \nabla)\mathbf{n} - (\nabla \cdot \mathbf{n})\mathbf{n}) \right] dV. \tag{14.1}$$

Tangent boundary conditions imply that the contribution from the K_4-term, which is a pure divergence, vanishes. We shall make use of the so-called one-constant approximation, in which the remaining elastic constants K_1, K_2, and K_3 are taken to be the same and set to unity. In this case, (14.1) becomes the Dirichlet energy:

$$E(\mathbf{n}) = \int_P (\nabla \mathbf{n})^2 \, dV. \tag{14.2}$$

Minimizers of the Dirichlet energy, which correspond to equilibrium configurations, are S^2-valued harmonic maps, as are local minimizers, which correspond to metastable configurations.

The homotopy classes of $C_T^0(P, S^2)$ are described in Section 14.2, and a lower bound for the infimum of the Dirichlet energy in each homotopy class is given in Section 14.3. The lower bound is expressed as a sum of minimal connections between fractional defects at the vertices of P, in analogy with the well-known result of Brezis, Coron, and Lieb (1986) for the infimum Dirichlet energy of a set of point defects in $\mathbb{R}^3$. For nonconformal homotopy classes, this bound can be improved by incorporating certain nonabelian homotopy invariants; this is shown explicitly for certain homotopy classes in a rectangular prism in Section 14.4. Unlike the case of point defects in $\mathbb{R}^3$, the lower bound of Section 14.3 is expected to be strictly less than the infimum; achieving the lower bound would require concentration along a minimal connection, which would be incompatible with tangent boundary conditions. However, for P a rectangular prism, we can construct trial configurations in each homotopy class whose energies differ from the lower bound by a factor which is bounded independently of h (Section 14.5). Generalizing the construction to arbitrary P requires finding conformal maps on S^2 which preserve a given set of geodesics.

It is an open question as to whether the infimum is achieved in a given homotopy class, as is the regularity of the local minimizers. Numerical results presented in Section 14.6 suggest that some homotopy classes always contain smooth minimizers, while others may or may not depending on the geometry of P. Numerical results for a model of a bistable display suggest that the observed nematic configurations are topologically distinct.

In addition to existence and regularity questions, it would be interesting to investigate dynamics under the influence of applied fields. Switching between configurations of different homotopy type requires the creation and destruction of defects, and one would like to understand this process in detail.

14.2 Homotopy classification

Given $\mathbf{n} \in C_T^0(P, S^2)$ we can identify a number of discrete-valued quantities which depend continuously on $\mathbf{n}$ and which are therefore homotopy invariants. (Details may be found in Robbins and Zyskin 2004 and Majumdar et al. 2007b). Along an edge of P, $\mathbf{n}$ must lie parallel to the edge, so its value there is determined up to a sign, which we call an *edge orientation* (see Fig. 14.2(a)).

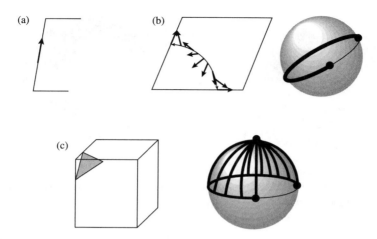

FIG. 14.2: Homotopy invariants. (a) Edge orientation (b) Kink number. **n** describes a 3/4-turn about the vertex (the image on S^2 is also shown), corresponding to kink number -1. (c) Trapped area. The image of the cleaved surface on S^2 has signed area $-3\pi/2$.

Next, along a path on a face of P between two edges, **n** must lie tangent to the face, and therefore describes a geodesic on S^2, i.e., an arc of a great circle (see Fig. 14.2(b)). As the endpoints of the geodesic are fixed by the edge orientations, the geodesic may be assigned an integer-valued relative winding number, or *kink number*. By convention, the shortest geodesic is assigned kink number zero. Another invariant is associated with a surface which separates one of the vertices of P from the other vertices—we call this a *cleaved surface* (see Fig. 14.2(c)). Along the boundary of a cleaved surface, **n** is determined up to homotopy by its edge orientations and kink numbers. Therefore, the signed area of the image of the cleaved surface itself, called the *trapped area* at the vertex, is determined up to an integer multiple of 4π (i.e., some number of whole coverings of the sphere).

Collectively, the edge orientations, kink numbers, and trapped areas constitute a complete set of homotopy invariants for $C_T^0(P, S^2)$; two configurations are homotopic if and only if their invariants are the same. We note that the invariants are not all independent—continuity of configurations on the faces of P implies that the kink numbers on each face satisfy a sum rule, while continuity on the interior of P implies that the trapped areas add up to zero. One can show that every set of invariants satisfying these sum rules can be realized.

From the preceding discussion, it is evident that the invariants of **n** can be determined from its values on a set of cleaved surfaces (the values of **n** on the corners and edges of the cleaved surfaces determine its edge orientations and kink numbers). Given a set of cleaved surfaces we can define an alternative set of invariants, the *wrapping numbers*, which will be used in subsequent sections.

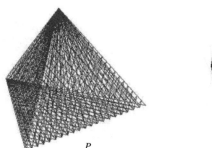 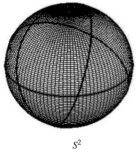

FIG. 14.3: Sectors for a tetrahedron. The great circles of directions tangent to the four faces partition the two-sphere into 14 sectors.

Let $T \subset S^2$ denote the set of directions which are tangent to one of the faces of P. Then T consists of a union of great circles. $S^2 - T$ consists of a union of disjoint open spherical polygons, which we call *sectors* (see Fig. 14.3). Let C^a denote a cleaved surface separating the ath vertex of P, say, from the others, and let $\mathbf{n}^a$ denote the restriction of $\mathbf{n}$ to C^a. Let Σ^σ denote the σth sector of S^2. The wrapping number $w^{a\sigma}$ is the number of times $\mathbf{n}^a$ covers Σ^σ, counted with orientation. For $\mathbf{n}$ differentiable, this is given by:

$$w^{a\sigma} = \frac{1}{A^\sigma} \int_{C^a} \mathbf{n}^*(\chi^\sigma \omega), \qquad (14.3)$$

where ω is the area two-form on S^2, normalised to have integral 4π, χ^σ is the characteristic function of Σ^σ, $\mathbf{n}^*$ denotes the pull-back, and $A^\sigma = \left|\int_{S^2} \chi^\sigma \omega\right|$ is the area of Σ^σ. Alternatively, $w^{a\sigma}$ can be expressed as the index of a regular value $\mathbf{s} \in \Sigma^\sigma$, i.e.:

$$w^{a\sigma} = \sum_{\mathbf{r} \mid \mathbf{n}^a(\mathbf{r}) = \mathbf{s}} \operatorname{sgn} \det(\mathbf{n}^a)'(\mathbf{r}). \qquad (14.4)$$

The wrapping numbers are homotopy invariants, and using Stokes' theorem can be expressed in terms of the edge orientations, kink numbers, and trapped areas. These relations can also be inverted to obtain the edge orientations, kink numbers, and trapped areas in terms of the wrapping numbers. Thus, the wrapping numbers constitute a complete (though redundant) set of homotopy invariants.

If the nonzero wrapping numbers at a given vertex are all negative, the homotopy class is said to be *conformal* with respect to that vertex, and if positive, *anticonformal* with respect to that vertex. A homotopy class is called *nonconformal* if there are vertices with wrapping numbers of different signs.

It is straightforward to count the number of invariants as well as the relations among them. Suppose that P has f faces, e edges, and v vertices (so that, from Euler's formula, $f - v + e = 2$). Then P has v trapped areas, which satisfy a single sum rule; $2e$ kink numbers, which satisfy f sum rules; and e edge orientations.

It also follows from Euler's formula applied to the set of tangent directions T, regarded as a graph on S^2, that there are generically (and at most) $f^2 - f + 2$ sectors ('generically' means that no direction is tangent to three or more faces of P). Thus, there are generically (and at most) $(f^2 - f + 2)v$ wrapping numbers, many more than the number of trapped areas and kink numbers. Among the constraints on the wrapping numbers, we point out that for a fixed sector Σ^σ, their sum over vertices must vanish, i.e.:

$$\sum_a w^{a\sigma} = 0. \tag{14.5}$$

We will use h to denote both an admissible set of values of the invariants as well as the homotopy class in $C_T^0(P, S^2)$ characterized by these values.

14.3 Lower bound: minimal connection

Brezis et al. (1986) established the infimum Dirichlet energy for unit-vector fields on $\mathbb{R}^3$ with point defects of specified position and degree. The result is expressed in terms of a *minimal connection* between the defects, defined below. A similar argument yields a lower bound for the infimum Dirichlet energy for tangent unit-vector fields on P of fixed homotopy type, in which the vertices of P play the role of defects, and the wrapping numbers that of generalized degrees. Details may be found in (Majumdar et al. 2004b), (Majumdar et al. 2004a), and (Majumdar et al. 2007b).

We first review the result of Brezis, Coron, and Lieb (1986). Let $\Omega = \mathbb{R}^3 - \{\mathbf{r}^1, \ldots, \mathbf{r}^n\}$, and let $\mathbf{n} : \Omega \to S^2$ denote a unit-vector field on Ω. Continuous unit-vector fields on Ω may be classified up to homotopy by their degrees, $d = (d^1, \ldots, d^n) \in \mathbb{Z}^n$, on spheres about each of the excluded points $\mathbf{r}_j$ (the restriction of $\mathbf{n}$ to such a sphere may be regarded as a map from S^2 into itself). For $\mathbf{n}$ smooth:

$$d^j = \frac{1}{4\pi} \int_{|\mathbf{r} - \mathbf{r}^j| = \epsilon} \mathbf{n}^* \omega \tag{14.6}$$

for small enough ϵ. For $\nabla \mathbf{n}$ square-integrable, the Dirichlet energy is given as in (14.2) by:

$$E(\mathbf{n}) = \int_\Omega (\nabla \mathbf{n})^2 dV. \tag{14.7}$$

In order for $E(\mathbf{n})$ to be finite, we require that:

$$\sum_j d^j = 0. \tag{14.8}$$

Let $C_\Omega^0(d)$ denote the homotopy class of continuous unit-vector fields with degrees d satisfying (14.8), and let:

$$E_\Omega^{\inf}(d) = \inf_{\mathbf{n} \in C_\Omega^0(d) \cap H^1(\Omega, S^2)} E(\mathbf{n}) \tag{14.9}$$

denote the infimum energy in $C_\Omega^0(d)$.

Given two m-tuples of points in $\mathbb{R}^3$, $\mathcal{P} = (\mathbf{a}^1, \ldots, \mathbf{a}^m)$ and $\mathcal{N} = (\mathbf{b}^1, \ldots, \mathbf{b}^m)$ (whose points need not be distinct), a *connection* is a pairing $(\mathbf{a}^j, \mathbf{b}^{\pi(j)})$ of points in $\mathcal{P}$ and $\mathcal{N}$, specified here in terms of a permutation $\pi \in S_m$ (S_m denotes the symmetric group). The length of a connection is the sum of the distances between the paired points, and a *minimal connection* is a connection of minimum length. Let:

$$L(\mathcal{P}, \mathcal{N}) = \min_{\pi \in S_m} \sum_{j=1}^m |\mathbf{a}^j - \mathbf{b}^{\pi(j)}| \qquad (14.10)$$

denote the length of a minimal connection, and let $|d| = \frac{1}{2}\sum_j |d^j|$.

Theorem 14.1 (Brezis et al. 1986) *The infimum $E_\Omega^{inf}(d)$ of the Dirichlet energy of continuous unit-vector fields on the domain $\Omega = \mathbb{R}^3 - \{\mathbf{r}^1, \ldots, \mathbf{r}^n\}$ of degrees d^j about the excluded points $\mathbf{r}^j$ is given by:*

$$E_\Omega^{inf}(d) = 8\pi L(\mathcal{P}(d), \mathcal{N}(d)), \qquad (14.11)$$

where $\mathcal{P}(d)$ is the $|d|$-tuple of excluded points of positive degree, with $\mathbf{r}^j$ included d^j times, and $N(d)$ is the $|d|$-tuple of excluded points of negative degree, with $\mathbf{r}^k$ included $|d^k|$ times.

In fact, the result of (Brezis et al. 1986) applies to more general domains with holes.

Here we sketch the argument that $8\pi L(\mathcal{P}(d), \mathcal{N}(d))$ is a lower bound for $E_\Omega^{\inf}(d)$. It suffices to consider smooth unit-vector fields on Ω, as these are dense in $C_\Omega^0(d) \cap H^1(\Omega, S^2)$. For any orthonormal frame $\mathbf{u}, \mathbf{v}, \mathbf{w}$, one has the inequality:

$$(\nabla \mathbf{n})^2 \geq 2|(d\xi \wedge \mathbf{n}^*\omega)(\mathbf{u}, \mathbf{v}, \mathbf{w})| \geq 2(d\xi \wedge \mathbf{n}^*\omega)(\mathbf{u}, \mathbf{v}, \mathbf{w}), \qquad (14.12)$$

where ξ is differentiable and $|d\xi| \leq 1$. (14.12) follows from the fact that, at every point, there is at least one direction (say $\mathbf{u}$) in which the directional derivative $\nabla_u \mathbf{n} := (\mathbf{u} \cdot \nabla)\mathbf{n}$ vanishes, while:

$$(\nabla_v \mathbf{n})^2 + (\nabla_w \mathbf{n})^2 \geq 2|\nabla_v \mathbf{n}||\nabla_w \mathbf{n}| \geq 2|\nabla_v \mathbf{n} \wedge \nabla_w \mathbf{n}|. \qquad (14.13)$$

Since $d\omega = 0$, it follows that $d\xi \wedge \mathbf{n}^*\omega = d(\xi \mathbf{n}^*\omega)$, so that:

$$(\nabla \mathbf{n})^2 \geq 2d(\xi \mathbf{n}^*\omega). \qquad (14.14)$$

Substituting (14.14) into (14.7) and applying Stokes' theorem, we get a lower bound:

$$E_\Omega^{\inf}(d) \geq 2\sum_j \xi^j d^j, \qquad (14.15)$$

which depends only on the values $\xi^j := \xi(\mathbf{r}^j)$ of ξ at the defects. Since $|d\xi| \leq 1$, these values are constrained by $|\xi^j - \xi^k| \leq |\mathbf{r}^j - \mathbf{r}^k|$. In fact, every set of ξ^j's satisfying these constraints can be realized by a piecewise-differentiable function ξ (eg, let $\xi(\mathbf{r}) = \max_j(\xi^j - |\mathbf{r} - \mathbf{r}^j|)$). Thus, one obtains a bound:

$$E_\Omega^{\inf}(d) \geq 2 \max_{\xi^j} \sum_j \xi^j d^j, \quad \text{where } \xi^j \geq 0, \ |\xi^j - \xi^k| \leq |\mathbf{r}^j - \mathbf{r}^k|, \quad (14.16)$$

in the form of a finite-dimensional linear optimization problem.

The dual formulation is given by:

$$E_\Omega^{\inf}(d) \geq 2 \min_{\eta_{jk}} \sum_{jk} \eta_{jk} |\mathbf{r}^j - \mathbf{r}^k|, \quad \text{where } \eta_{jk} \geq 0, \sum_k (\eta_{jk} - \eta_{kj}) \geq d^j. \quad (14.17)$$

This is a sort of transport problem, in which the degrees are the quantities to be transported and the costs are the distances between defects. We can take η_{jk} to be 0 unless $d^j > 0$ and $d^k < 0$. Without loss of generality, we may also assume that the degrees are either $+1$ or -1, so that there are an equal number, $m := n/2$, of each, with positions $(\mathbf{a}^1, \ldots, \mathbf{a}^m)$ and $(\mathbf{b}^1, \ldots, \mathbf{b}^m)$ respectively (if not, repeat each defect according to its multiplicity). (14.17) becomes:

$$E_\Omega^{\inf}(d) \geq 2 \min_{M_{pq}} \sum_{p,q=1}^{m} M_{pq} |\mathbf{a}^p - \mathbf{b}^q|, \quad \text{where } M_{pq} \geq 0, \sum_p M_{pq} = \sum_q M_{qp} = 1. \quad (14.18)$$

As M is constrained to be doubly stochastic, a theorem of Birkhoff (Birkhoff 1946) implies that it lies in the convex hull of the set of m-dimensional permutation matrices. The optimal solution will be amongst the permutation matrices themselves, leading to $E_\Omega^{\inf}(d) \geq 8\pi L(\mathcal{P}(d), \mathcal{N}(d))$.

A similar argument leads to a lower bound for the infimum Dirichlet energy for tangent unit-vector fields on P of given homotopy type.

Theorem 14.2 (Majumdar et al. 2007b) *Let $h = \{w^{a\sigma}\}$ be an admissible topology for continuous tangent unit-vector fields on a polyhedron P. The infimum $E^{\inf}(h)$ of the Dirichlet energy of continuous tangent unit-vector fields on P with invariants h is bounded below by:*

$$E_P^{\inf}(h) \geq \sum_\sigma 2A^\sigma L(\mathcal{P}^\sigma(h), \mathcal{N}^\sigma(h)), \quad (14.19)$$

where $\mathcal{P}^\sigma$ (resp. $\mathcal{N}^\sigma$) contains the vertices of P for which $w^{a\sigma}$ is positive (resp. negative), each such vertex included with multiplicity $|w^{a\sigma}|$.

Thus, to each sector σ may be associated a constellation of point defects at the vertices $\mathbf{v}^a$ with degrees $w^{a\sigma}$. The lower bound of (14.19) is a sum of the lengths of minimal connections for these constellations, weighted by the sector areas A^σ.

14.4 Lower bound: nonabelian invariants

For nonconformal homotopy classes, the lower bound of Theorem 14.2 can be improved by incorporating certain nonabelian invariants. These invariants, and the sense in which they are nonabelian, are introduced in Section 14.4.1 in a two-dimensional setting. For tangent unit-vector fields on $P \subset \mathbb{R}^3$ we describe this phenomenon in a particular case (Section 14.4.2), reflection-symmetric homotopy classes in a rectangular prism. Details will be given in (Majumdar et al. 2008).

14.4.1 Absolute degree and spelling length

Let $\phi : D^2 \to \mathbb{R}^2$ be a smooth map of the two-disk into the plane. We recall that $x \in D^2$ is a regular point of ϕ if x is in the interior of D^2 and $\det \phi'(x) \neq 0$, $y \in \mathbb{R}^2$ is regular value of ϕ if all of its pre-images are regular points, and a regular value has a finite number of pre-images. Let $\mathcal{R}(\phi) \subset \mathbb{R}^2$ denote the set of regular values of ϕ. From Sard's theorem, $\operatorname{Im} \phi - \mathcal{R}(\phi)$ is of zero measure.

Given $y \in \mathcal{R}(\phi)$, the algebraic degree of y (or degree, for short) is given by:

$$d_\phi(y) = \sum_{x \in \phi^{-1}(y)} \operatorname{sgn} \det \phi'(x), \qquad (14.20)$$

and is invariant under smooth deformations of ϕ which preserve the boundary map $\partial \phi$. We define the *absolute degree* of y by:

$$D_\phi(y) = \sum_{x \in \phi^{-1}(y)} 1. \qquad (14.21)$$

Clearly $D_\phi(y)$ is not invariant under all deformations which preserve $\partial \phi$, and:

$$D_\phi(y) \geq |d_\phi(y)|. \qquad (14.22)$$

Let $R = \{y_1, \ldots, y_n\} \subset \mathcal{R}(\phi)$ denote a set of n regular values of ϕ. We may regard the boundary $\partial \phi$ as a map $\partial \phi : S^1 \to \mathbb{R}^2 - R$ from the circle to the n-times-punctured plane. Let $\pi_1(\mathbb{R}^2 - R, p)$ denote the fundamental group of $\mathbb{R}^2 - R$, based at a point p, and let $[\partial \phi] \in \pi_1(\mathbb{R}^2 - R, p)$ denote the homotopy class of $\partial \phi$.

The fundamental group $\pi_1(\mathbb{R}^2 - R, p)$ may be identified with the free group on n generators, $F(c_1, \ldots, c_n)$ (see, e.g., Magnus et al. 1976). Let us specify that the generator c_j corresponds to a loop which encircles y_j once with positive orientation but contains no other y_k's. This determines the c_js up to conjugacy. Given $g \in \pi_1(\mathbb{R}^2 - R, p)$, we define a *spelling* to be a factorization of g into a product of conjugated generators and inverse generators, e.g.,

$$g = h_1 c_{i_1}^{\epsilon_1} h_1^{-1} \cdots h_r c_{i_r}^{\epsilon_r} h_r^{-1}, \qquad (14.23)$$

where $h_s \in \pi_1(\mathbb{R}^2 - R, p)$ and $\epsilon_s = \pm 1$. The length of a spelling is the number of factors (i.e., r in (14.23)). Define the *spelling length*, denoted $\Lambda_n(g)$, to be the shortest possible length of a spelling of g (e.g., $\Lambda_n(c_1c_2c_1^{-1}c_2^{-1}) = 2$). The spelling length of the identity, e, is taken to be zero. It turns out that the spelling length of $[\partial\phi]$ gives a lower bound on the sum of the absolute degrees of points in R.

Proposition 14.1 *Given $\phi : D^2 \to \mathbb{R}^2$ smooth, $R = \{y_1, \ldots, y_n\} \subset \mathcal{R}(\phi)$, and $\pi_1(\mathbb{R}^2 - R, p) \simeq F_n(c_1, \ldots, c_n)$, with generators c_j as above. Then:*

$$\sum_{j=1}^{n} D_\phi(y_j) \geq \Lambda_n([\partial\phi]). \tag{14.24}$$

Let $\bar{F}_n(c_1, \ldots, c_n)$ denote the abelianization of $F_n(c_1, \ldots, c_n)$, obtained by taking all of the c_js to commute, and given $g \in F_n(c_1, \ldots, c_n)$, let $\bar{g}$ denote the corresponding element of $\bar{F}_n(c_1, \ldots, c_n)$. Then $\bar{g}$ can be written as $c_1^{\delta_1} \cdots c_n^{\delta_n}$ for some integers δ_j. Let $\bar{\Lambda}_n(g) = \sum_{j=1}^{n} |\delta_j|$. Clearly $\Lambda_n(g) \geq \bar{\Lambda}_n(g)$ (e.g., $\bar{\Lambda}_n(c_1c_2c_1^{-1}c_2^{-1}) = 0$). It is readily seen that:

$$\sum_{j=1}^{n} |d_\phi(y_j)| = \bar{\Lambda}_n([\partial\phi]). \tag{14.25}$$

Thus, Proposition 14.1 implies that $\sum_j D_\phi(y_j)$ is strictly greater than $\sum_j |d_\phi(y_j)|$ provided that $\Lambda_n([\partial\phi])$ is strictly greater than $\bar{\Lambda}_n([\partial\phi])$. For example, if $[\partial\phi] = c_1c_2c_1^{-1}c_2^{-1}$, then ϕ takes values y_1 or y_2 at least twice, even though y_1 and y_2 are of degree zero.

For our applications we shall want to consider maps $\nu : D^2 \to S^2$ from the two-disk into the two-sphere. Let $R = \{\mathbf{e}_0, \ldots, \mathbf{e}_n\} \in \mathcal{R}(\nu)$ denote a set of $n+1$ regular values of ν. In contrast to the case of maps to the plane, the algebraic degrees $d(\mathbf{e}_j)$ are not determined by $\partial\nu$, since the image of ν itself is determined only up to whole coverings of S^2. We can remove this ambiguity by specifying the degree at one of the regular values, e.g., $d_\nu(\mathbf{e}_0) = d_0$. We may identify the fundamental group $\pi_1(S^2 - R, \mathbf{q})$ with the free group $F_n(c_1, \ldots, c_n)$ on n generators. As above, we specify that the generator c_j corresponds to a closed loop which encircles $\mathbf{e}_j$ once with positive orientation but contains no other $\mathbf{e}_k$s, which determines the c_js up to conjugacy. c_0, which corresponds to a loop about $\mathbf{e}_0$, may be expressed as a product of the generators c_1 through c_n and their inverses. In what follows, we write $b \sim c$ to denote that b and c are conjugate. In analogy with Proposition 14.1, we have the following:

Proposition 14.2 *Given $\nu : D^2 \to S^2$ smooth, $R = \{\mathbf{e}_0, \ldots, \mathbf{e}_n\} \subset \mathcal{R}(\nu)$, and $\pi_1(\mathbb{R}^2 - R, p) \simeq F_n(c_1, \ldots, c_n)$ with generators c_j as above, such that $c_0 \in F_n(c_1, \ldots, c_n)$. Suppose that $d_\nu(\mathbf{e}_0) = d_0$. Then:*

$$\sum_{j=0}^{n} D_\nu(\mathbf{e}_j) \geq |d_0| + \min_{\substack{g_1,\ldots,g_{r+|d_0|} \sim c_0 \\ h_1,\ldots,h_r \sim c_0^{-1}}} \Lambda_n([\partial\nu]g_1 \cdots g_{r+|d_0|}h_1 \cdots h_r). \tag{14.26}$$

While it is straightforward to compute the spelling length of a given element g, evaluating (14.26) may not be as straightforward.

14.4.2 Reflection-symmetric homotopy classes in a prism

A crude way to obtain a lower bound for the Dirichlet energy of tangent unit-vector fields on P is to estimate the contributions from nonoverlapping balls centred on each vertex. Let $\mathbf{n}$ denote a smooth tangent unit-vector field on P with invariants h. Let $\mathbf{v}^a$ denote the ath vertex of P, and let $O^a \subset S^2$ denote the set of directions about $\mathbf{v}^a$ subtended by P. For r less than the length of any of the edges coincident at $\mathbf{v}^a$, define $\boldsymbol{\nu}_r^a : O^a \to S^2$ by:

$$\boldsymbol{\nu}_r^a(\mathbf{e}) = \mathbf{n}(\mathbf{v}^a + r\mathbf{e}). \tag{14.27}$$

Up to parameterization, $\boldsymbol{\nu}_r^a$ describes the restriction of $\mathbf{n}$ to a spherical cleaved surface of radius r. We have that:

$$|\nabla \mathbf{n}(\mathbf{v}^a + r\mathbf{e})|^2 \geq \frac{1}{r^2}|(\boldsymbol{\nu}_r^a)'(\mathbf{e})|^2 \geq 2\frac{1}{r^2}|\det(\boldsymbol{\nu}_r^a)'(\mathbf{e})|, \tag{14.28}$$

where the last inequality follows from the same reasoning as in (14.12). Then:

$$E(\mathbf{n}) \geq \sum_a \int_0^{R^a} W_r^a \, dr, \quad \text{where } W_r^a = \int_{O^a} |\det(\boldsymbol{\nu}_r^a)'| \, d\mathbf{e} \tag{14.29}$$

and the R_a's are chosen so that $R_a + R_b \leq |\mathbf{v}_a - \mathbf{v}_b|$.

The quantity W_r^a is just the unsigned area of Im $\boldsymbol{\nu}_r^a$. The unsigned area of Im $\boldsymbol{\nu}_r^a \cap \Sigma^\sigma$ is at least the area of Σ^σ times the minimal absolute degree of the regular values in Σ^σ. Thus we have that:

$$W_r^a \geq \sum_\sigma \min_{\mathbf{e} \in \mathcal{R}(\boldsymbol{\nu}_r^a) \cap \Sigma^\sigma} D_{\boldsymbol{\nu}_r^a}(\mathbf{e}) A^\sigma. \tag{14.30}$$

Noting that $D_{\boldsymbol{\nu}_r^a}(\mathbf{e}_0) \geq |w^{a\sigma_0}|$ for all $\mathbf{e}_0 \in \mathcal{R}(\boldsymbol{\nu}_r^a) \cap \Sigma^{\sigma_0}$, we may apply Proposition 14.2 to (14.30) to obtain:

$$W_r^a \geq |w^{a\sigma_0}|A^{\sigma_0} + \sum_{\sigma \neq \sigma_0} \min_{\substack{g_1,\ldots,g_{r+|w^{a\sigma_0}|} \sim c_{\sigma_0}, \\ h_1,\ldots,h_r \sim c_{\sigma_0}^{-1}}} \Lambda_{s-1}([\partial \nu]g_1 \cdots g_{r+|w^{a\sigma_0}|}h_1 \cdots h_r) A^\sigma \tag{14.31}$$

(s in (14.31) is the number of sectors). We note that it follows from (14.30) and (14.22) that

$$W_r^a \geq \sum_\sigma |w^{a\sigma}| A^\sigma. \tag{14.32}$$

For certain homotopy classes of tangent unit-vector fields on a rectangular prism, R, one can show that the estimate (14.32) based on spelling lengths leads to an improvement of the lower bound of Theorem 14.2. Let:

$$R = \{\mathbf{r} \mid 0 \leq r_j \leq L_j, j = x, y, z\}, \tag{14.33}$$

where for convenience we have chosen coordinates with the origin at one of the vertices and axes parallel to the edges. By convention, we take $L_x \geq L_y \geq L_z$. In this case, the sectors are the coordinate octants of S^2 with area $A^\sigma = \pi/2$.

Reflection-symmetric homotopy classes on R are the homotopy classes of tangent unit-vector fields which are invariant under reflections through the mid-planes of the prism:

$$\mathbf{n}(x, y, z) = \mathbf{n}(L_x - x, y, z) = \mathbf{n}(x, L_y - y, z) = \mathbf{n}(x, y, L_z - z). \tag{14.34}$$

In this case, the wrapping numbers at two vertices a and $\bar{a}$ related by a single reflection differ by a sign:

$$w^{a\sigma} = -w^{\bar{a}\sigma}. \tag{14.35}$$

Thus, the wrapping numbers about the origin determine all the rest, and for simplicity we denote these by w^σ. The prism, and reflection-symmetric configurations in particular, will also feature in Sections 14.5 and 14.6.

To estimate $E_R^{\inf}(h)$ for reflection-symmetric h, it suffices to consider tangent unit-vector fields $\mathbf{n}$ which are themselves reflection symmetric. From (14.29) and (14.32) it follows that:

$$E_R^{\inf}(h) \geq 4\pi \sum_\sigma |w^\sigma| L_z, \tag{14.36}$$

which coincides with the lower bound (14.19) of Theorem 14.2 (a minimal connection in this case is obtained by pairing vertices at the endpoints of the (shortest) L_z-edges of R). However, by using the estimate (14.31) instead of (14.32), we get the following.

Theorem 14.3 (Majumdar et al. 2008) *Let R be a nonconformal reflection-symmetric homotopy class in R. Let σ^+ denote the sector with largest positive wrapping number, denoted W^+, and let σ^- denote the sector with largest (in magnitude) negative wrapping number, denoted W^-. Let:*

$$\Delta(h) = \max\left(W^+ - \sum_{\sigma \in adj(\sigma^+) \mid w^\sigma > 0} w^\sigma - \chi, |W^-| - \sum_{\sigma \in adj(\sigma^-) \mid w^\sigma < 0} |w^\sigma| - \chi, 0\right), \tag{14.37}$$

where $adj(\sigma)$ denotes the set of (three) octants adjacent to (i.e., sharing an edge with) σ and χ is equal to 0 or 1 depending on the signs of the edge orientations and kink numbers. Then:

$$E_R^{inf}(h) \geq 4\pi \left(\sum_\sigma |w^\sigma| + 2\Delta(h)\right) L_z. \tag{14.38}$$

For typical nonconformal homotopy classes, $\Delta(h) > 0$.

14.5 Upper bound in a prism

In Theorem 14.1, one obtains an equality for the infimum Dirichlet energy for a prescribed set of point defects, rather than just a lower bound, by constructing a sequence $\mathbf{n}^{(j)}$ whose energies approach $8\pi L(\mathcal{P}(d), \mathcal{N}(d))$. It can be shown that a subsequence $\mathbf{n}^{(k)}$ approaches a constant away from lines joining the paired defects in a minimal connection (here assumed unique), while $|\nabla \mathbf{n}^{(k)}|^2$ approaches a singular measure supported on these lines (Brezis et al. 1986). For tangent unit-vector fields on P, the boundary conditions preclude such a construction; $\mathbf{n}$ is required to vary across the faces of P, and therefore throughout its interior. However, by constructing tangent unit-vector fields which saturate the local inequality (14.12) over most of P, we can produce upper bounds for the Dirichlet energy with the same scaling with homotopy invariants as the lower bound of Theorem 14.2. Details are given in (Majumdar et al. 2004a; Majumdar 2006; Majumdar et al. 2008).

Here in outline is a procedure for constructing such configurations. Fix a set of values h of the homotopy invariants. As in Section 14.4.2, let $O^a \subset S^2$ denote the set of directions about the vertex $\mathbf{v}^a$ subtended by P. Define spherical cleaved surfaces:

$$C^a = \{\mathbf{v}^a + r^a \mathbf{e} \mid \mathbf{e} \in S^2\}, \tag{14.39}$$

where r^a is taken to be less than half the length of the smallest edge coincident at $\mathbf{v}^a$ (so that the C^as do not intersect). Specify $\mathbf{n}$ on C^a so as to satisfy tangent boundary conditions with wrapping numbers given by h, and take $\mathbf{n}$ to be constant along rays from C^a to $\mathbf{v}^a$. It remains to define $\mathbf{n}$ on $\hat{P}$, the (closed) domain obtained by excising the cones between the C^as and $\mathbf{v}^a$s. The boundary of $\hat{P}$ is composed of i) the C^as and ii) the faces of P truncated by the C^as. Extend $\mathbf{n}$ smoothly to these truncated faces so as to satisfy tangent boundary conditions. Choose a point $\mathbf{p}$ in the interior of $\hat{P}$. Along rays from C^a to $\mathbf{p}$, take $\mathbf{n}$ to be constant. Along rays from each truncated face to $\mathbf{p}$, rotate the values of $\mathbf{n}$ out of the tangent plane to the outward normal. There emerges a discontinuity at $\mathbf{p}$, but this is easily removed. If $\mathbf{n}$ is specified on the C^as to be conformal or anticonformal, except possibly on a small subset where its derivative is suitably controlled, then the local inequality (14.12) is saturated throughout most of P, and the Dirichlet energy can be shown to be proportionate to the lower bound of Theorem 14.2 independently of h.

The main difficulty in carrying out this procedure is in defining $\mathbf{n}$ on the C^as. Let $\boldsymbol{\nu}^a : O^a \to S^2$ be given by $\boldsymbol{\nu}^a(\mathbf{e}) = \mathbf{n}(\mathbf{v}^a + r^a \mathbf{e})$ (similarly to (14.27)). We note that O^a is a geodesic polygon on S^2; its sides are arcs of the great circles of directions tangent to the faces of P which are coincident at $\mathbf{v}^a$. Tangent boundary conditions require that $\boldsymbol{\nu}^a$ maps each side of O^a into the great circle containing it. If h is conformal with respect to $\mathbf{v}^a$ (the anticonformal and nonconformal cases are discussed below), we are led to the following:

Problem 14.1 *Find conformal maps on S^2 which preserve a given set of geodesics.*

Restricting the domain of such a map to O^a yields a candidate for ν^a.

In the case of the rectangular prism R, Problem 14.1 is readily solved. There are three geodesics which meet at right angles, and which may be taken to be the great circles about $\hat{x}$, $\hat{y}$, and $\hat{z}$. Under the stereographic projection $\mathbf{e} \mapsto w = (e_x + ie_y)/(1+e_z)$ from S^2 to the extended complex plane, these are mapped to the real axis, imaginary axis and unit circle respectively. Problem 14.1 becomes one of finding locally analytic functions $f(w)$ such that i) $f(w)$ is real when w is real, ii) $f(w)$ is imaginary when w is imaginary, and iii) $|f(w)| = 1$ when $|w| = 1$. Property i) implies that f is real; ii) then implies that f is odd; iii) then implies that $f(1/w) = 1/f(w)$. Therefore, if w_* is a zero of f, then $-w_*$ and $\bar{w}_*$ are zeros, while $1/\bar{w}_*$ is a pole. Restricting to f to be meromorphic, we may conclude that f is rational of the form:

$$f(w) = \pm w^{2m+1} \prod_{j=1}^{a} \left(\frac{w^2 - r_j^2}{r_j^2 w^2 - 1}\right)^{\rho_j} \prod_{k=1}^{b} \left(\frac{w^2 + s_k^2}{s_k^2 w^2 + 1}\right)^{\sigma_k} \times$$
$$\times \prod_{l=1}^{c} \left(\frac{(w^2 - t_l^2)(w^2 - \bar{t}_l^2)}{(t_l^2 w^2 - 1)(\bar{t}_l^2 w^2 - 1)}\right)^{\tau_l}. \tag{14.40}$$

The r_js denote the real zeros ($\rho_j = 1$) and poles ($\rho_j = -1$) of f between 0 and 1; the s_ks, the imaginary zeros and poles of f (according to $\sigma_k = \pm 1$) between 0 and i ; and the t_ls, the complex zeros and poles of f (according to $\tau_l = \pm 1$) with modulus less than one and argument between 0 and $\pi/2$.

The parameters in (14.40) can be chosen to realize any admissible set of conformal (i.e., nonpositive) wrapping numbers. Anticonformal topologies can be realised by replacing w with $\bar{w}$. Nonconformal topologies can be produced by modifying f in a small neighbourhood to be anticonformal and smoothly interpolating between the conformal and anticonformal domains.

Let:

$$E_P^-(h) = \sum_\sigma 2A^\sigma L(\mathcal{P}^\sigma(h), \mathcal{N}^\sigma(h)) \tag{14.41}$$

denote the lower bound of Theorem 14.2.

Theorem 14.4 (Majumdar et al. 2007b) *Let R denote a rectangular prism with sides of length $L_x \geq L_y \geq L_z$ and largest aspect ratio $\kappa = L_x/L_z$. Then:*

$$E_R^{inf}(h) \leq C\kappa^3 E_P^-(h) \tag{14.42}$$

for some constant C independent of h and L_x, L_y, L_z.

In the proof of Theorem 14.4, the positions of the zeros and poles of the conformal map (14.40) must be chosen carefully to ensure that the bound is achieved.

For reflection-symmetric conformal homotopy classes (cf (14.34)), a simpler construction leads to an improved result, in which $C = 1$ and κ^3 is replaced by $(L_x^2 + L_y^2 + L_z^2)^{1/2}/L_z$.

Theorem 14.5 (Majumdar et al. 2004a) *Let R denote a rectangular prism with sides of length $L_x \geq L_y \geq L_z$ and h a reflection-symmetric homotopy class which is conformal about one of the vertices. Then:*

$$E_R^{inf}(h) \leq \frac{(L_x^2 + L_y^2 + L_z^2)^{1/2}}{L_z} E_P^-(h). \qquad (14.43)$$

Theorem 14.5 extends to nonconformal reflection-symmetric homotopy classes, provided $E_P^-(h)$ is replaced by the lower bound given by Theorem 14.5 (Majumdar et al. 2008).

14.6 Existence and regularity of local minimisers: numerical results

Using direct methods, one might expect to establish the existence of a global minimizer of the Dirichlet energy for tangent unit-vector fields on P. The existence of (continuous) local minimizers in a given homotopy class h, however, is more difficult to address. The homotopy classes are not weakly closed, so that the existence of such local minimizers is not guaranteed; $E_P^{\inf}(h)$ may not be realized (just as the infimum energy for a prescribed set of point defects in $\mathbb{R}^3$ is not realized). We also recall the Hardt–Lin phenomenon (Hardt and Lin 1989) – global minimizers of the Dirichlet energy may have interior singularities, even when continuous unit-vector fields are admissible. If continuous local minimizers exist, then one would like to analyse their regularity (Schoen and Uhlenbeck 1982, 1983; Duzaar and Mingione 2004; Moser 2005).

Questions about the existence and regularity of continuous local minimizers of given homotopy type appear to be open for the problems we are considering. Below we describe some numerical results which suggest that, for some homotopy classes, smooth minimizers always exist, while for others, they may exist or not depending on the geometry of P.

The first examples concern two reflection-symmetric homotopy classes in a rectangular prism, denoted here by h_0 and h_1, which are both conformal with respect to one of the vertices (and, therefore, conformal or anticonformal with respect to the others). Details are given in (Majumdar et al. 2004a). h_0 is the simplest possible, in which there is a single nonzero wrapping number equal to -1, so that $\mathbf{n}$ takes values in a single octant of S^2. The restriction of $\mathbf{n}$ to a spherical cleaved surface corresponds to the conformal map given by $f_0(w) = w$ (cf (14.40)). Such a configuration is shown in Fig. 14.4. The lower bound for the infimum energy of Theorem 14.2 is $4\pi L_z$. The upper bound of Theorem 14.3 can

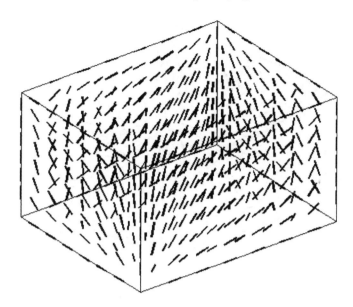

FIG. 14.4: Unwrapped configuration in class h_0.

be improved in this case by explicit evaluation of the Dirichlet energy for a trial configuration, yielding:

$$E_R^{\text{inf}}(h_0) < 8\frac{L_y L_z}{L_x^2} F_2\left(1, \frac{1}{2}, \frac{1}{2}, \frac{1}{2}, \frac{1}{2}, -\frac{L_y^2}{L_x^2}, -\frac{L_z^2}{L_x^2}\right)$$
$$+ (x \to y \to z) + (x \to z \to y), \tag{14.44}$$

where $F_2(\alpha, \beta, \beta', \gamma, \gamma'; s, t)$ is the Appell hypergeometric function (Gradshteyn and Ryzhik 1980). For a unit cube, we get the bounds:

$$12.5 \lesssim E_R^{\text{inf}}(h_0) \lesssim 15.3 \tag{14.45}$$

We computed minimizers numerically using two methods, namely solution of the Euler–Lagrange equation (using FEMLAB, a commercial PDE solver) and gradient descent. The converged energies from both methods agree, giving approximately 14.8. The converged unit-vector field is indistinguishable from Fig. 14.4 at the resolution shown, and appears to be regular away from the vertices.

The homotopy class h_1 is the next simplest among the reflection-symmetric conformal classes. There are three nonzero wrapping numbers equal to -1 in contiguous octants, so that **n** takes values in three-quarters of a hemisphere. The restriction of **n** to a spherical cleaved surface corresponds to the conformal map:

$$f_1(w) = w\frac{w^2 + s^2}{s^2 w^2 + 1}. \tag{14.46}$$

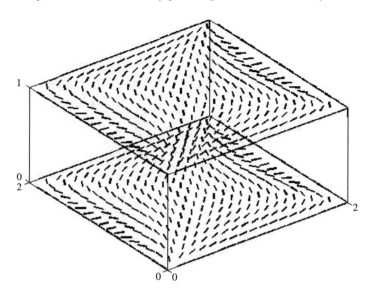

FIG. 14.5: A configuration **n** in h_1, generated by the conformal map $f(w) = w(w^2 + s^2)/(s^2w^2 + 1)$ for $s = .5$. **n** describes a 3/4-turn about each vertex in the top and bottom face.

Such a configuration is shown in Fig. 14.5. On the xy-faces of the prism, **n** executes a three-quarter turn about each vertex, corresponding to kink numbers of ± 1; the kink numbers on the other faces all vanish. As the parameter s approaches 1 from below, half-turns becomes concentrated along the y-edges, while f_1 approaches f_0 away from the y-edges. The corresponding family of configurations $\mathbf{n}_s$ is weakly but not strongly continuous with respect to s (an example of the fact that h_1 is not weakly closed).

Both numerical methods indicate that h_1 supports a smooth local minimizer for sufficiently thin slabs ($L_x/L_y, L_x/L_z \lesssim 1/10$), while for aspect ratios closer to unity, the numerical solution converges to the minimizer in h_0. Some insight into this behaviour is provided by computing the Dirichlet energy of trial configurations characterized by the one-parameter family (14.46), as shown in Fig. 14.6. For a cube (dashed curve), the energy approaches a minimum as s approaches 1, corresponding to a configuration in which half-turns concentrate along the y-edges. For $L_x = 20$, $L_y = 10$, $L_z = 1$ (solid curve), the energy has a minimum for s between 0 and 1, corresponding to a smooth configuration. Note that concentration along the shortest (L_z)-edges (which support the minimal connection) is not compatible with the topology, as the nonzero kink numbers lie in the xy-faces. Analogous arguments suggest that reflection-symmetric homotopy classes with two or more nonzero kink numbers do not contain smooth minimizers (for these classes, concentration along the shortest edge is compatible with the topology). However, it is conceivable that more nonreflection-symmetric

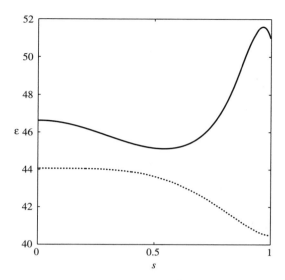

FIG. 14.6: Scaled energy $\epsilon = E/(L_x L_y L_z)^{1/3}$ of the conformal configuration $w(w^2 + s^2)/(s^2 w^2 + 1)$. Solid curve: $L_x = 20$, $L_y = 10$, $L_z = 1$. Dashed curve: $L_x = L_y = L_z = 1$.

homotopy classes (for which minimal connections do not necessarily pair vertices along edges) support smooth local minimizers.

The last numerical example is an idealized model of the PABN device. Details are given in (Majumdar et al. 2007). In fact, the model lies outside the class of problems we have considered so far; the domain is not a polyhedron, and the boundary conditions are not purely tangent. We take the PABN to consist of a rectangular post of square cross-section centred on the bottom surface of a rectangular cell of square cross-section, as in Fig. 14.7. In keeping with the device dimensions, the cell height is taken to be three times the cell width, and the cell width to be twice the post width. The height of the post is variable. Boundary conditions are dictated by material characteristics of the substrates. Tangent boundary conditions apply on the bottom substrate and on the post, while normal boundary conditions are appropriate for the top substrate. Periodic boundary conditions are imposed on the vertical sides of the cell, simulating a two-dimensional array of cells supporting the same nematic configuration (at a given time) and comprising a single pixel.

We consider four simple homotopy classes, in which the kink numbers are zero and the trapped areas taken to have their minimal allowed values. The orientation of **n** on the horizontal edges of the post are fixed, as in Fig. 14.7(a). The classes are distinguished by the relative orientations of **n** on the vertical edges of the post. Up to symmetry, there are four distinct possibilities. For the tilted class T, the orientation on all four vertical edges is the same. The other

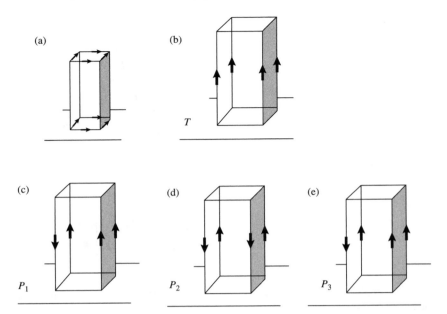

FIG. 14.7: Edge orientations for the four PABN configurations. (a) Orientations on the horizontal edges are the same for all. (b) Tilted profile T. **n** points up on all vertical edges. (c) Planar profile P_1. **n** points down on a single vertical edge (d) Planar profile P_2. **n** points down on a pair of adjacent vertical edges (e) Planar profile P_3. **n** points down on a pair of opposite vertical edges.

three classes, called planar, are obtained by taking the orientation to be opposite on, respectively, one of the vertical edges (the P_1 class), two adjacent vertical edges (P_2), and two opposing vertical edges (P_3). Configurations in T exihibit a large vertical component n_z in the region around the post. In configurations in P_1 through P_3, n_z is suppressed by the change in orientation between the vertical edges.

Local minimizers for each of these homotopy classes were computed using FEMLAB for a range of post heights. The converged configurations appear to be smooth away from the vertices of the post. In Fig. 14.8 we plot the converged energies of the local minimizers as a function of post height. The tilted class has the lowest energy, which is consistent with experimental observations which show that the liquid crystal always relaxes into the high-tilt state when cooled from the isotropic state (Kitson and Geisow 2002). The computations support the hypothesis that the bistable states of the PABN are topologically distinct.

Acknowledgments

We thank C.J.P. Newton, our co-author on (Majumdar et al. 2007), for many helpful discussions and, along with A. Geisow, for stimulating our interest in

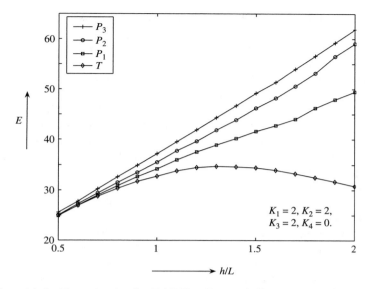

FIG. 14.8: Energies in the PABN cell as a function of post height.

these problems. AM was partially supported by an EPSRC/Hewlett-Packard Industrial CASE Studentship. A.M. and M.Z. were partially supported by EPSRC grant EP/C519620/1.

References

Birkhoff, G. (1946). Tres observaciones sobre el algebra lineal. *Univ. Nac. Tacumán Rev, Ser. A* (5), 147–51.

Brezis, H., Coron, J.-M., and Lieb, E. (1986). Harmonic maps with defects. *Comm. Math. Phys.* **107**, 649–705.

de Gennes, P.-G. and Prost, J. (1995). *The Physics of Liquid Crystals* (2nd ed.). Oxford University Press: Oxford.

Duzaar, F. and Mingione, G. (2004). The p-harmonic approximation and the regularity of p-harmonic maps. *Calc. Var.* **20**, 235–56.

Gradshteyn, I. and Ryzhik, I. (1980). *Tables of Integrals, Series and Products.* Academic Press: London.

Hardt, R. and Lin, F. (1989). Stability of singularities of minimizing harmonic maps. *Journal of Differential Geometry* **29**, 113–23.

Jones, J., Hughes, J., Graham, A., Brett, P., Bryan-Brown, G., and Wood, E. (2000). Zenithal bistable devices: Towards the electronic book with a simple LCD. In *Proc IDW*, pp. 301–304.

Kitson, S. and Geisow, A. (2002). Controllable alignment of nematic liquid crystals around microscopic posts: Stabilization of multiple states. *Appl. Phys. Lett.* **80**, 3635–37.

Kleman, M. and Lavrentovich, O. (2002). *Soft Condensed Matter.* Springer: New York.

Magnus, W., Karras, A., and Solitar, D. (1976). *Combinatorial Group Theory.* Dover: New York.

Majumdar, A. (2006). *Liquid crystals and tangent unit-vector fields in polyhedral geometries.* Ph. D. thesis, University of Bristol.

Majumdar, A., Newton, C., Robbins, J., and Zyskin, M. (2007). Topology and bistability in liquid crystal devices. *Phys. Rev. E* **75**, 051703 (11 pages).

Majumdar, A., Robbins, J., and Zyskin, M. (2004a). Elastic energy of liquid crystals in convex polyhedra. *J. Phys. A* **37**, L573–L580. ; *J. Phys. A* **38** (2005) 7595–595.

Majumdar, A., Robbins, J., and Zyskin, M. (2004b). Lower bound for energies of harmonic tangent unit-vector fields on convex polyhedra. *Lett. Math. Phys.* **70**, 169–83.

Majumdar, A., Robbins, J., and Zyskin, M. (2008a). In preparation.

Majumdar, A., Robbins, J., and Zyskin, M. (2008b). Energies of S^2-valued harmonic maps on polyhedra with tangent boundary conditions. *Annales de l'Institut Henri Poincaré. Analyse non linéaire.* **25(1)**, 77–103.

Moser, R. (2005). *Partial Regularity for Harmonic Maps and Related Problems.* World Scientific: Singapore.

Robbins, J. and Zyskin, M. (2004). Classification of unit-vector fields in convex polyhedra with tangent boundary conditions. *J. Phys. A* **37**, 10609–623.

Schoen, R. and Uhlenbeck, K. (1982). A regularity theory for harmonic maps. *J. Diff. Geom.* **17**, 307–35.

Schoen, R. and Uhlenbeck, K. (1983). Boundary regularity and the Dirichlet problem for harmonic maps. *J. Diff. Geom.* **18**, 253–68.

Stewart, I. W. (2004). *The Static and Dynamic Continuum Theory of Liquid Crystals.* Taylor and Francis: London.

Tsakonas, C., Davidson, A., Brown, C., and Mottram, N. (2007). Multistable alignment states in nematic liquid crystal filled wells. *App. Phys. Lett.* **90**, 111913 (3 pages).

Virga, E. (1994). *Variational Theories for Liquid Crystals.* Chapman and Hall: London.

INDEX

Note: page numbers in *italics* refer to Figures

absolute degree 314
action, Freidlin–Wentzell theory 270, *274*
action function, reduced 289–91
action functional, Ising model 250–1
 reversibility 255–6, 258
Adams, S., 184
Adams, S. and Bru, J.-B., 150, 174
Adams, S. and Dorlas, T., 159, 161, 175, 184
Adams, S. and König, W., 148, 150, 151, 159, 161, 175, 184
Adams, S. et al., 175, 182
adhesion model 281, 299–302
admissible sets of points 94
Airy processes 12
Aizenman, M. and Graf, G. M., 201
Aizenman, M. and Molchanov, S., 204
Aizenman, M. and Newman, C. M., 209
Aizenman, M. et al., 202, 204, 205, 207, 210
Aizenman–Molchanov criterion 195, 200, 201, 205, 214
Alexander, K., 43
Allaire, G., 89
Allen–Cahn equation 249, 254, 259
Anderson, M. et al., 174
Anderson localization 2, 194, 196–9, 201–2
 finite-volume criteria 205
 analogy with percolation 205–9
 consequences 209–10
 localization for large disorder 210–11
 proof, 214–16
 rigorous results 199–201
 use of fractional moments 202–4
Anderson model 194–6
annealed random walks in random potential 58
anticonformal homotopy classes 310
antisymmetric wave functions 149
Appell hypergeometric function 321
Arnol'd, V. I., 281, 299
Arratia, R. et al., 154
asymmetric simple exclusion process (ASEP), 22, 24, 25
 diffusive fluctuations 29

asymptotic shape, Richardson model 43–4
atomistic scale 244
 Ising model 247–8

backward irreducibility, ballistic paths 67
Baik, J. et al., 34
ballistic annihilation 2, 103, 104
ballistic paths 56, 59–60
 class of models 56–8
 coarse graining 60–4
 connectivity constants, Lyapunov exponents and Wulff shapes 58–9
 irreducible decomposition 64, 67–*8*
  cone points of paths 67
  cone points of skeletons 66–7
  cone points of trunks 66
  probabilistic structure 68–70
  surcharge function and surcharge inequality 65–6
 large deviation rate function 70
 local limit result 72
 small perturbations 72–6
Ball, J. M. et al., 267
Banach spaces 85
BBGKY hierarchy 102, 226, 238–9
Bec, J. and Khanin, K., 281, 300
Becker–Döring model 266–8
Bellettini, G. et al., 250, 255, 256, 259, 261
Bellissard, J. et al., 201
Benfatto, G. et al., 151
Benjamini, I., 43, 52
Birkhoff's ergodic theorem 14
bistable cells 307
BK inequality 136–7, 145
blocked cone points 66
Bodineau, T. and Ioffe, D., 257
Bogoliubov, N., 150, 173
Bolthausen, E. et al., 185
Boltzmann equation 2, 101–4
 justification of gainless homogeneous Boltzmann equation, 105
 non-validity 113–18
 validity 118

Boltzmann factor traces 150, 175, 183, 184
Boltzmann–Grad relation 102, 108
Boltzmann particles 152
bond configurations, notation 138–9
bonds 124
Borel–Cantelli lemma 301
Bose condensate 174
Bose–Einstein condensation 2, 149, 150–1, 153, 159, 173–4, 181
  path measure interpretation 168–9
Bose functions 157, 169–70
Bose gas, specific free energy 156–7
Bose statistics 149
Bosons 149
  low temperature studies 150
Boson systems, probabilistic models 151–3, 183–5
boundary layers, LSW-theory 239–40
Bourgain, J. and Kenig, C. E., 202
Bradley, C. et al., 174
Bratteli, O. and Robinson, D., 151
Brezis, H. et al., 308, 311, 312, 318
bricklayer processes 21
Brownian bridge measures 152, 155
Brownian bridge probability measure 158–9, 162
Brownian last-passage model 34
Brownian motions
  interacting 2, 151
  probabilistic models 183–5
Brownian spheres 103
Brydges, D. and Spencer, T., 128
bubble condition 127–8
Burgers equation 3, 281, 285–6
  geometric results 291–5
  inviscid limit 287, 289
  reduced action function 289–91
  stochastic forcing 286–9, 296–7
Burgers fluid mass density 285
Burgers fluid velocity 298
Busemann function 46
Buttà, P. et al., 256

cadlag 10
Cahn–Hilliard theory 227, 228
Campanino, M. et al., 56
canonical ensemble 150
canonical ensemble model 184
  large deviation results, vanishing temperature 186
canonical partition function 155
Carr, J. and Penrose, O., 227
Cassandro, M. et al., 245, 268
caustics 281, 282, 288, 290, 291
  geometric results 293–5

central limit theorem 11, 269
chaos propagation 101–2, 104–5
  hierarchy of evolutions 106–7
  marked trees 107–10
    convergence of empirical distribution to mean-field distribution 113–18
    empirical distribution 111–12
    mean-field distribution 110–11
  mean-field theory
    non-validity 118
    validity 113–18
Chayes, L., 55
Cioranescu, D. et al., 89
classical nucleation theory 268, 275–6
cleaved surfaces, 309
coagulation 102
coagulation term, LSW-theory 241
coarse graining 248, 255, 257
  self-interacting random walks 60–4
  in time 259–60
coarsening rates 227–8, 231
collisions, inclusion in LSW-theory 240–2
collision trees 107–10, 108
  empirical distribution 111–12
  convergence to mean-field distribution 113–18
  mean-field distribution 110–11
combinatorial studies 151
  Brownian bridge probability measures 159
Comets action functional 251, 255
comparison principle 261
competition models 40
complex turbulence 296
cone points 66–7
confluent Brownian motion 188
conformal homotopy classes 310
connection 257, 258
connectivity, percolation 206
  exponential decay 207, 209–10
connectivity constants, self-interacting random walks 58
continuum models 40
contours 255
convex analysis, large deviation theory 12
cool part of Maxwell set
  adhesion model 299–302
  vortex filament structure 298–9
Cornell, E. A., 174
corner growth model 12–13
  fluctuations 23–4, 25
corrector problem 90
correlation effects, coarsening processes 238–40

correlation functions 151
cost of switching 252
coupling
  exclusion processes 25
  Markov jump processes 259
  two-type Richardson model 45, 48
Cramér's theorem 162, 167, 188
crashing droplets 260–1
critical behaviour 123
  mean-field behaviour 127–8
critical droplet size 265–6, 267, 274
critical exponents 126–7
critical points 125–6
critical radius, Ostwald ripening 226
cusps, Maxwell sets 292–5
cycle counts 153
  discrete empirical shape measure 154–5
  Ewens sampling formula 154
  large deviations 155–8
cycle structure, empirical path measures 165–9
Cycon, H. L. et al., 200

Dalfovo, F. et al., 181
Davies, I. M. et al., 281, 289, 291
Davies, I. M. and Truman, A., 288
Davis, K. et al., 174
de Broglie wavelength 176
de Gennes, P.-G. and Prost, J., 306
Deijfen, M. and Häggström, O., 41, 45, 50, 51
del Rio, R. et al., 198, 199
De Masi, A. et al., 247, 248, 254, 255, 258
Dembo, A. et al., 151, 155
den Hollander, F., 268
determinantal point processes 12, 24
de Witt, C. and Storaeds, R., 151
Dickhoff, W. and Van Neck, D., 178
diffusion limited aggregation (DLA), 12
dilute interacting systems 151
dilute quantum gases 175–6
  Gross-Pitaevskii approximation 178–83
  potentials and scattering length 176–8
Dirac notation 195
directed bonds, notation 139
directed last-passage percolation models 13
Dirichlet energy 308
  infimum 312–13
  local minimizers 320–4
  lower bound
    minimal connection 311–13
    nonabelian invariants 314–18
    upper bound in a prism 318–20
discrete empirical shape measures (empirical cycle counts), 154–5

div–curl lemma 87–8
Domb–Joyce model 57
Donsker, M. and Varadhan, S., 183, 185
Donsker–Varadhan rate function 164, 186
Dorlas, T. et al., 151
double-well potential, 249
drift velocities 269–70
droplet growth 265–6
droplet size, Becker–Döring model 266–8
Droz, M. et al., 104
Duzaar, F. and Mingione, G., 320
dynamical environment 32

early universe, adhesion model 299–302
edge orientation
  PABN configurations, 324
  unit-vector fields 308, 309
effective evolution law, mesoscopic scale 248
elastic (Oseen–Frank) energy 307–8
elliptic PDEs, stochastic homogenization 83
  general theory of homogenization 87–8
  periodic case 88–9
  stationary ergodic case 89–91
Elskens, Y. and Frisch, H., 104
Elworthy, K. et al., 281, 302
empirical cycle count (discrete empirical shape measure), 154–5
empirical path measures 158
  cycle structure 165–9
  spatial structure 158–65
encounters, inclusion in LSW-theory 240–2
ends, Richardson models 43–4
envelope property, interfaces on plane, 19, 22
ergodic theorem 90
Euler–Lagrange equation 271, 321
Euler's formula 310–11
Ewald summation method 235
Ewens sampling formula 154
exclusion processes, growth processes and interface models 23–6
exponential spectral localization 197–9

far-field term, LSW-theory 236
Faris, W. G. and Jona-Lasinio, G., 250
Fatou's lemma 214
Feng, C. and Zhao, H., 302
Fermions 149
Fermi projection kernel, exponential decay 201
Ferrari, P. A. and Fontes, L. R. G., 29
Ferrer diagram 154
Fetter, A. and Walecka, J., 178

Feynman, R., 151, 164, 174
Feynman–Kac formula 152, 175
Fichtner, K., 162
finite patterns, self-interacting random walks 75–6
finite-volume criteria 205
  consequences 209–10
  percolation 205–9
finite volume instanton 250
first-passage percolation 12, 40
flip rate 247
fluctuations 11
  exclusion processes 22–6
  Hammersley process 26–9
  linear models 29–32
Flury, M., 55
Föllmer, H., 159
forgetfulness property, exponential distribution 10, 44
forward irreducibility, ballistic paths 67
fractional Brownian motion 31
fractional moments, Anderson localization 200, 202–4
free energy per particle 190
Freidlin, M. I. and Wentzell, A. D., 251, 287, 301
Freidlin–Wentzell theory 270–5
Frisch, U. and Bec, J., 281
Froese, R. et al., 202
Fröhlich, J. and Spencer, T., 199, 201

gainless homogeneous Boltzmann equation 102–3
  validity 105
galaxy formation, adhesion model 299–302
Garet, O. and Marchand, R., 46
Gärtner–Ellis theorem 161, 167
Gaussian Unitary Ensemble (GUE), 23
Geman, D. et al., 188
generalized cusps, Maxwell sets, 292–5
generalized exclusion 21
generic cusp 293, *294*
geodesics, Richardson model 43
Germinet, F. and Klein, A., 201
Germinet, F. et al., 202
Gibbs–Thomson law 224, 231
Giron, B. et al., 227
Girsanov theorem 284
Glauber dynamics 247
global inverse function theorem 283
Goncharov, V., 154
good time intervals 259–60
'good' trees 113–14
gradient flow models
  motion on a stationary manifold, *246*

multiple scales 244–5
  switching 245, *246*
Gradshteyn, I. and Ryzhik, I., 321
grandcanonical ensemble 150, 151
Green's function 124, 127, 128, 196
  use of fractional moments 202
Griffin, A. et al., 150
Griffiths, R. B. et al., 141
Gronwall lemma 301
Gross, E., 181
Gross–Pitaevskii approximation 2, 174, 175, 178–83
growth processes
  fluctuations 22–32
    exclusion processes 22–6
    Hammersley process 26–9
    linear models 29–32
  large deviations 32–4
  limit shape and evolution 12–22
  stochastic processes 9–12

Häggström, O. and Pemantle, R., 40, 41, 43, 45, 46, 48, 52
Hall effect 201
Hambly, B. et al., 34
Hamilton–Jacobi equations 21, 22, 24
Hamilton–Jacobi level surface 288, 290, 291
Hamilton–Jacobi theory 281–6
Hamilton operator 149–50, 174
  ground product states 179–80
  ground-state energy per particle 179
Hammersley, J. and Welsh, D., 40
Hammersley process 26–9
  large deviation 32–4
Hara, T. and Slade, G., 136, 138, 141
hard ball dynamics 101–2
  collision trees 107–10
hard-core potentials 177
Hardt–Lin phenomenon 320
Hartree–Fock ansatz 178, 179, 185
Hartree formula 179
Hartree model 175, 182, 185
  large deviation results
    large systems at positive temperature 187–90
    vanishing temperature 187
  many-particle limit 189–90
head of directed bond 139
'healing length', 176
heat equations, semi-classical solution 283–5
height fluctuations, interface models 24–5
height functions 9
hierarchy of evolutions, interacting particle systems 106–7

Hilbert D., sixth problem 102
Hilbert spaces 85
  decomposition 196–7
Hoffman, C., 46
homogenization of elliptic PDEs 83
  general theory 87–8
  link with thermodynamic limits 84, 97–8
  periodic case 88–9
  stationary ergodic case 89–91
homotopy classification, unit-vector fields 308–11
Hönig, A. et al., 232, 238
Hopf–Cole transformation 285
Hopf–Lax formula 20–1
Hunziker, W. and Sigal, I. M., 197
hydrodynamic limits 20, 21
  Hammersley process 28
  random average process 30
hypersurfaces, minimal 257

Illner, R. and Pulvirenti, M., 102
independence-principle, Poisson-point processes 114
independent walks 32
infection propagation, Richardson model 39–42
  one-type 42–4
  two-type 44–52
infinite coexistence, two-type Richardson models 41, 43, 45–50
infinite sets of points, average energies 83–4, 93–4
  deterministic case 94–5
  random deformation of periodic lattices 95–8
instantons 249–50
  linear stability 254–5, 258
  optimal displacement 253–4, 259–62
integral kernel 151
interacting particle systems 10
interaction kernel 247
interaction potentials 177, 178
interface models
  fluctuations 22–32
    exclusion processes 22–6
    Hammersley process 26–9
    linear models 29–32
  large deviations 32–4
  limit shape and evolution 12–22
  stochastic processes 9–12
interfacial energy 231–2
invariant manifolds, *250*, 256, 258
inverse temperature 248

inviscid classical mechanical flow map 281
inviscid limit
  Burgers equation 287, 289
  of a vorticity 298
irreducible decomposition, ballistic paths 64, 67–*8*
  cone points 65–7
  probabilistic structure 68–70
  surcharge function and surcharge inequality 65–6
irreversibility of macroscopic systems 101
Ising correlation functions 55
Ising model 123, 125, 200
  critical exponents 126
  critical point 125–6
  lace-expansion 141–6
  multiscale 247–50
iterated logarithm, law of 297
Ito processes 283

Jikov V. V. et al., 90
Johansson, K., 34
Jones, J. et al., 307
jump rates, height variables 9–10

Kac potential 247
Kallenberg, O., 90
Kesten, H., 43
Kesten's pattern theorem 56
Ketterle, W., 174
K-exclusion process 21
Khinchin, A., 150
kinetic annihilation 102–4
Kingman, J., 154
  subadditive ergodic theorem 40
kink number, unit-vector fields, *309*, 310, 311
Kirsch, W., 196
Kirsch, W. and Metzger, B., 196
Kitson, S. and Geisow, A., 307, 324
Klein, A., 202
Kleman, M. and Lavrentovich, O., 306
Kohn, R. V. and Otto, F., 227
Kolchin, V., 154
Kolokol'tsov, V. N. et al., 290
K-skeleton, self-interacting random walks 60–1
  attractive case 62–4, *63*
  cone points 66–7
  repulsive case, *61*–2
  surcharge function and surcharge inequality 65–6
Kunita, H., 289
Kunz, H. and Souillard, B., 199, 200

lace-expansion 2, 128–30
  idea of the proof 130–1
  for Ising model 141–6
  for percolation 136–41
  for self-avoiding walk 132–6
Lalley, S., 44
Landau, L., 150, 173, 183
Landau, L. and Lifshitz, E., 174
Lanford, O., 102
Lang, R. and Nguyen, X., 102
Laplace operator 149
large deviation principle (LDP), 11, 12, 151, 190–1
large deviation rate function, self-interacting random walks 70
large deviation results, canonical ensemble and Hartree models, 186–90
large deviations
  cycle counts 153–8
  empirical path measures 158–69
  growth processes 32–4
large-N behaviours, Gross–Pitaevskii formula 181–3
large numbers, laws of 10–11, 163, 168
last-passage models 13
  Brownian 34
Laurençot, P., 229
leading behaviour 298
Ledoux, M., 34
Lib, E. and Seiringer, R., 174
Licea, C. and Newman, C., 44
Lieb, E. and Seiringer, R., 181
Lieb, E. and Yngvason, J., 177, 181
Lieb, E. et al., 149, 150, 174, 181
Lifshitz, I. M. and Slyozov, V. V., 223, 226, 240, 241, 267
Lifshitz, I. M. et al., 202
Lions–Nisio theorem 22
Liouville equation 226, 238
liquid crystals 306–8
local limit theory, ballistic paths 72, 73–5
London, F., 150, 173
lower bound, LSW-theory 235
LSW-theory (leading order theory) 223–4, 226–7
  extension 237
  BBGKY hierarchy to capture correlations 238–9
  boundary layers due to fluctuations 239–40
  model including encounters 240–2
  rigorous derivation 229–30
  scaling of first order correction 230
  proof 233–7
  result 232–3
  set-up and assumptions 231–2
  shortcomings 228
Lyapunov exponents 58, 76–8
Lyapunov functionals 249

macroscale, Ising model 250
Majumdar, A. et al., 308, 311, 313, 317, 318, 319–20, 323
Manhattan norm 194
many-body wave function 149
many-particle limit, Hartree model 189–90
Marder, M., 238
marked trees 103, 107–10
  convergence of empirical distribution to mean-field distribution 113–18
  empirical distribution 111–12
  mean-field distribution 110–11
Markov jump processes 258–9
Marqusee, J. A. and Ross, J., 238
Martinelli, F. and Scoppola, E., 199
martingales 19
Maslov, V. P. and Fedoriuk, M. V., 281
Massari, U. and Miranda, M., 257
Matthies, K. and Theil, F., 111, 118
Maxwell–Klein set 291
Maxwell set 281–2, 288–9, 290, 291
  adhesion model 299–302
  geometric results 292–5
  vortex filament structure 298–9
mean-field behaviour 127–8
mean-field distribution 110–11
  non-validity 118
  validity 113–18
mean-field nature, LSW-theory 228
mean-field theory 104
measurability 196
mesoscopic scale 244
  Ising model 248
  deviations 250–1
metastability 265–6
  application of Freidlin–Wentzell theory 270–5
  Becker–Döring model 267–8
  pathwise approach 268
  proposed SD 268–70, 276
Minami, N., 200
minimal connection 312
minimal hypersurfaces 257
Molchanov, S. A., 200
momentum 271
monomers 266
monopole approximation 231

monotonicity, two-type Richardson model 50
Moser, R., 320
Mullins–Sekerka evolution 224–5, 227, 230
multi-instanton manifold 260
multiple scale approach, Anderson model 201, 202
multiscale Ising model 247–50
  deviations from mesoscopic equation 250–1
  optimal displacement of instanton on diffusive scale 253–4, 259–62
  switching 252–3, 256–9

n-connection 138
near-field term, LSW-theory 236–7
Neate, A. D. and Truman, A., 281, 291, 292, 296, 299
Nelson diffusion process 283, 300
nematic liquid crystals 306–8
Neumann boundary conditions 246
Newman, C., 44
Newman, C. and Piza, M., 43
Nguetseng, G., 89, 98
Niethammer, B., 229
Niethammer, B. and Otto, F., 229
Niethammer, B. and Pego, R. L., 227, 229
Niethammer, B. and Valásquez, J. J. L., 228, 229, 230, 232, 240
no level repulsion 200
nonconformal homotopy classes 310
non-diagonal long-range order 174
normalized integrals 86
normalized occupation measures 184, 185
Novikov condition 283
nucleation 261, 265–6
  application of Freidlin–Wentzell theory 270–5
  Becker–Döring theory 266–8, 275
  cost 253
  pathwise approach 268
  proposed SD 268–70, 276
nucleation rate 268

occupation numbers see cycle counts
off-diagonal Green's function, self-avoiding walk representation 211–14
Olivieri, E. and Vares, M., 245, 251, 268
one-particle number density, Ostwald ripening 226–7
one-type Richardson model 42–4
Onsager, L. and Penrose, O., 151, 159, 174
Ornstein–Zernike renewal techniques 55

Oseen–Frank energy 307–8
Ostwald ripening 3, 223–4
  basic model
    dynamic scaling and coarsening rates 227–8
    leading order theory (LSW-theory), 226–7, 228, 229–30
    Mullins–Sekerka evolution 224–5
    screening effect 228–9
  see also LSW-theory (leading order theory)

PABN (post-aligned bistable nematic cell), *307*
  edge orientations, *324*
  energies as a function of post height, *325*
  local minimizers of Dirichlet energy 323–4
pair-interaction potentials 177, 178
pair measures 159
pair potential 149
pair probability measures, relative entropy 160
partially asymmetric interface models 22
particle flux 19
particle growth rate 225
Pastur, L. A., 196
pathwise approach, metastability 268
Pauli's exclusion principle 149
Penrose, O., 151, 174
percolation 124–5
  exponential decay of connectivity 20–109, 207
  finite volume criteria 205–9
  lace-expansion 136–41
percolation probability 125–6
periodic homogenization theory 88–9
periodic lattices, random deformation 95–8
periodic problems, random deformations 91–3
perturbations, self-interacting random walks 72–6
perturbed mean-field model 151
phase boundaries, *246*
  multiple scales 244–5
phase functions 195
phase transition 125–6
  nucleation 265–6
Pitaevskii, L., 181
Pitaevskii, L. and Stringari, S., 150, 174, 176
Pitman, J., 154, 155
pivotal directed bonds 139
Poisson clocks 9–10, 17–18

polynomial swallowtail 293–5, *294*
polynuclear growth model, (PNG) 29
pre-Maxwell set 291
  geometric results 292–5
Presutti, E., 255
prism
  local mimimizers of Dirichlet
    energy 320–3
  reflection-symmetric homotopy
    classes 317
  upper bound for Dirichlet
    energy 318–20
pure point spectrum 198

quantized vortices 174
quantum canonical partition function 150
quantum models 2

RAGE (Ruelle, Amrein, Georcescu, Enss)
    theorem 197, 199, 201
random average process (RAP), 29–30
random-current representation 141–2
random growth models 1
random Schrödinger operator (random
    Hamiltonian), 195, 202
random stationary diffeomorphisms,
    91–2
random walk in random environment
    (RWRE), 31–2
random walk representation, off-diagonal
    Green's function 211–14
rate functions 160–3, 166, 167
  Donsker–Varadhan rate function 164
  Freidlin–Wentzell theory 270, *274*
  large deviations principles 187–8
real turbulent times 296
recollisions of particles 113
reduced action function 289–91
reflecting boundary conditions 246
reflection-symmetric homotopy
    classes 317
  local minimizers of Dirichlet
    energy 320–3
relative entropy, pair probability
    measures 160
reversibility, action functional 255–6, 258
Reynolds, C. N., 297
Richardson model 39–40
  one-type 42–4
  two-type 40–2, 44–5
    infinite coexistence 45–50
    unbounded initial
      configurations 51–2
Robbins, J. and Zyskin, M., 308
Ros, A., 257
Rost, H., 21

RSK correspondence 24
Ruelle, D., 149, 150

Sakai, A., 136, 142, 145
Sanov's theorem 159
Sard's theorem 314
scales
  dilute quantum gas systems 176
  phase boundaries 244–5
scattering lengths, dilute quantum
    gases 177–8
  link to ground states 183
scattering states 104, 106–7
  collision trees 107–10
Schoen, R. and Uhlenbeck, K., 320
Schrödinger, E., 159–60
Schwartz alternating method 240
screening effect, Ostwald ripening 225,
    228–9
screening length, Ostwald ripening,
    230, 231
second class particles 25, 26
sectors, *310*
self-avoiding walk (SAW), 57, 124
  lace-expansion 132–6
  mean-field behaviour 127
  representation of off-diagonal Green's
    function 211–14
self-energy formula 202, 203
self-interacting polymers 55–6
self-intersections of particle path
    cylinders 113
self-similarity, Ostwald ripening 223, 227
shape theorem 40, 43
Simon, B., 198
Simon, B. and Wolff, T., 200
Simon–Lieb inequalities 205
simple symmetric random walk 123–4
skeletons, self-interacting random
    walks 60–1
  attractive case 62–4, *63*
  repulsive case, *61–2*
smooth subsets, definition 84
'Snapshot' analysis 231
Sobolev spaces 85
soft-core potentials 177
source-switching lemma 145
space-time process 30–2
spatial structure, empirical path
    measures 158–65
specific free energy, non-interacting Bose
    gas 156–7
spectral gap 257–8
spectral localization 197–9, 200, 201
spelling length 314–16
Spohn, H., 101

standing waves 249–50
stationarity 89
stationary behaviour, interfaces on a plane 18–19
stationary diffeomorphisms 91, 97
stationary homogenization theory 89–91
Stewart, I. W., 306
stochastic differential equation (SDE), nucleation 268–70, 276
 application of Freidlin–Wentzell formula 270–5
stochastic homogenization of elliptic PDEs 83
 general theory of homogenization 87–8
 link with thermodynamic limits 84, 97–8
 periodic case 88–9
 stationary ergodic case 89–91
stochastic turbulence 296–7
Stollman, P., 202
stopping times 16–17
strangled sets 45
Strassen form, law of iterated logarithm 297
Stratonovich heat equation 286
strong dynamical localization, 200, 201, 205
strong law of large numbers 11
subadditive ergodic theorem, in analysis of Richardson model 40, 42
subcritical regimes, LSW-theory 231, 233
subharmonic functions 208–9
supercritical droplets 265–6, 274
supercritical regimes, LSW-theory 231, 233, 235
superfluidity 150, 173–4
surcharge functions 65
surcharge inequality 65–6
surface tension, relationship to droplet size 265
swallowtails 292–5
switching 245, *246*, 252–3, 256–9
Sylvester determinants 291
symmetric wave functions 149
symmetrization 153
symmetrization correlations 151
synergetics 123
Sznitman, A., 103, 152

tagged Poisson point processes 105
tail of directed bond 139
tangent boundary conditions 307–8
thermodynamic equilibrium states 175
thermodynamic limits 83–4, 150
 link to homogenization 84, 97–8
Thirring, W. 150, 178

tight binding approximation 194
time, coarse graining 259–60
time evolution equations, nucleation 275
time-minimizing paths, Richardson model 43
totally asymmetric simple exclusion process (TASEP), 14–15, 18, 19, 29
 large deviations 34
totally asymmetric K-exclusion 21
traces of Boltzman factor 150, 151, 175, 183, 184
 Feynman–Kac formula 152
Tracy–Widom fluctuations 12
Tracy–Widom GUE distribution 23
trapped areas, *309*, 310, 311
trap potentials 149, 176–7
triangle condition 127–8
Truman, A. and Zhao, H. Z., 281, 287, 288
Truman, A. et al., 282, 290, 296, 297
Tsakonas, C. et al., 307
turbulent times 296
two-body interaction energy 176
two-phase material, multiscale model 247–50
 deviations from mesoscopic equation 250–1
 optimal displacement of instanton 253–4, 259–62
 switching 252–3, 256–9
two-scale convergence, periodic homogenization theory 89
two-type Richardson model 1, 39, 40–2
 asymmetric 44–5
 infinite coexistence 45–50
 symmetric 44
 unbounded initial configurations 51–2

unbounded initial configurations, two-type Richardson model, 51–2
unit-vector fields
 homotopy classification 308–11
 local minimizers of Dirichlet energy 320–4
 lower bound for Dirichlet energy minimal connection 311–14
 nonabelian invariants 314–18
 upper bound for Dirichlet energy in a prism 318–20
universality, critical exponents 126–7
upper bound, LSW-theory 235
Ushiyama, K., 104

van den Berg–Kesten inequalities 209
van der Hofstad, R. and Holmes, M., 56
van der Waerden, B. L., 291

Varadhan's lemma 189
variational formulation, LSW-theory 234
Velásquez, J. J. L., 228
Vershik, A., 151, 155
Virga, E., 306
vortex filament structure, Maxwell
    set 298–9

Wagner, C., 223, 226
Wasserstein distance 259
Watterson, G., 154
weak convergence 86, 186

weal selection criterion, LSW-theory 228
wedge process 15–16
Wieman, C. E., 174
Williams, D., 297
wrapping numbers 309, 310, 311, 317
Wulff shape 59, 253, 256–7

Young tableaux 24, 154–5

Zeldovich approximation 281, 299
zero range processes 21
zeta process 296, 297